# The Victorian & Edwardian
# House Manual

**SECOND EDITION**

First published in September 2005
Reprinted 2006, 2007 and 2011
Second edition published in May 2015

British Library Cataloguing in Publication Data:
A catalogue record for this book is available
from the British Library

ISBN 978 0 85733 284 4

Published by Haynes Publishing,
Sparkford, Yeovil, Somerset BA22 7JJ, UK
Tel: 01963 442030  Fax: 01963 440001
Int. tel: +44 1963 442030  Int. fax: +44 1963 440001
E-mail: sales@haynes.co.uk
Website: www.haynes.co.uk

Haynes North America Inc.
861 Lawrence Drive, Newbury Park, California 91320, USA

Printed in the USA by Odcombe Press LP,
1299 Bridgestone Parkway, La Vergne, TN 37086

# Acknowledgements

Photos: Ian Rock and Ian MacMillan, except where credited or listed below.
Editing, technical advice and step by steps Ch 6 & Ch 13: Ian MacMillan
Thanks also to: Louise McIntyre, project manager and Ian Heath, copy editor

For assistance with photos and research; Jerry Blears at DIY sashwindow.co.uk; Judith and
Sarah Ferrier; Jason Perry; Ray Axtell; Mike Parrett at dampbuster.com; Stephen Cobb at
AngelPlastics.co.uk;  Mathew Newton; Debbie Taylor at CWS; Howard Joynes and Phil
Parnham at Sheffield Hallam; Lafarge Plasterboard; Jenny Alderman; Alex Dover BA
(hons); Russell and Maria Bell; Reena Suleman at Linley Sambourne House; E.G. Books.

Many thanks to the following chartered surveyors: Steve Lees, Jeff Stott, Andrew Lister,
and also Chris Bone, Kevin Stevens, Bill Greenfield, Sam Holton, David Mitchell, Rob
Stambrey, Graham Andrews, Keith Woodbine, Alan Woodward, David Buchanan,
Andrew Walmsley.

# The Victorian & Edwardian

## House Manual

**SECOND EDITION**

**Ian Alistair Rock** MRICS

*Edited by* **Ian R. MacMillan** FRICS

# CONTENTS

# INTRODUCTION

Looking for trouble? Well, you've chosen the right house. Despite their many excellent features, when you buy a typical Victorian or Edwardian house there's a good chance it'll be crying out for some serious TLC. Well-known problems such as timber decay, beetle infestation, damp and subsidence can sometimes afflict properties of this age. But in recent years, a lot of unnecessary and expensive remedial work has been carried out because common defects have been misdiagnosed. Much harm has also been done as a result of well-intentioned but misguided home improvements and modernisation – and sometimes because of inappropriate 'essential repairs' stipulated by mortgage lenders.

So how can you tell if your house genuinely has significant defects or is just typical for its age? This Haynes manual shows you what to check and where to look; what's 'normal' and what isn't; and how to remedy common defects. In many cases the correct solution involves nothing more than some fairly straightforward maintenance work. You may still need to call in the professionals, but there's a lot you can do yourself.

# About this book

This Haynes manual is designed to be easy to read and is written broadly in the format of a building survey. However, each chapter could easily have made an entire book in itself, so, inevitably, some aspects – like architecture and history – are only touched on very briefly.

The 'Defect/Cause/ Solution' format makes it quick and easy to look up the causes of technical problems and see how to fix them. And the classic Haynes step-by-step photo guides illustrate how to repair and maintain your house – in easy stages. There are also a number of useful projects showing how to add value to your home with popular improvements. This Haynes manual should tell you considerably more than the average house survey. To find out more about Building Regulations, planning issues and sourcing materials for period houses visit the book's website www.victorian-house.co.uk.

**Above:** Victorian villas with skilful contrasting 'polychrome' brickwork

**Below left:** Survivor! – despite deliberate damage by misguided developer

**Below right:** Late Victorian exuberance, built with quality materials and workmanship

# Before you start

There are two key things to consider before embarking on the projects described on the following pages: first, that you're not putting yourself at risk of injury or death; and second, that you are not about to unwittingly commit a criminal offence, *eg* by carrying out work without the necessary Planning or Building Regulations consents, or in contravention of legislation such as the Party Wall Act.

## HEALTH AND SAFETY

Serious injury is a real risk when working on buildings –

Scaffolding + high voltage supply cables – a potentially fatal combination!

especially when using power tools, ladders and electrical equipment, as well as from less well-known causes such as trench collapse. So before you begin it makes sense to reduce the risks by taking sensible precautions:

■ Wear eye protectors, dust masks, and gloves, particularly when using power tools or chemical sprays.
■ For roof work and any work at a high level use access towers or scaffolding. Where ladders are used, they must be well secured so that they're stable and level. Roof ladders or 'wheel ridge hook' ladder adaptors can be used by those with suitable experience.
■ Use an RCD electrical safety cut-out adaptor.
■ Angle grinders and electric saws are notorious for causing nasty injuries. A disk under load may shatter, or a blade may snap, so always wear protective safety clothing and a mask when using power tools.
■ Keep hands well behind the edges of cutting tools.
■ If you're unfamiliar with hired equipment or any tool always read the instructions.
■ Skips parked on roads must have nightlights.
■ Never rush a job, and don't be too proud to call in a tradesman if a job turns out to be harder than anticipated. If in doubt always seek professional advice.

*NB Every year about 70 fatalities and 5,000 serious injuries are caused by falls from a height. Health and Safety legislation now requires anyone who could potentially fall more than 2m to use a fall-protection system, so it's worth investing in a full body harness and safety helmet (available from most DIY stores).*

# How Victorian houses work

Old houses work in a totally different way from modern buildings. Today's homes are built with super-deep, rock-solid foundations designed to inhibit even the slightest movement.

This permits the use of strong modern bricks, hard inflexible cement and impervious paints, which are very effective at repelling rain and sealing out moisture.

In contrast, old houses tend to shuffle about on their shallow foundations in tune with seasonal ground changes, almost like living beings. And because traditional materials such as brick, stone and soft lime mortar are relatively porous, rather than repelling rainwater they work by temporarily absorbing moisture

until it can evaporate away in dry weather. This natural cycle is known as 'breathing'.

Problems really begin when these two very different philosophies get mixed up. Whereas traditional lime mortars, renders and paints had sufficient flexibility to cope with the natural movement in old houses, brittle modern cement-based materials simply can't tolerate it. The resulting cracking in re-pointed or cement-rendered walls can allow rain to penetrate. But instead of naturally evaporating away via lime mortar joints, the hard cement traps it. Modern paints add to the problem by sealing in damp, which has little chance to dry out screened from the sun and wind behind layers of waterproof paint. Of course damp doesn't just get in from rain; moisture from indoor condensation also needs to be allowed to disperse, something that well ventilated old buildings managed very ably.

Today, many Victorian houses have suffered from well intentioned repairs that have unwittingly damaged their natural breathing cycle resulting in consequential problems, from eroded masonry to decayed timber.

The aim of this book is to help restore the health of Britain's wonderful Victorian housing heritage.

## HOW VICTORIAN IS YOUR HOUSE?

Strictly speaking, a Victorian house is one constructed during the reign of Queen Victoria – between 1837 and 1901; but the vast majority were built after 1860, and the methods and materials used in mainstream housing remained largely

# Remedying past mistakes

Many repairs and improvements carried out to Victorian houses in the past have subsequently turned out to be damaging and later need to be 'unpicked'.

It's a sad fact that people sometimes lavish large sums of money on 'improvements' that actually reduce the value of the property. Popular 'negative value' improvements include gluing artificial stone cladding to the walls, widening original window openings to fit ugly plastic or aluminium double-glazed windows, and re-cladding roofs with clumsy modern interlocking concrete tiles. Other botches include roofs sprayed internally with expanding foam, or painted externally, and meter boxes stuck on historic brickwork. So making good past inappropriate work can add value to a period property.

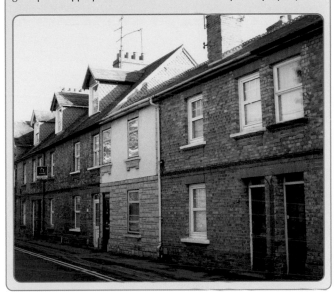

unchanged up until after the First World War. Of course builders didn't suddenly adopt a radical new Victorian architectural style in 1837, and by the same token, designs didn't change overnight in 1901. So the advice in this book is also relevant to Edwardian properties built between 1901 and 1914 (King Edward VII actually died in 1910 but to simplify matters the 'Edwardian era' is usually defined as embracing the first four years of George V's reign).

# GETTING TO KNOW YOUR HOUSE

Everybody can recognise a 'typical' Victorian house – and yet the sheer variety of styles and materials used in 19th and early 20th century buildings around the country was truly phenomenal. Here we look at architectural clues to help date the age of your house, the four most common layouts are identified, and we consider why the terrace design was so popular.

Victorian and Edwardian homes today comprise around one in six of all UK houses – an incredible 4 million plus properties. Estate agents never fail to highlight original features and enthuse about 'period charm', and for many aspiring buyers owning a 'house with history' has a certain appeal, compared to the equivalent bland modern box. So before we look at how best to care for houses of this vintage, it's worth taking a few moments to step back in time and explore how they were built, and what features were important to their original occupants.

# Plain or posh?

One of the joys of owning a Victorian house is that even the humblest terrace was built as a cheaper replica of grander dwellings, mimicking as far as possible their wealthier cousin's layouts and features. Houses from this era can be roughly divided into middle-class and working-class types, although both are similar. The biggest difference between them is their size – the middle classes needed more space for their servants.

The design of the Victorian home was very much concerned with reflecting status. The front was designed to impress the visitor, so entrance halls would often boast highly ornate ceilings, joinery and floors. These features could be admired by guests after being greeted by a maid and ushered into the hall while being 'announced'. Where the main flight of stairs could be seen by guests they were wider and more ornate than the upper flights. The hall also facilitated separate access to individual rooms, a status-enhancing factor that differentiated a family house from a mere cottage.

The most prestigious room was the drawing room or 'front parlour' (where the ladies would 'withdraw' after dinner and *parlez*) which housed the family's best furniture (also known as the sitting room, living room, or best room – the word 'parlour' being considered slightly 'common'). Here, the provision of a bay window was a much-prized feature and status symbol. Yet this was the least used room in the house, reserved for the receiving of guests on Sundays, formal family gatherings and occasional use as a study or reading room. The earlier Georgian custom of placing the main reception room upstairs (away from kitchen smells and servants) had now largely been abandoned.

A second reception room, the back parlour, was the one used most for family recreation. In well-to-do families it was known as the morning room, where breakfast was taken, as there would be a separate dining room and study or smoking room. Despite marked differences in the way these adjoining rooms were used, the front and rear parlours were frequently separated only by a set of double doors.

Space was also defined according to sex, particularly in larger houses with the 'feminine' drawing room and 'masculine' dining room and study. But the most basic distinction was between 'staff' and 'family' with strict demarcation lines to separate service areas from the rest of the house.

A desire for better hygiene and the realisation that damp basements were not conducive to good health saw the phasing out of traditional basements. From the mid century, the earlier 'upstairs/downstairs' division with basement kitchens gradually evolved to 'front/back', the rear of the property being the domain

**Above left:** Knowing your place! Status mattered
**Above right:** Imposing stone & brick Gothic semi in Cardiff
**Below:** Classic mid Victorian stucco town houses with semi-basements, accessed by separate front steps (with coal store below main steps)

**Above:** Plain but well-built workers' cottages (note lack of bays)
**Below:** More expensive houses had attics for live-in servants and nurseries

of servants in all but the poorest houses. Here the scullery was used for doing the laundry, preparing vegetables, and washing up after meals, with a separate kitchen for cooking.

Upstairs, two good sized bedrooms would be sufficient to accommodate even quite large families (each child having their own room is a relatively recent concept). The majority of small houses were built with just two bedrooms until quite late in the century, by which time a third bedroom above the rear addition had become the norm, many having since been converted to bathrooms. Because kitchen floors and ceiling heights were lower to the rear of the property, the layout had to be 'split-level', with steps down into the back rooms.

A larger four- or five-bedroom house from the mid Victorian period would typically have a small side bathroom off the landing with the back bedrooms for the servants (over the kitchen and away from the family bedrooms), with additional servants' bedrooms and a nursery sometimes built into the attic.

Another measure of 'plain or posh' was where you stored your coal. Kitchen ranges and coal fires were the only source of heat in the house, consuming staggering quantities of fuel. In larger townhouses with external front steps the space underneath would be used for coal storage, sometimes extending into dank cellars under the pavement. Coal was delivered into a chute connected to the cellar via a circular hole in the front garden path or pavement, or via a cast-iron plate covering a front step. Smaller houses without cellars had a coal bin in the back yard, or a cupboard under the stairs would suffice as a 'coal-hole'.

# Victorian builders

The inner suburbs of just about every major town and city in Britain were constructed during a wave of 19th century building booms, the biggest taking place in the 1880s. With no real planning restrictions, farmland around the edge of a growing town would become ripe for development once it was more valuable as a building plot than for agriculture. The landlord would then divide it up into leasehold parcels for sale to local builders. Unlike today's giant corporate housebuilding PLCs, Victorian builders were mostly individual tradesmen, ambitious joiners or bricklayers who could borrow sufficient capital to buy a building lease and pay for materials and labour. To generate enough profit to fund the next project and repay the loan, they had to complete the new houses as quickly as possible and sell them off to investors who in turn would rent them out. This meant that streets tended to be built a handful of houses at a time, and it's often possible to detect subtle breaks in terraces where one phase ended and another began as new funds were raised.

But building in this way was risky. Narrow profit margins combined with unpredictable markets meant that bankruptcy was common. This discouraged risk-taking, with most builders sticking to well-tried formats, relying on established designs in pattern books without any professional architectural input.

Only on the larger and more prestigious estates would the landowner lay down rules about the size and appearance of the houses to be built on his land. This was particularly common where contracts provided for the land and the houses erected on it to revert to the landowner upon the expiry of the lease.

This was also the era when charitable organisations (eg The Guinness Trust) became involved in housing 'working class people of good character'. Trusts erected solidly built 'no frills' blocks of flats. Some more enlightened employers also provided workers' housing, and around the turn of the century some of the earliest Council housing appeared.

## Location location

In stark contrast to the present day, the concept of home ownership was irrelevant to the Victorians – nearly everyone rented their homes, even the aristocratic occupants of London mansions. This meant that builders went to great lengths to ensure their developments would appeal to 'respectable' middle-class tenants. As we shall see, this directly affected the type of features that were incorporated into houses, such as bay windows, front and rear reception rooms, and decorative joinery and cornicing helping to boost 'snob value'. To convey an impression of respectability (and thereby generate higher rents), the first thing many larger developers erected on a new estate was a church. Paradoxically, the second edifice appearing on a new estate was often a pub, as a way of generating additional income by taking the first bite of workers' pay packets.

Once established, the housing market was swift to react to the subtlest changes in the social standing of an area. So it was a developer's nightmare that the seeds of a slum would take root, blighting the chances of earning a good return on their investment. To pre-empt this, Victorian house deeds commonly contained restrictive covenants listing prohibited 'dishonourable' trades that could lower the tone of the area, such as selling liquor, bone-boiling, manure-making and the training of fighting dogs.

Historically, the 'posher' parts of town tended to be located on the west side, and the reason for this was well known to Victorian builders; thanks to the prevailing wind direction in much of Britain, smoke and smells from industry generally blow eastward. And with every house burning huge amounts coal this was an important smog-inducing factor even in non-industrial towns. In London, matters were compounded by the eastward flow of the giant sewer that the Victorian Thames had become.

Slum areas were more common where the inhabitants lived cheek by jowl with industry, which by nature was often smelly and noisy. It therefore became a mark of respectability to live as far away from your work as possible, and as transport improved it allowed the middle classes to live further out in new suburbs on the edge of town.

# The four most common layouts

The vast majority of Victorian and Edwardian houses were designed with one of the following ground floor layouts – or a very similar variation. Terraced houses were especially consistent in their basic plan, being relatively narrow in width, with a frontage often of no more than 4 metres. Wide 'double-fronted' houses, with the front door in the middle and rooms either side of the hall, were the exception, being built on plots where there was not enough depth for the regular 'three-room-deep' design.

The classic 'deep corridor' hallway was a function of the constraints imposed by the limited width, and also reflected the social desirability of having both a front and a rear parlour. A separate entrance hall where guests could be greeted before being shown into the front drawing room mimicked the grander homes of the well-to-do. In terraces with adjoining front entrances the hallways helped insulate the reception rooms from noise in the neighbouring property.

Houses were typically two or three storeys high, although some urban townhouses extended another floor or two.

Upstairs, the layout of the bedrooms echoed the pattern of the ground floor, with a larger 'master bedroom' usually spanning the full width of the house over the entrance hall. More expensive homes might have a secondary staircase to the rear for servants. In earlier townhouses with basements (or semi-basements) the reception rooms would be on the elevated main entrance floor, but in most houses built after about 1870 they occupied the ground floor.

a) The 'classic' layout found in more expensive houses, as well as in many smaller suburban villas, would typically start with the front door opening into the main entrance hall, via a small recessed porch or 'vestibule'. The front parlour ('drawing room') was entered via a doorway from the hall. Directly to the rear there would be a back parlour (often linked via double doors) with its own independent access from the hall. The hallway would continue past the staircase to the rear addition, with a step down to a lower level, both in terms of floor height and status. In larger houses this might contain a third reception 'breakfast room', but more commonly would comprise the kitchen area, divided into a kitchen proper for cooking and a scullery for washing-up and cleaning. The better-off might have a pantry (a large store room/larder) for household goods and cutlery. A typical rear addition would terminate with a small single storey lean-to that might incorporate the scullery and various storage areas, plus an outside toilet.

b) Some less expensive houses used a variation on the 'classic' theme, retaining the socially important separate entrance hall and front parlour. But space would be saved, and costs reduced, by combining the rear hall and back parlour leading to the kitchen with the scullery located in a smaller rear addition. Access to the outdoor toilet was via the garden.

c) The popular small 'cottage' design often dispensed entirely with the separate entrance hall. Here, the front door would open straight into the front parlour, or in end terrace or semi-detached houses from the side to a small stair lobby. To the rear there is a back parlour leading to the kitchen and scullery, with an outside 'privy'. In cheaper two-bedroom houses the kitchen replaced the rear parlour, the scullery occupying a small single story lean-to addition.

Staircases were sometimes positioned alongside the party wall in the rear parlour, but in the majority of small houses the stairs formed a space-saving partition between the front and back rooms, running from side to side in the middle of the house. The main drawback with this design could be seen upstairs, with the small third bedroom to the rear accessed through the larger bedroom. However, where space permitted, a full length landing might be incorporated.

d) The later 'Edwardian square' design had a more modern 20th-century layout, although open-plan living was still a long way off. A wider main entrance hall led to the kitchen, which was now tucked into the main house, or within a less-cramped, broader rear addition. For the first time there was no separate scullery, and indoor WCs and first-floor bathrooms off the landing were commonly installed.

# The Victorian terrace

Of the five million plus Victorian houses that were built, most were terraced, and the vast majority of these are still standing. The popularity of the terraced house design with Victorian builders was simply down to the fact that it made the optimum economic use of land, with guaranteed big savings on bricks and labour. Furthermore, it was a strong structure with each house supported by its neighbour. But that doesn't mean that all Victorian terraces were well built. By now the worst examples have either fallen down, or been demolished or rebuilt.

Even cheaper to build were primitive back-to-back houses (the original 'cluster homes'). Although increasingly prohibited by law after the 1860s due to unsanitary conditions and poor ventilation, they continued to be built in many large industrial towns well into the 1890s. Other variations on the terraced theme included 'cottage flats', popular in Tyneside and East London, and well-appointed Edwardian maisonettes, which look almost identical to terraced houses except for the telltale twin front doors.

**Top:** Cheap back-to-back housing with courtyard
**Above:** Classic mid Victorian terrace
**Below:** Substantial 4 storey townhouses c 1870s

# House history: Architectural clues

The Victorian house managed to be distinctive and instantly recognisable, yet at the same time enormously varied and eclectic. Loosely based on a classical Georgian template, Victorian architecture was variously influenced by Italianate and Gothic styles, followed later in the century by fashionable 'Old English' Queen Anne revival and 'cottagey' Arts & Crafts designs. Fantasy-like 'Scottish baronial' features found on many grander houses of the period – corner turrets, stepped gables and castellated 'battlements' – were adopted to embellish humble suburban villas. Take a closer look and you may notice some quite bizarre combinations: windows with classic Roman or Greek columns, Medieval cathedral-style roofs, Tudor chimneys, Dutch gables, Swiss bargeboards and folksy front porches.

Around the turn of the century two conflicting fashions could be seen. In some more affluent architect-designed homes there was a trend towards increased simplicity with traditional 'aesthetic' architecture. Meanwhile, automation was bringing down the price of a wide range of decorative manufactured products such as fancy woodwork and ornate cast iron for mainstream housing.

The immense variety of styles can make 'house-spotting' a tricky business – for instance, some features on a more fashionable property of the 1870s may still be seen on conservative houses built in the early 1900s. There are plenty of books that explore architectural history extensively, but here are some key design features that should help identify the approximate age of your house:

### EARLY VICTORIAN

Georgian influence is still evident in with elegant, symmetrical terraces with slate-clad shallow pitched roofs hidden from the street behind low parapet walls and smooth well-proportioned facades with symmetrically arranged windows and doors. Plain brickwork walls were commonly rendered with 'stucco', but with some classical 'Italian villa' features, such as string courses, arches and cornerstones. Some more expensive houses have well-proportioned classical facades made from smooth ashlar stonework. Sash windows still have multiple panes. The standard plan comprises two main rooms over two or three

floors. Traditional basement kitchens and first-floor living rooms remained in larger town houses, with coal cellars and stores extending under the front pavement. Imposing flights of stone steps led to front doors, with separate steep steps providing access for servants.

### MID-VICTORIAN

Italianate classical styles compete with increasing Gothic influences. Traditional slated pitched roofs with eaves supersede

hidden parapet roofs – usually sporting ornamental features (finials, crestings and decorative ridge tiles). Facades of exposed brick or stonework start to eclipse big fully stuccoed frontages. Facades were becoming important as an indicator of social status, featuring contrasting colours of polychrome brickwork, with string course bands and arches of mainly red or yellow brick. Doors and window surrounds became more ornamental.

Manufactured stone lintels and sills featured along with sash windows with just one or two larger panes. One of the most popular ways of ornamenting the facade was to add a bay window. Formerly associated with seaside towns, by the mid 1860s polygonal 'splayed' or 'canted' bay windows were almost ubiquitous. From the 1870s two-storey bay windows were common with sashes grouped in threes – two small side windows and a larger central one. Later, roof gables and porches started to feature elaborate bargeboards.

Town houses with semi-basements were still common by 1870, with large flights of steps up to the front door. The standard layout evolved into the classic three-rooms-over-two-floors format. Basements had largely disappeared by the 1880s, superseded by deeper layouts with long, corridor hallways leading to rear addition kitchens and sculleries. Single-storey rear additions sprouted another floor (or two) occupied by one or more bedrooms. More expensive houses might incorporate an internal bathroom and WC, but mainstream housing had an outdoor 'privy'.

## LATE VICTORIAN AND EDWARDIAN
Layouts become wider with squarer hallways while rear additions were less deep or phased out altogether. Roofs were built to a steeper pitch and clad with fashionable manufactured clay tiles.

**Above:** Late Victorian mansion block c 1880–1900

Queen Anne revival was a major influence, impacting on suburban house design from the 1890s: walls of red brick with white stone dressings, or plain brick walls contrasted with white painted joinery; fine cut-and-rubbed brickwork and gables sometimes of curved Dutch or crow stepped designs.

Later Arts & Crafts influence saw white painted roughcast, half-timbering, tile-hanging and pebbledash, large overhanging gables over stout square bay windows. Upper sashes divided into multiple panes and broader casement windows started to supersede sashes, some of cast iron with traditional leaded lights. Edwardian features include elaborate fretwork timber porches, coloured glass in front doors and mock Tudor half-timbering to gables. Bathrooms and WCs were indoors, except in the poorest housing.

**Left:** Contrasting styles: the shape of things to come c.1910 (top) compared with lavish Edwardian exuberance of the same era (below).

## HINTS & TIPS

To discover more about the history of your house, and to see who lived there, and in your street, over a century ago, try the 1901 and 1911 census link at www.victorian-house.co.uk. Also, old Ordnance Survey maps are widely available, showing the rapid urbanisation of the time. Parish records in local libraries are worth delving into, and original house deeds can reveal all sorts of interesting facts about previous owners and prohibited domestic activities.

# Style wars: Gothic vs Classicism

In the mid-Victorian period there were two powerful architectural forces at work. Blood ran hot on both sides – you were either a 'Goth' or an Italianate neo-classicist. Classical Georgian architecture provided the basic template from which Victorian house designs evolved. But in contrast to the Georgian emphasis on symmetry and the unity of whole terraces, the Victorians were more concerned with promoting individuality, breaking up terraced frontages with prominent bays and porches. But the inspiration for devotees of the neo-classical Italianate style was Queen Victoria's Osborne House (below), completed in 1851, which was strongly influenced by the architecture of medieval Italian villas. In contrast, Gothic inspiration was derived from medieval cathedrals with 'pointed arches', the style manifesting itself in the new Houses of Parliament (completed in 1860). Inevitably, speculative builders pinched ideas from both camps, and by the 1850s elements of Italianate and Gothic had crept into many street fronts. Here are some of the chief characteristics:

## Italianate features
- Shallow-pitched roofs, often hipped, with overhanging eaves supported on brackets
- Showy polychrome brickwork with bands and arches of coloured brick (chiefly red and yellow)
- Brickwork set off by stuccoed quoins, carved capitols to bays (often in mass-produced artificial stone), and featuring string courses delineating each storey
- Rounded 'Romanesque' arches typically set above pairs of double arcaded windows (but gently curved 'segmented' arches were more common)
- Grander houses might display a tower or 'campanile' emulating Osborne House

**Left:** Gothic vicarage c1860 with stone porch

## Gothic features

- Prominent pointed arches to windows and door openings

- Prominent pointed gables adorned with large, conspicuous carved or moulded bargeboards

- Steeper-pitched slate roofs, some sprouting 'church belfry' towers

- Ecclesiastical pointed apex roofs to front porches

- A preference for brickwork in 'English bond' but also contrasting polychrome colours

- Gothic capitol columns with carved foliage to bays and porches

- Coloured glass in doors and windows evocative of churches

- Rib-vaulted cellars

# So what was new?

No other period in British house building has witnessed a greater variety of materials with an enormous range of bricks, stone, terracotta, tiles, decorative timber and iron available to local builders. The contrasting colours and textures of these different materials were brought alive by skilled workmanship.

Technical improvements in mass housing included the provision of suspended timber ground floors, damp proof courses (from 1875), and the development of effective plumbing. Larger windows, improved ventilation and higher ceilings helped create a lighter, more airy indoor environment. The widespread use of architectural pattern books helped perpetuate tried-and-tested designs, while new legislation drove improvements in building standards. This combination of tradition and progress may explain why so many Victorian and Edwardian houses have not only survived but are generally much sought-after in the 21st century.

# Building control and legislation

The Victorians may not have invented jerry-building, but 19th century builders had plenty of tricks up their sleeves when it came to cutting corners and economising on materials. As the century progressed, various local Building Acts and By-laws were introduced, with fines for non-compliance, and new construction was increasingly inspected by official District Surveyors (although negligence and corruption was not unknown). Plans for new streets and drainage had to be submitted to Local Authorities prior to work commencing, and rudimentary local planning laws were introduced to control the height of buildings in relation to street width, the space at the back of dwellings and the building line. Minimum distances between dwellings were stipulated and even the front garden was controlled: gates had to open inward, and fences could not exceed a certain height.

Towards the end of the century, large 'garden suburb' developments such as Bedford Park (pictured below) in Chiswick (from 1875), Port Sunlight on Merseyside (1888) and Bournville in Birmingham (1895), followed by Letchworth (1903) and Hampstead Garden Suburb (1907) pioneered radical improvements in suburban house design and town planning.

But the area where greatest progress was achieved during the Victorian

era was in domestic sanitation. The motivation of the Health of Towns Act of 1848 was to control contagious disease, with the realisation that it did not respect class barriers – it was only the foul stench of the sewage-infested River Thames that finally alerted Parliament to the dangers of infection caused by poor drainage. From the mid century, there was rapid development in domestic piped water supplies, flushing toilets and latterly public sewers, spurred on by further public health legislation in 1858. This was followed by the 1875 Public Health Act, which made all local authorities responsible for providing mains sewers and collecting refuse, transforming sanitary conditions in major towns and cities.

# Materials

Traditionally, many building materials (such as bricks and lime) were produced on or near the site.

But from the mid-Victorian period onwards, modern methods were applied to the manufacture of key building materials such as brick, iron, tiles and glass, whilst steam power facilitated the excavation and processing of natural resources such as slate, timber and stone. The spread of the railways in the 1850s made it possible to deliver manufactured materials at affordable prices across the country. So builders increasingly had access to cheaper, standardised materials, resulting in more consistent standards of construction. However, the perpetuation of local skills and traditions made this a uniquely rich and varied period – with the best of both worlds.

## BRICKS

Bricks were traditionally made from clay dug out of the ground for the foundations, or close to the building site. So the first stage in a new development was often to erect a temporary kiln or 'clamp' for baking the clay, some of which were large multi-storey affairs. This was often done by gangs of specialist brick-makers travelling from site to site.

But making bricks by hand from clays of variable quality using makeshift kilns made it impossible to achieve a consistent, reliable product, with the size, evenness and quality of bricks varying considerably. So in the earlier years of the 19th century, brick was regarded as an inferior material to be avoided by those who could afford to build in stone, or else hidden behind a coat of stucco. And of course in many parts of the country there was no readily available clay subsoil.

From the mid-Victorian period, the new technology of coal-

fired kilns (eg the Hoffman kiln) and the introduction of special presses allowed bricks to be manufactured continuously along a primitive production line. The popularity of brick was helped by the removal of excise duty in 1850, which reduced prices. Brick-making

machines were in widespread use by the 1870s with consequent improvements in evenness, colour and appearance. Manufactured bricks only became widely available by the later Victorian period, with the advent of the railways. Mass-production offered the tremendous advantages of being both cheap and of a predictable quality.

The best local 'facing bricks' (sometimes called 'stocks') were used for the visually prominent front elevations facing the street. But money could be saved by employing lower quality 'common bricks' for less noticeable side and rear walls. Cheapest of all were the

underburned 'rejects' known as 'place bricks'. These were uneven and soft and widely used for internal partition walls; however, it was not unusual to find them also in load-bearing party walls – in fact anywhere they could be

covered over with plaster and swiftly forgotten (including the load-bearing inner faces of some main walls).

After about 1900 low-priced, mass-produced (but relatively soft) commons, such as 'Flettons' from the East Midlands, and blotchy pink 'Cheshires', were increasingly used for internal walls and less-visible rendered walls.

## Brick sizes

Standardisation of brick sizes is a relatively modern concept. Despite enormous progress in manufacturing, much of production served local markets and brick sizes remained fairly inconsistent until the 20th century. Today trying to precisely match original brickwork can be challenging for this reason and also because it can take considerable skill to replicate lime mortar joints, which were generally thinner than today. The nearest to a standard size was probably 9in long x 4in wide x 2.5in high (229mm x 100mm x 60mm), or the larger imperial 9in x 4½in x 3in (229 x 114 x 76mm), compared to today's standard metric size of 215 x 102.5 x 65mm.

## STONE

In many areas, stone was the traditional building material. Smooth dressed ashlar stonework was reserved for the most expensive houses. Cheaper squared rubble blocks or rough rubble stone was common in stone districts such as Cardiff, Bristol, York, Harrogate and much of Scotland.

But as manufactured brick became more widely available it started to replace stone as the main building material, being both cheaper and more practical to use. However, new technology was also applied to quarrying and stonecutting so that buying ready-cut sills, lintels and copings became much cheaper than building them on site, and hence suburban builders incorporated large

amounts of pre-cut stone into speculative housing. So even in dedicated brick regions, stone was widely used for decorative door and window surrounds, piers of bays, corner quoins, carved cornices and balcony railings. The appeal of certain types of stone was also driven by snob value, because smooth dressed ashlar was the aristocrat of building materials used in expensive mansions. Hence incorporating decorative stone detailing and dressed stone components added a certain cachet to houses lower down the social scale. Smooth dressed ashlar blocks could be employed in small amounts as posh looking quoins to contrast with brickwork. By the close of

Scottish Baronial style

work than sandstone and was therefore more widely used for decorative facade features, such as elaborately carved Gothic porches. Garlands of flowers can often be seen carved on window lintels, or strawberry leaf capitals on porch pillars.

However, decorative components such as ornate columns to bay windows were increasingly made from artificial stone, moulded from cement mixed with stone or brick dust, the mixture pressed into moulds to cast an imitation of expensive carved or dressed stone.

## MORTAR

Victorian builders traditionally used mortars made with lime mixed with sand which were made up on site. Sometimes ash was added to the mix for decorative effect (and was a cheap material for bulking up the mix). By the late 19th century cement was slowly becoming more widely used, often with a little lime in the mix to improve elasticity. It's relative strength and quick setting properties made it particularly useful for jointing pipes, or mixed with 'coke breeze' (waste from blast furnaces) to make concrete for foundations, as well as in the manufacture of some early reinforced concrete lintels. But lime was not generally superseded by cement-based mortars until well into the 20th century.

## RENDER

The Georgian love of stucco (a hard external plaster) persisted into the early Victorian period. Stucco was a cheap way of disguising rough brickwork and could be decorated to look like expensive real stone, but it also had the practical purpose of helping resist rain penetration. Imposing white stuccoed front elevations with their mighty classical columned porches can be seen on numerous houses from the earlier Victorian period in areas such as Bayswater and Kensington in London, and ports and seaside resorts like Bristol and Brighton. Coastal towns in particular adopted stucco because of the extra protection it afforded from sea winds.

Traditionally based on lime putty and sand to which animal hair and other fibres were added for reinforcement, stucco could be built up around thin timber battens, which were later removed once dry to leave grooves or 'incised' lines to replicate blocks of stone. Stucco could also be scored to replicate the appearance of finely jointed ashlar, or moulded to form string courses and cornices.

From the 1860s mechanisation meant builders could obtain real stone, good-quality brick and fashionable terracotta, and

the century, the use of dressed stone components had become universal even for small houses.

But it wasn't all about image. The practical, hard-wearing nature of stone made it especially useful for heavy-duty parts of the building – hence it's not unusual to find lintels of yellow Bath limestone, window sills of Derbyshire Grit, and steps of hard-wearing brown York stone.

Limestone was easier to

## TYPES OF STONE

There were three main types of stone used for 19th century wall construction:

### Sandstone

Quarried extensively in central Scotland, Wales and much of northern, western and southern England, colours range from yellow or grey through to browns and reds. The stone's strength and colour are determined by the extent of local impurities, such as clays, iron and chalks. Many Glasgow tenements were built of an especially distinctive warm red stone. Hard-wearing 'York stone' was a very popular Yorkshire sandstone, widely used for steps and paving slabs.

### Limestone

The 'limestone belt' sweeps across the country in a broad arc from Dorset on the south coast to the Cleveland Hills of Yorkshire. Colours range from white through to honey yellows and browns. Well-known varieties include whitish grey Portland stone and warm, mellow Bath or Cotswold stone.

### Granite

The hard-wearing qualities of granite made it ideal for buildings in very exposed locations, famously being used to clad Tower Bridge, as well as for kerbs and paving. Aberdeen is well known as 'Granite City', but it was also quarried in the Highlands and south-west Scotland, as well as in Cumbria, Leicestershire, Devon and Cornwall.

once these 'real' materials became so cheap there was no longer any need to resort to fakery. However, that wasn't the end of the story. Stucco found a new less-prominent role in the form of smaller complementary architectural details such as ornamental columns, parapets and veranda balusters in place of stone. This appealed to middle-class professionals would not have bought houses made entirely from humble unadorned brick.

It also retained a discreet presence fronting some front door vestibules and bays, presumably as a way of adding a touch of nobility to otherwise plain designs (or concealing cheap brickwork).

After a period of being deeply unfashionable it was later revived as a finish for some grander late Victorian and Edwardian 'wedding cake' houses, hymns to the glory of stucco. The imitation of expensive painted stonework could be remarkably authentic.

Stucco was later superseded by more modern smooth cement/sand render mixes, commonly seen in fashionable 'Queen Anne' architecture contrasted with tile-hung walls, or complementing fine rubbed brickwork. Roughcast and pebbledash became enormously popular in later Arts & Crafts buildings remaining in vogue right through to the mid 20th century.

## SLATES AND TILES

Slate was by far the most common Victorian roof covering as it was relatively cheap and widely available. It also had a some appealing practical and aesthetic qualities, such as being fashionably grey, pleasingly smooth and non-absorbent. And because it was such a light material, it facilitated simpler, cheaper roof structures and could be used to clad roofs built to a relatively shallow pitch. Most popular was the thin, regular-sized Welsh slate, with its distinctive grey-blue, purple, or greenish hues. Thicker, rougher Cumbrian green- or blue-tinged slate was also widely used, as were Highland, West Scottish and Cornish varieties.

But slate didn't rule the roost in all parts of the country. Traditional large wavy pantiles were no dearer, and remained popular in some tile strongholds such as East Anglia. The demise of slate as the dominant roofing material came towards the end of the century as manufactured clay tiles started to become widely available. These were cheaper, lighter and thinner than traditional handmade tiles, and more fashionable than slate. 'Broseley' plain clay tiles from Shropshire became enormously popular. Other notable tile-making regions were Staffordshire, Berkshire, Somerset and Leicestershire.

On some better quality houses, tiles or slates were laid in decorative patterns, such as diamonds or fishtails. Cladding walls with tiles or cut slate was a traditional and economical method of providing enhanced protection in exposed coastal locations,

becoming very fashionable later in the century. By the 1880s no high-class suburban development was without its tile-hung gables or bays.

## TIMBER

Victorian buildings were constructed using high-quality, well-seasoned wood that has lasted considerably longer than the timber in many modern houses. Structural timber was commonly obtained from softwood species such as Baltic fir, whitewood spruce, and Scandinavian yellow pine or red deal. Mechanisation in timber yards meant that sizes of joists and rafters became fairly standardised. Items like sash windows were mass-produced ready for fitting. Hardwoods like oak and mahogany were used for joinery in more expensive homes.

Decorative timberwork flourished – some areas like Hull and Sunderland excelled in exuberantly designed woodwork, and bargeboarded decorative gables later became a prized architectural feature. Later 'Queen Anne' designs introduced decorative 'half-timbering' to resemble traditional timber-frame houses.

## GLASS

Glass had previously been a luxury material, but excise duty and window tax were finally abolished by 1851, just as new manufacturing techniques were making sheet glass widely available. Health legislation now stipulated minimum sizes of windows and the sash window became almost ubiquitous. The 12-pane Georgian window was out of fashion – four- or two-paned vertical sliding sash windows were now the style, with rolled plated glass, or, in some later houses, more durable sheet glass facilitated single-paned sashes. From the 1860s, decorative acid-etched patterned glass became widely used in more expensive homes, for doors and bathroom windows, superseded a decade or so later by cheaper sandblasting methods used to create a frosted appearance.

However, just as technological progress was making larger expanses of sheet glass affordable for mass housing, fashions changed. Traditional glazing bars, smaller panes and leaded lights made a re-appearance in 'Queen Anne' style houses. Towards the turn of the century, small vividly coloured blue or red panes became popular in fanlights and windows adjoining doors.

## TERRACOTTA

Terracotta was an important Victorian substitute for brick or stone, and was manufactured in a similar process to brickmaking, but fired from higher quality fireclays mixed with fine sand.

It was mainly produced in distinctive reddish-brown and sometimes buff colours (although after 1875 a wider range of colours became available such as pinks and greys).

Three varieties were manufactured – *ordinary* with a plain, unglazed surface (which is slightly porous), *vitreous* which is lightly glazed and impervious to moisture, and full-glazed *faience* with a smooth shiny surface, mainly used internally (although it was a popular cladding for the exterior of Victorian pubs in prominent corner locations, resplendent in colourful bright green, turquoise or maroon glazes).

As well as providing architectural decoration, glazed terracotta offered the practical advantage of being fireproof with a hard skin that resisted damp and pollution. Being washable, the impervious finish made it an ideal facing material for buildings in sooty industrial towns where stonework would quickly become tarnished.

Terracotta became very fashionable from the 1860s in more expensive homes, but decorative ridge tiles, finials and crestings were applied in enormous numbers to enhance rooflines on all types of houses. A diverse range of architectural ceramics could be ordered from catalogues including whole doorcase structures and mullioned window frames. Later in the century, patterned terracotta tiles with sunflower motifs, or moulded with dates or coats of arms, were commonly embedded in brickwork above front doors and bay windows.

## IRON

Early Victorian traditional wrought iron was made by hand, heating up the raw material and hammering it into shape, often with incredibly fine detailing. Manufactured cast iron was formed from the molten ore cast into moulded components that could be bolted together. Mass production inevitably meant that cast-iron products were considerably cheaper.

Structural ironwork is rarely found in Victorian houses (being largely reserved for industrial, commercial or grand public buildings like stations). However, one area where iron

components play a structural role is where expensive dressed stone was secured in the traditional manner using embedded iron cramps. The two main areas where iron was widely used in mainstream housing were for cast-iron rainwater fittings and for railings to gardens and balconies. Large conservatories of glass and iron were the preserve of the well-to-do. But cast iron was also widely used for making innumerable small components, such as drain covers, pipes, doorknockers, letterboxes, bootscrapers, window-box rails and iron finials used to enliven many a roof ridge.

## DECORATION

Think original Victorian house. Think glossy black railings, creamy white stucco, and white window frames. It may look that way in black-and-white photographs, but research tells us that railings were often painted a bright bluish green. The new Palace of Westminster (built between 1840 and 1870) had window frames specified in bronze. So it became fashionable to mimic this by painting residential window frames the colour of expensive patinated bronze. Alternatively they could be grained to simulate oak, the most expensive wood. Maroon and brown were other popular colours for external woodwork.

Victorian stucco would have been painted to match local stone, *eg* the honey colour of Bath stone or the creamy white of London Portland stone, perhaps with a contrasting colour in the joints to represent mortar. The patchiness of the limewash finish helped add to the illusion.

# Regional variations

One of the great delights of Victorian houses is the local variation in materials. Despite increasing standardisation, local skills and traditions with materials like stucco and tiling, finely carved joinery and preferences for elaborate ironwork or terracotta all made for tremendous diversity. But above all it was the gloriously colourful regional variations in brick and stone that are so striking by today's uniform standards. Almost every town had its local brickworks and there were distinct contrasts in colours between different towns and regions. The precise colour and texture of the locally produced bricks was dependent on the method of burning, and differences in minerals and impurities in the local clay (such as chalk, sand, flint or iron) plus special ingredients added to the recipe such as salt or ashes.

The predominance of iron in most soil types accounts for the disproportionate weighting towards 'natural' reds. These dominated in the north, with bricks such as 'Accrington bloods' and Leicestershire reds, but elsewhere there was a riot of colour; 'whites' from East Anglia (today often a sooty grey but whitish yellow when cleaned); from Cambridgeshire and Suffolk came hard chalky white Gaults, with red and yellow Bedfords, the Luton purple, Wrexham's red ruabons, flint silver greys from Reading and Sussex, reds and whites from Maidstone, West Country reds and yellows, and of course London's famous yellow stocks made from clay rich in magnesium and iron mixed with waste ash from coal fires. Staffordshire blues ('engineering bricks') were actually a deep purple-black colour on account of their high iron content; because of their strength and density they were used in footings and damp-proof courses, for copings on garden walls, as well as enlivening walls with decorative bands of colour.

Brickwork combining different coloured bricks in intricate 'polychrome' patterns was soon being used for all kinds of housing, but it was generally reserved for more visible front elevations, the rear of the houses commonly being built of plain brickwork. A traditional decorative technique, common in the South Midlands and Oxfordshire, was to create a chequerboard pattern of reds with contrasting yellow, white or black-burnt headers.

To enhance the appearance of a wall, local bricks could also be contrasted with pressed smooth facing bricks. Clay could be moulded into shapes, such as 'bullnose' bricks with rounded shoulders, used for wall corners and around window openings. Moulded ornamental bricks, in patterns such as nailheads and rosettes were widely used to enliven house facades in areas such as Liverpool and Manchester. Special red bricks called 'red rubbers' that were soft enough to be rubbed into tapered wedge shapes ('voussoirs') were widely used to form traditional decorative arches. Moulded, carved and rubbed bricks laid with very fine joints of pure lime putty were very much in vogue in 'Queen Anne' style houses built in the 1870s and 1880s.

Sounding rather like a selection of fine cheeses, many regional varieties of sandstone, from quarries long gone, can still be widely seen, among them Runcorn Red (Cheshire), Darley Dale (Derbyshire), Craigleith (Edinburgh), Bolton Wood, Millstone Grit (Yorkshire) and Pennant sandstone (Bristol). Sandstone was used extensively to build Victorian millworkers houses in Lancashire and Yorkshire.

Limestone areas of distinction included the famous honey-yellow Bath and Cotswolds stone, White Portland (Dorset), and Hopton Wood and Bolsover Moor (Derbyshire). Other 19th century favourites included Kentish Rag, White Magnesian (Nottinghamshire) and Ketton (Leicestershire).

Toughest of all, granite was widely used in north-east Scotland and Cornwall, while some houses in chalk districts, such as parts of Sussex and Hampshire, continued the tradition of flint work as a panel infill between bricks.

So much for history and architecture. The object of this book is to assess how these buildings are faring well over a century after their construction. So let's now take a look at each part of building in turn to see what commonly goes wrong, what causes common defects and how to fix them.

Luton purple and red

Reading silver and cream

Checkerboard pattern

Edinburgh sandstone

Cotswold limestone

# ROOFS

The crowning glory of many a Victorian house was its skillfully decorated roof with delicately patterned ridge tiles and fancy crestings or finials – the icing on the cake. Original roof coverings may look nice, but this is about the last place you want to find serious defects. So here we explain all the component parts of traditional roofs of this era, identify the typical weak points, and summarise what is important and what isn't.

**Four in a row:** left to right – standard pitched, butterfly roof behind parapet, gabled front, hipped bay

To most of us, roofs can be a little scary. When you're standing high up on a ladder, or the wind is howling around the scaffolding, you tend to develop a certain respect for the job in hand. And having to withstand the full force of the weather means that roofs are more likely to develop faults than just about any other part of the property.

So it's not uncommon to find the occasional broken or missing slate or tile, the prompt replacement can be the vital 'stitch in time' that halts water penetration, thereby preventing damp and rot.

# Main and subsidiary roofs

The predominant form of Victorian roof was the traditional pitched variety of simple 'cut timber' construction, clad in natural slates or clay tiles. The structure follows the layout of the house, which in a classic 'three-box' design would comprise:

**MAIN ROOF:** Over the main bedrooms and landing, the roof slopes meet centrally at the ridge. At the base of each slope the

rafters rest on the main walls. On terraces, the roofs of each house are sometimes separated from those of neighbouring properties by the party wall being built up into a parapet, but most just have a continuous roofline. To the front there is often a bay window with its own small roof. Some earlier properties instead had 'butterfly roofs' concealed behind a front parapet wall comprising a pair of small lean-to roofs resting on the party walls either side – see below.

**REAR ROOF:** Above the small back room in the back addition there is often a separate roof abutting the main house at the rear wall or roof (usually at right angles). Most are semi-detached, combined with next-door's rear addition, with the ridge located over the party wall. Roofs are often gabled, sometimes hipped or may simply comprise a single lean-to roof slope. Others take the form of stand-alone structures with flat roofs.

**LEAN-TO ROOF:** At the back of the house you may find a small, single-storey outside WC or store area with a primitive lean-to roof propped up against the back wall.

The major weak points are at the junctions, where one roof meets another or meets an adjacent wall or chimney stack.

Classic hipped slate roof

Main roofs and rear additions (re-clad with concrete tiles)

Rafter feet and wall plate structure visible

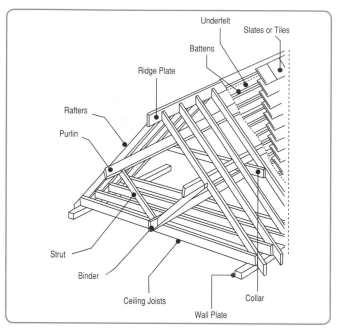

# Roof structure

As everyone knows, a basic roof structure takes the form of a triangle. The front and rear rafters meet at the top on the ridge, and the triangle is coupled together at its base by ceiling joists acting as 'collars'.

In most houses the roof timbers were made from slow-grown, naturally seasoned softwood, such as Baltic pine, which was of considerably better quality and more resistant to decay than much of the building timber available today. In some early Victorian roofs you may see unplaned 'rough hewn' rafters, still looking remarkably like tree branches.

Access to the roof space in an unmodernised house may not be possible without cutting a new opening, in which case you probably have an original loft unseen for well over a century! If there is a loft hatch, it normally opens into the large main roof, usually with an adjoining small roof space over the rear bedroom or bathroom visible (but not easily accessible!).

## RAFTERS

The roof slopes are constructed of timber rafters, which, depending on the span of the roof, are typically sized about 100 x 50mm and spaced at 400mm centres. The rafters meet at the top of the roof at a horizontal timber 'ridge plate' running along the apex. In terraces this typically runs from side to side resting on the party walls (or side walls usually at the pinnacle of a gable end). At their base, the rafter feet sit on a 'wall plate' (a wooden beam running along the top of the wall). To stop the rafters pushing outwards, the ceiling joists form a strong and secure 'triangle', which relies on adequate nailing at the feet of the rafters with the wall plate and the ceiling joists.

## PURLINS AND STRUTS

The rafters are usually supported by a large timber beam known as a purlin, running underneath, typically measuring about 200 x 75mm, depending on the span. Larger, steeper roofs sometimes have two purlins, but on lightweight slate roofs with shorter rafter spans there may be no need for any additional support. At each side the purlins normally meet the firebreak party wall (in a terrace) or the side walls, where they are either embedded in the masonry or rest on protruding bricks called corbels. For all but the smallest houses, each purlin is supported in turn by a central strut, with the loading ultimately taken by a bedroom wall, usually via horizontal timber ceiling 'binders' that spread the load (binders may also distribute the load to the side walls).

**Right:** Weak and weedy – early Victorian roof with branch-like 'rough hewn' timbers – recently underfelted

**Below:** Purlins normally need support from struts (slight bowing evident here)

Looking back – view from main roof to small rear addition loft. Roofs were not originally lined with underlay

Queenposts

Mid-Victorian Mansard roofs

## HANGERS AND COLLARS

Although not so common on simple Victorian roofs, you may find the roof structure has been strengthened with additional timbers known as hangers or collars. Hangers are central vertical timber posts below the ridge (traditionally known as 'Kingposts'). Collars are horizontal timber beams joining the two main roof slopes, and are more structurally effective at lower level (hence the importance of ceiling joists, which act as collars); they are sometimes used as an alternative to struts to support the purlins.

For extra strength, bigger roofs in grander houses (where the roof space was designed for use as accommodation) sometimes have 'Queenpost' supports. Attic rooms were also facilitated by using the split-level Parisian mansard design, where the upper roof slopes are shallow pitched but the lower roof is at a much steeper angle.

## BATTENS

The slates or tiles were fixed to rough-sawn softwood battens running horizontally across the rafters. These typically measured 50 x 25mm or 38 x 25mm, spaced apart according to the size of the slates or tiles and the pitch of the roof.

A superior alternative was a layer of timber 'sarking' boards laid over the rafters, with slates fixed either directly to the boards or to additional battens on top.

**Above:** Stripped slates reveal battens – not generous overlap of slates

**Right:** Rafters lined with timber sarking boards under slates

## IRON FITTINGS

Iron straps, clamps, wedges and bolts are rare but were occasionally used to join timbers in larger roof structures to save on skilled carpentry. Cast and wrought iron is far more rust-resistant than modern steel, so if fittings have a little surface rust, it is not normally significant and best left well alone.

## FIREBREAK WALLS

In the roof space of terraced or semi-detached houses there should be original party walls of brick or stone. Usually one side will incorporate the front and rear chimney breasts which combine into the stack above. However, many terraced houses were built without firebreak walls – see 'Defects' below. If the party walls are of blockwork or plasterboard, then the house was built without them and has since been upgraded. Reinstatement of missing firebreak walls can also be beneficial from a structural perspective, providing additional support to the timber purlins that in turn strengthen the rafters that form the main roof slopes. Where the party walls project up above the roof (as required by the Building Acts in London, to prevent skimping on firewalls) they are typically capped with coping stones, hard engineering bricks, or have a course or two of 'creasing tiles' projecting out either side to disperse rainwater.

**Right:** Recent blockwork firebreak wall where original omitted

**Right:** Many party walls were originally built up above roof level forming parapets as stipulated by London byelaws. Interlocking concrete tiles were widely used for re cladding in the 1980s but are heavy and look inappropriate (even without the roof garden!)

New roof with breather membrane and high quality reclaimed Victorian slates

### UNDERFELT

Since the 1950s, roofs have been built with a layer of underfelt beneath the tiles. This provides a secondary line of defence against wind-driven rain, and typically comprises black, bitumenised, hessian-based 'sarking felt'. Victorian roofs didn't originally have underlay, so if yours does it probably means the roof has subsequently been reclad. So if you can see the underside of the slates or tiles in the loft then the roof is likely to be original. Whilst there is more risk of rainwater or snow seeping through in severe weather conditions, the inherent draughtiness of such lofts allowed moisture to swiftly evaporate.

Before the days of underfelt, some roofs were 'torched' underneath with 'hair mortar' for added weatherproofing. Original lime mortars are acceptable for this purpose as they allow moisture to evaporate, but modern cement or foam materials retain water and restrict ventilation, which can lead to timber decay in the battens. As noted earlier some roofs in more expensive later Victorian and Edwardian houses have timber boarding under the tiles, a standard method of construction in Scotland.

Sarking felt underlay should sag slightly between the rafters and be carried down into the guttering to discharge any water that may have been driven under the tiles. Older underlay was impermeable, which could trap moisture in the loft, and it can suffer from damp and mould and from being torn when installed.

Today's new generation of vapour-permeable 'breathable' underlays are a major improvement over old sarking felt. Modern 'breather membranes' prevent the ingress of rain but allow air from outside to pass through, ventilating the roof space. Rather like hi-tech mountaineering clothing they simultaneously allow any internal moisture to escape, thereby reducing the risk of condensation forming on cold undersides, and are therefore worth incorporating when re-roofing old buildings. However, the cost of stripping the slates and battens and fitting new breather membrane underlay is not normally warranted, unless complete recladding is required anyway.

### INSULATION AND VENTILATION

Old roofs had lots of air passing through them, which was a good thing – it helped maintain ventilation around the battens and roof timbers, reducing the risk of decay by allowing any moisture that got in to dry out again and evaporate. Problems can therefore occur where roofs have been reclad and underfelted, as they can suffer from condensation and damp unless ventilation elsewhere is improved.

Ventilation can be aided by clearing any insulation blocking air paths at the edges of the loft by the eaves or by fitting special vents at the ridge tiles, to roof slopes, or to gable ends.

Originally lofts were not insulated, but as the roof accounts for at least 25% of heat loss in uninsulated houses most will by now have had some loft insulation fitted, although probably to a lesser depth than is currently recommended. Loft insulation should be at least 270mm (11in) deep laid over the ceiling joists. Where large batts of loft insulation have instead been stuck between the roof rafters, there is a risk that they can stop air circulating and trap moisture. See chapter 15.

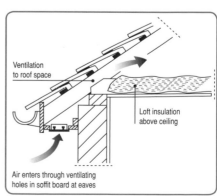

Ventilation to roof space

Loft insulation above ceiling

Air enters through ventilating holes in soffit board at eaves

## Avoiding quick fix botches

Some old roofs have suffered from modern 'quick fix' treatments that actually damage them further. For example, polyurethane spray foam is widely advertised as a miracle cure that can rejuvenate decrepit old roofs. In effect the foam glues the undersides of tiles or slates to the battens and rafters to stop them slipping. This is a classic false economy because any moisture in the roof timbers will then be sealed in, hastening decay. The foam also has the unfortunate effect of blocking essential ventilation to lofts, and in any case is often misapplied and gaps are left in areas where access is restricted. It is sometimes even applied to underfelt, which defeats the whole point of doing it anyway! Despite bogus claims in press advertisements, spray foam offers virtually zero insulation qualities. And to add insult to injury spraying old slates or tiles prevents them from being salvaged and re-used.

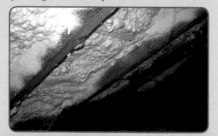

Another well-known quick fix botch is 'turnerising'. Here old roofs are painted with a thick layer of bituminous paint, sometimes applied over a layer

of mesh. Alternatively an unholy combination of roofing felt and tar is sometimes used. In time the coating becomes rock hard, making it impossible to later recycle the slates or tiles. Because of the cost of scaffolding and labour, turnerising can cost as much as half the price of a proper roof overhaul. A classic false economy that only lasts a few years. Surveyors will immediately 'downvalue' properties subjected to such disastrous botches.

# Early Victorian 'Butterfly' roofs

As we saw earlier, the Georgian fondness for concealing roofs behind elegant facades was perpetuated during the early Victorian period. From ground level the house can appear to have a flat roof, because the front wall is built up in the form of a parapet that hides its slopes. See chapter 4.

Looking at a terraced house from the front, the 'butterfly roof' (aka 'London' roof) is M-shaped, with its rafters running sideways to the street. The ridges are at the highest points on each side directly above the party walls, from where the rafters slope downwards to meet in the middle. Rainwater runs into the hidden central valley gutter, which normally feeds it to a hopper and downpipe at the back of the house.

Being out of sight, roofs of this type are generally forgotten about and neglected. Which is rather unfortunate, because unless they're well maintained butterfly roofs can

**Left:** Parapet gutter behind balustrade

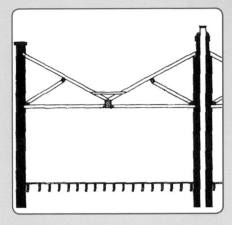

spell trouble. Valley gutters are always at risk of becoming blocked with leaves or debris, which causes water to build up until it seeps under the slates and into the roof space, wetting ceilings in the bedrooms below. Worse, the big timber beam (roof plate) underneath that supports the gutter can sag with age, especially if it is undersized, as Victorian timbers sometimes are. This causes water to pond in the middle of the gutter where, over time, frost expansion can cause cracks, allowing water to seep into the timbers below, which will eventually rot and collapse. This gutter beam can also fail where the main front or rear walls bulge out (eg if they're not tied in above bedroom ceiling level), or if settlement of the overloaded internal spine wall causes the rafters to slip. But the scariest defect that can develop with roofs of this type is where the entire parapet wall collapses – see boxout overleaf.

On the bright side, there may be scope for converting the loft by infilling the valley with a new mansard roof structure – subject to obtaining the necessary planning consents – overcoming the drawback of the small amount of loft space.

'Parapet' roofs are a similar beast, with the main front wall built up as a parapet, sometimes designed in the form of a classical balustrade. The difference with these roofs is that the main gutter runs directly behind the parapet. The roof itself is either a conventional pitched design or a mansard. Again, hidden gutters of all types tend to be 'out of sight, out of mind', which can lead to similar problems to butterfly roofs. See chapter 4.

**Left:** Classic early Victorian butterfly roof with central valley gutter running front-to-rear, concealed by front parapet

**Right:** Double pile farmhouse roof with central valley

## Danger! parapet collapse

In June 2012, the roof parapet wall to a row of four houses in Stockwell, South London collapsed, scattering masonry across the street and leaving the lofts exposed. Miraculously, no one was injured. The insurers initially refused to pay, citing 'lack of maintenance'. But the fact that one house belonged to the novelist Will Self and *Guardian* columnist Deborah Orr resulted in extensive publicity and the insurers eventually paid for remedial works.

### Cause

Parapet walls depend for their stability on being tied to the party walls at roof level between the properties, and at bedroom ceiling level. However, there are likely to be a combination of factors at work here:

■ **Poor build quality**. The fact that the parapet walls sheared away from the party walls to all four houses suggests they may have never been fully bonded, or tied in.

■ **Lack of maintenance**: Blocked valley gutters behind parapets can leak unseen for years, causing ingress of water and decay to bonding timbers. Flashings are also prone to leakage, This can be checked from within the loft spaces.

■ **Wind direction** – the predominant wind direction in the UK is from the SW, so open valleys can sometimes channel wind forces to exposed walls.

### Solutions

i  Fit stainless-steel wall ties or helical bars to tie the parapet walls to the party walls on both sides (note Party Wall Act). Fit steel ties at rafter level, with the ties run across at least four rafters, and similarly to ceiling joists. Consult Building Control and a structural engineer.

ii  Check the condition of valleys and flashings for watertightness and fit new leadwork as necessary. Replace any rotted timbers.

iii  Overhaul parapet masonry – eg the pointing and coping stones. Parapet walls are highly exposed, and over time the masonry can become soaked suffering spalling and erosion due to repeated freeze-thaw cycles.

# Roof coverings and detailing

Sagging slate roof needs strengthening

A typical Victorian roof has a covering either of natural slate or clay tiles. However, because the Victorian house was an economically built but finely balanced structure, roof timbers were used sparingly – no more than required – and are sometimes undersized by modern standards. Slight bowing is often evident where timbers have settled even under the weight of light slates, which is why serious problems occurred when many original slate roofs were replaced with large, heavy concrete tiles in the 1980s.

Different coverings have different recommended pitches (angles of slope). Slates and pantiles could be laid at shallower angles than small plain tiles, so the pitch of a slate roof would often be as low as 25°, compared to typically 40° or more for plain tiles.

Slates and tiles are laid in courses (rows), starting upwards from the eaves. The first row should consist of a special double course made with under-eaves slates or tiles, which are shorter than those used elsewhere. Each course is laid so that their tails 'lap' (cover) the heads of the row below, the vertical joints being staggered – like the bonding of brickwork – to protect the joints. The depth of the lap can be less for a steep roof (of say 50°), as the rain runs off quickly. On shallow roofs the rainwater runs slower and tends to 'fan out', getting under the edges, so a greater lap is needed. The area most at risk is lower down near the eaves, where the most water accumulates and where the pitch is sometimes slightly shallower in order to slow the rainwater for a safe landing in the gutter.

### NATURAL SLATES

Slate is a type of rock that splits along the grain to produce a highly durable roofing material and was by far the most common Victorian roof covering. Until later in the 19th century slate was both readily available and cheaper than tiles. It is lightweight, frost resistant, and extremely hard-wearing, with a lifespan normally well in excess of a hundred years.

# Roofing work

Surveyors sometimes have a tendency to be over-cautious and pass premature death sentences, condemning old roofs that may still have a good 10 or 20 years' life left in them. It's also not unknown for some less ethical roofers to succumb to the temptation to drum up business by scaring homeowners into unnecessary major works.

There's an old rule of thumb that if the number of slipped slates or tiles exceeds about a quarter of the total then complete stripping and re-cladding a roof is more economical than spot repairs. But today's 'Working At Height Regulations' mean that scaffolding or platform towers are required even for minor roof repairs, adding to the cost. Although hydraulic platform 'cherry pickers' can often make access easier for minor repairs, they are still expensive. Nonetheless, re-fixing or renewing the odd slipped or missing tile should be a fairly straightforward job for a competent roofing contractor.

When roofs have to be re-covered it is sometimes possible to salvage as much as 75% of the original slates or tiles for reuse with new fixings, to achieve that authentic look. If using new materials, it is important to ensure that the correct size of slate or tile has been selected for the angle of pitch and that they are laid to the correct lap, otherwise rainwater may rise back underneath. Before the job starts it's a good idea to take some photos to record the condition of the old roof, to help reinstate the detailing correctly.

A change in the type of covering is sometimes worth considering for aesthetic reasons – to 'right an old wrong'. For example where an old roof has been inappropriately re-clad with ugly, modern, concrete interlocking tiles.

**NB** Before undertaking any external roof work you must first be certain that safe access has been provided – purpose-made roof ladders and platform towers or scaffolding are essential – and care must be taken to avoid damaging the slates or tiles further. If you have any doubts, do not attempt it.

Vast quantities of roofing slate were once produced in many parts of Britain. Welsh slate is probably the best known and most widely used across the UK, recognisable by its subtle grey-purple hue (but appearing uniformly black when wet). However, not all slate was as thin and lightweight as the widely used Welsh variety – for example the distinctive bluish-green Westmorland slate from Cumbria is thicker and heavier with a rougher, more grainy texture and less uniformly flat, but it is very robust and long lasting. The quality and thickness of different types of slate varies in different parts of the country, and was also sourced in large quantities from quarries in Scotland (*eg* Easdale and Ballachulish) as well as Cornwall (*eg* Delabole) and Lincolnshire (*eg* Collyweston).

The world of slate has aristocratic pretensions. You may well have some *Countesses* sitting on your roof, perhaps a few *Duchesses*, a *Wide Viscountess*, or even some *Broad* and *Narrow Ladies*. These are all different sizes of slate, the commonly used standard sizes being 24 x 12in *Duchesses* (610 x 300mm) and 20 x 10in *Countesses* (500 x 250mm).

Traditionally slates were laid in diminishing sizes and thicknesses with the largest and heaviest on the bottom courses, getting smaller and thinner towards top. However, later 'mass-

Slates laid in uniform sizes to late Victorian roof

Slipped slates due to nail fatigue – note fixing holes; roof needs to be stripped and re-clad with re-fixed original slates

Slates are female – and aristocratic!

produced' slates from larger quarries were generally laid in uniform sizes. Depending on the pitch, slates would typically be laid to a lap of about 75mm. They were usually nailed – through two holes in their centres or tops (heads) – to timber battens running across the rafters, or sometimes fixed directly to timber boarding. The nails were of copper, zinc, or galvanised iron and failure is usually due to 'nail sickness' where the fixings become so corroded over the years that the slates start to slip and fall loose. The cheaper, widely used, galvanised iron nails are especially prone to corrosion. Slates are generally less affected by water penetration than manufactured tiles of the same era. However, they can eventually start to delaminate ('exfoliate'), absorbing water and becoming soft. Occasionally they need replacing because the nail holes are cracked.

New Welsh slate is the ideal covering if you are recladding your roof. New natural Cornish slate can also still be obtained, as can the 'Lake District' varieties. But quality of this type costs money, so recycled Welsh slate is a good compromise, plus you should be able to salvage some of the sound old slates from your roof. Alternatively, it may be tempting to use cheaper foreign imports such as Brazilian, Chinese, Canadian or Spanish varieties. Although these may look authentic, they are generally far less durable and have a shorter lifespan, being softer than the native equivalent.

When patching an old roof with new slates, as well as selecting new ones of the right size, colour and texture, it's particularly important to match the thickness, as slates that are much thinner or thicker than their neighbours will not lie flat and will be prone to lifting in strong gusts of wind.

## ARTIFICIAL SLATES

Modern composite slates have become a popular, cheaper alternative to natural slate for re-roofing. These are manufactured from artificial materials such as concrete, fibre cement or glass-fibre resin mixed with slate dust, and moulded to *imitate* the colour, form and texture of the original. They can normally be laid to a very shallow pitch (some as low as 17.5°). Slates manufactured with moulded side grooves are easy to lay, as they can be overlapped at the side like single-lap interlocking tiles. Unlike concrete tiles, however, they are lightweight and look reasonably authentic, if a little smooth and shiny. Most are much lighter even than real slate, so to prevent wind damage their tails sometimes have to be laid with special metal rivets or ties. Early artificial slates contained asbestos fibre reinforcement, and were prone to discoloration and warping. Superseded in the 1980s by non-asbestos composite-fibre slates, they are now the budget

**Left:** Modern artificial slate – uniformly smooth and even

**Right:** Delightful late Victorian hipped roof clad with plain clay tiles

material of choice. But although at first glance these can appear similar to traditional materials when first laid, after a few years they will weather differently and don't have the enduring beauty of natural materials, and their lifespan may prove comparatively short.

## STONE 'SLATES'

Rare, but found in some traditional stone districts such as Yorkshire, the Cotswolds and parts of Sussex, these are small, very heavy, irregular slabs of sandstone or limestone that can last more than 150 years. However, they are often secured to the battens by wooden pegs, which are prone to decay, or are simply bedded down on hair mortar over timber boarding or lath. Because they shed water more slowly than natural slates, a steep pitch is needed, and their weight means that roof structures need to be extremely robust. They are now very expensive, and require specialist skills to repair or renew.

## PLAIN TILES

Plain clay tiles were originally made with small holes in one end so they could be nailed to timber battens, or fixed with wooden pegs ('peg tiles'). However, in the Victorian period they were increasingly produced with projecting 'nibs' so they could be hung from the battens (like modern tiles) with only every fourth or fifth course nailed (though additional nailing was required in windswept coastal locations).

Plain tiles are usually rectangular, though they were also made to various ornate designs such as arrowhead, bullnose, fishtail, and club, for use in decorative roofing. They are typically about 13–24mm thick, and slightly cambered, with a standard size of approximately 265 x 165mm. Being fairly small, the pitch would need to be greater than 35° with the lap as deep as 88mm.

Early Victorian tiles were handmade, with a characteristic slight irregularity in shape and an uneven surface. They were sometimes known as 'sand-faced' tiles because the moulds were sanded to stop the clay sticking. By the late Victorian and Edwardian periods, cheaper manufactured tiles became fashionable and were widely available in standard sizes. Machine-made tiles were smoother and flatter than the traditional handmade variety, and despite not being as tough should still last a century or more. But both types

Handmade plain clay tiles, some patterned, have a pleasing unevenness

View from the loft: tiles and nibs show light salt staining but are still sound

Leaks commonly develop at joints with parapet walls and stacks; hence importance of lead flashings (but plant life suggests a risk of water ingress)

can eventually become unduly porous, due to under-burning or salts in the clay, and can suffer from 'spalling', where they break down into flakes as a result of frost damage or thermal movement. They will then absorb moisture from rain (or from condensation in poorly ventilated lofts) leaving them vulnerable to frost action causing delamination. It can also hasten the onset of decay in the battens, eventually resulting in the tiles slipping, leaving a hole in the roof. If there are several slipped or missing tiles, and extensive white powdery deposits can be seen around the nibs when the tiles are viewed from the roof space, the most likely solution is complete re-tiling.

## PANTILES

These are traditional large tiles of Dutch origin with a wavy S-shaped profile, but are rare outside parts of East Anglia, Lincolnshire, and Nottinghamshire. They are hung by their nibs and traditionally were often pointed up with mortar. They have similar properties to the smaller flat plain clay tiles, but overlap each other at the sides, are lighter (per m$^2$), and can be laid to a shallower pitch. By the mid 19$^{th}$ century double 'Roman tiles' were introduced. These are similar to pantiles but are slightly larger and typically have two rolls rising from a flat surface. Roman tiles provide good weather resistance but are considered less visually appealing.

## CONCRETE TILES

Plain concrete tiles manufactured in sizes and styles that matched smaller traditional clay varieties were introduced as a cheaper alternative from the 1930s. Some original clay tile roofs have been patched with concrete tiles, which have a darker colour (when new) and a rougher surface. However, relatively few seem to have been used as replacement coverings for Victorian houses.

The main drawback with plain concrete tiles is that they are relatively heavy and have a shorter lifespan of only around 50–75 years. Some modern varieties have an almost identical appearance to clay tiles when new, but they do not usually weather 'naturally' over the years, becoming pale and blotchy with age.

Large interlocking concrete tiles, on the other hand, have had a dramatic effect on the nation's Victorian housing stock. In the 1980s many houses had their roofs reclad with these, much of the work being funded with the help of council grants. They were quick and easy to lay, needing fewer battens than traditional coverings, but proved much heavier than the original slates and imposed substantially higher loads on the roof structure. Without additional support, such roofs commonly show signs of distress and in extreme cases may ultimately collapse. Large concrete tiles are also difficult to adapt to angles, such as hips over bay windows, but they can be laid to a pitch as low as 17.5°. As a result of their weight and clumsy appearance they are currently unpopular as a replacement roofing material.
Nevertheless, if manufactured and fitted properly they should last well over 50 years.

## FLASHINGS

The most significant weak point on most roofs is where the roof meets a wall, such as where the party walls of a terraced house project above roof level as parapet walls. At the junctions with these or with chimney stacks you will find metal flashings or mortar fillets. Cement mortar fillets have a short life and tend to crack – a very common cause of leaks. They should be replaced with lead flashings.

House on right: original slate, but next door re-clad with inappropriate concrete interlocking tiles

Mortar fillets are prone to cracking – rain will penetrate even very thin hairline cracks

**Top row:** Stepped lead flashings suit brickwork (left) or regular stone – otherwise a continuous strip can suffice (right).

**Lower row:** Mortar fillets (traditionally tiled, left pic) without lead soakers underneath are at risk of cracking

But as well as specifying the correct thickness of lead, you also need to be clear that you want the new flashing fixed in a 'stepped' configuration (as opposed to a continuous strip cut into the wall which can look clumsy in brickwork).

## RIDGE TILES

Probably the most exposed part of the roof is its very top – the apex or 'ridge' – which therefore needs to be well waterproofed.

Classic Victorian detailing – finials, crests and delightful patterned ridge tiles; be sure to retain them

This was normally done in Victorian times by capping the ridge either with lead strips (see below) or with terracotta ridge tiles. These V-shaped angle tiles were bedded in place with hair mortar and the joints pointed. Some designs have a thin roll along the top, but conspicuous ridge tiles with elaborate designs became increasingly fashionable, with ornamental styles such as the 'fleur de lys crest', the 'double six point star', the 'toothed cogged crest' and the 'punched three toothed crest'.

But you didn't have to stop there. To really impress the neighbours required exotic iron or timber finials, weathervanes and exuberant terracotta crests on gable ends – until around 1900, when the excitement began to wear off and simpler ridge designs were considered more tasteful.

Storm damage to ridge tiles is quite common, the end ones often coming loose first. Most rely on no more than their own weight and a bed of mortar to hold them down, in which cracks can occur for a number of reasons (*eg* the mortar mix was excessively strong or dried too quickly as a result of dry tiles sucking out the moisture; or due to different rates of expansion between the mortar and the tiles). It is common to find that the mortar pointing has eroded. This should be pointed up before damp can penetrate leaving the tiles vulnerable to frost action, which will weaken the bond so that a strong gust of wind could dislodge them. Loose ridge tiles can normally be fixed by re-bedding them in a mortar mix of 1:1:6 cement/lime/sand.

Modern 'dry' ridge tiles have overcome these problems by being secured with special screws or a wire tied to the ridge timber below but tend to look inappropriate on roofs of this

3 Salvaged ridge tiles – Left to right: 2 hole crest, hexagonal crest, and toothed crest

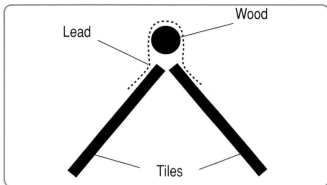

Above: Valley formed in purpose made tiles
Right: Typical lead lined valley gutter and hopper to party wall – commonly neglected, blocked and overflowing into bedrooms below
Right: Hip rolls and open valley (with slipped slate)

Above: Lead hip roll (left). Joint with dormer cheeks sealed with soaker extending under slates (right)

age. Fortunately, the original ridge tiles can normally be re-used once the old lime mortar has been cleaned off. Crested Victorian terracotta ridge tiles are particularly appealing and can be repaired or replicas specially made using broken old pieces as a template.

## RIDGE ROLLS, HIP ROLLS AND SOAKERS

Where ridge tiles were inappropriate (eg on very shallow-pitched roofs or on small bay or porch roofs) a ridge or hip could instead be sealed with a 'ridge roll' or 'hip roll', a strip of lead or zinc flashing wrapped round a wooden 'broom-handle' pole with side wings. Most rolls had a wood core, but some were hollow and supposedly more durable, as timber can rot, but they can be damaged by being stepped on (eg by roofers or TV aerial installers).

Alternatively, metal soakers comprising strips of lead or zinc sheet were sometimes placed under the slates or tiles at a join, eg to smaller hipped roofs and dormers. Where lead is used, water run-off combined with lead oxide can eventually cause pale streaks to slate roofs below ridge or hip rolls or flashings. This can be prevented with by applying a coat of patination oil prior to fixing new leadwork.

## HIPS

Whereas most 'standard' roof structures have a triangular gable wall providing the 'infill' between the two main slopes, hipped roofs instead have a third slope. Looking rather like elongated pyramids, hipped roofs are found on some semi-detached and end terrace houses of this era, as well as a large number of Victorian bay windows.

The vertical corners where the three roof slopes meet are constructed with timber 'hip rafters'. They are usually covered with zinc or lead strips (hip rolls), or with ridge tiles (which require a 'hip iron' protruding at the base for stability) or sometimes with

upturned 'bonnet tiles'. Alternatively, hips could be mitred with slates cut to meet neatly over hidden lead or zinc soakers, thereby achieving 'invisible' weather protection as described above.

## VALLEYS

Valleys are found where one roof joins another at an angle, eg where a bay window roof meets the main roof. They are often formed by a strip of lead or zinc sheeting (minimum width 100mm/4in) over a timber board base. Such 'open valleys' leave the metal strip exposed, for easy maintenance, with any gaps at the sides pointed up with mortar.

Alternatively, 'mitred' valleys have slates neatly cut along the join at the valley edges and fixed over soakers. But corrosion of the metal (especially zinc) by acidic urban rainwater over time can result in leaks where joins are 'secretly' waterproofed in this way. Later, ready-made 'swept' valley tiles became available, providing a robust, watertight solution. Valleys are a common weak point and cause of water penetration and should be periodically cleared of moss and leaves to prevent rainwater overflowing into the roof space.

Note that modern purpose-made plastic (GRP) linings are a cheaper but less durable alternative to lead for valleys, mitred hips, secret gutters and flashings.

Right: Typical open valley lined with overlapping lead sheet – can be prone to blockage and overflow

## BAYS

Bay windows are a major architectural feature of Victorian and Edwardian houses but can be problematic. On full-height bays, the valley where the bay roof meets the main roof is a common weak point. Worse, on single-storey (or lower level) bays with small flat 'balcony' roofs, the rainwater discharge pipes can become blocked or the flat roof may perish, causing damp to the main front wall. Unfortunately, there is a large timber lintel known as a 'bressummer' just where the living room ceiling opens into the bay. Damp seeping through the wall over time can result in serious hidden rot in this lintel, ultimately causing the collapse of the masonry above. Replacing bressummer beams is a major structural job. See chapter 7.

Bays on large, rambling, turn-of-the-century houses sometimes exhibit fabulous domed roofs, turrets and towers of the *Chateau Disney* school of architecture. However, exotically tiled or slated cones and pyramids and metal-clad bell-shaped roofs were the preserve of the rich, and are rare on mainstream houses of the period.

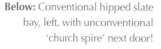

**Right:** Single storey flat roofs can be prone to blockage with rot developing to 'bressummer' beams in the front wall

**Below:** Conventional hipped slate bay, left, with unconventional 'church spire' next door!

## EAVES

There are various kinds of external detailing where the rafter feet meet the main walls to create a roof overhang that helps protect the upper walls from the weather. These include box eaves with a ventilated soffit underneath; bare projecting rafter ends; a fascia board over the rafter ends; and brickwork that is stepped outwards (corbelled). At the eaves, the lower edge of the roof should generally overhang the guttering by between 25 and 50mm to ensure rainwater discharges into the gutter and doesn't blow back under the slates.

## GABLES AND VERGES

The verges are the side edges of a roof slope, typically visible at gables. The tiles at the side of a roof were traditionally raised up or tilted slightly to keep rainwater away from the edges, by fixing a course of slate 'undercloak' bedded in mortar under the batten ends prior to the slates or tiles being laid. The verges were pointed up with mortar and often decorated with wooden bargeboards underneath. Modern replacement 'cloaked' verge tiles, or verge coverings, are a 'maintenance-free' alternative, but are not architecturally compatible with older buildings.

Large overhanging verges on big front elevation gables are a powerful architectural feature on late Victorian and Edwardian houses, especially over bays. Later models sprouted ever more elaborate bargeboards and ornate fascias, but the masonry to gable walls at this level may be only of thin, single thickness brick (125mm), often rendered or tiled for protection. Sometimes render was applied over traditional thin timber laths, which can be prone to damp.

## DORMER WINDOWS

Rooms in the roof space often have dormer windows that project through the roof slope. Like bays, they normally have small gabled or flat roofs, or very rarely may be hipped or arched. The fronts are often built straight up in brick or stone from the wall below as 'half dormers', with a side framework of upright timber studwork from the rafters or floor joists.

The dormer sides ('cheeks') were commonly clad in zinc or lead sheet, or hung with tiles or slates. At the junctions to the roof there may be 'secret gutters' – soakers of sheet metal obscured by the lap of the slates.

Common failings with dormers are thin and poorly insulated walls and ceilings, prone to condensation and damp; leaks to defective coverings at cheeks; and leaks and blockages at roof joints. Brick or stone fronted half dormers built up from the front wall would commonly settle inwards in the years following construction, adopting a weary backward slant, before achieving stability. It is very unusual for such dormers to lean forwards, but any signs of outward movement should be professionally checked to pre-empt any risk of collapse from unrestrained masonry.

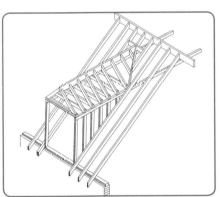

# Metal claddings and flat roofs

Original flat roofs ('flat' meaning any slope less than 12°) are commonly found on bays, porches and dormers, and more rarely on some small rear additions. Construction was equivalent to a timber floor made of joists and boarding. For most houses, the choice of weatherproof covering was between sheets of either lead or zinc (also used to clad many small hipped roofs to bays). Some grander houses benefited from flat roofs clad in superior, expensive copper sheet. Asphalt (bitumen) was less common as a flat-roof covering but is found on some bays and porches, especially where surfaces were subject to foot traffic. A feature found on some more upmarket flat roofs was the 'lantern light', a kind of small greenhouse structure that allowed light through to stairwells underneath.

## ZINC

Zinc was a cheaper, thinner substitute for lead, and was the most common flat-roof covering on less expensive Victorian houses, where it was normally of 0.8mm (14 gauge) thickness. Though its lifespan is normally well over 40 years, zinc has a major weakness – it is corroded by cats' urine, and cats just love warm tin roofs! It is also badly affected by acid pollution, becoming brittle and crusted with carbonate. Consequently any surviving Victorian zinc-covered roofs are likely to be overdue for replacement. Look for splits and pitting to the surface, or past attempts at coating with bitumen. Zinc-working is now a fairly rare skill, so it is best replaced with more durable lead.

## LEAD

Durable cast lead was the preferred material for better quality houses. However, it was the cheaper milled variety that was most widely used, though even this was equivalent to a good quality modern thickness. Today, lead is still specified in terms of 'Codes' based on weight in pounds per square foot – varying between 4lb and 8lb psf, corresponding to Code 4 and Code 8. The higher the Code, the thicker and heavier the lead and hence the larger the size of sheets that can be laid, but Code 4 lead is the one that builders tend to use unless instructed otherwise. Flashings should typically be made of Code 4 or 5 lead (1.8mm or 2.24mm thick respectively), and valley gutters of Code 5 or 6 lead (2.24mm or

**Above:** Intricate leadwork dome

**Below:** Nicely re-leaded flat roof to bay

**Below:** Lead roll expansion joints run parallel to the water flow. Stepped 'drips' are also needed for longer lengths

**Below:** Classic Victorian flat roof to rear addition

'I never promised you a roof garden'

# TECHNICAL DATA

Typical terraced house original roof timber dimensions are:

COMMON RAFTERS 4 x 2in (100 x 50mm) or 3 x 2.5in (75 x 63mm)

PURLINS 7 x 2.5in (175 x 63mm) or 8 x 3in (200 x 75mm)

RIDGE PLATE 6 x 1.25in (150 x 31mm)

WALL PLATES 4.5 x 3in (112 x 75mm)

Standard plain tile sizes varied between 250 x 150mm (10 x 6in) and 280 x 175mm (11 x 7in).

NB All stated laps and pitches are rough guides only.

See www.Lafarge-roofing.co.uk for specific figures.

See www.victorian-house.co.uk for further advice on party walls, specialist materials, and contractors.

2.65mm). Thickest of all is sheet roofing, of Code 6, 7, or 8 lead (2.65mm, 3.15mm, or 3.55mm thick).

Lead is susceptible to two forms of corrosion: from alkaline conditions, typically caused by eroding cement mortars; and from acidic conditions, such as rainwater coming from roofs covered with algae, moss, or lichen. In fact, acid rain and moss are the enemies of both lead and zinc, so watch for rainwater that runs off nearby roofs of different materials. Lead and zinc mustn't be used together, as they can react and corrode. Despite this, the lifespan should be at least 80–100 years – so it's possible that some original coverings may need replacement, which isn't likely to be cheap. If you can see old splits, patch repairs, surface ripples, or impressions of the boarding below, then it's on its way out.

Lead is notoriously prone to expansion in hot weather, so any large areas of lead without expansion joints, known as 'rolls' or 'drips', will eventually split and buckle. Rolls are run parallel to the water flow and are similar to the 'hip rolls' described earlier. Drips are basically steps with overlapping sheets. Sheets should therefore not exceed a length of about 2.5m (for Code 7 thick lead). Lead is also prone to 'creeping' downhill, so the roof pitch should not be too steep – between 25–75mm per 2.5m run (1–3in per 8ft) is best. At joints with walls, the sheets need to be turned up and lapped over by a separate lead flashing. Sheet leadwork is a specialist skill, and not something most roofers are familiar with other than for flashings. Fortunately the Lead Sheet Association has diagrams detailing leadwork, via the website Period-house.com.

**Left:** Zinc and lead were sometimes used for small bay roofs.

**Right:** Modern flat roof clad with short-life felt.

## MODERN FLAT ROOFS

Many older houses have modern extensions with felted flat roofs on timber decks. These are often problematic, not least because of a complete lack of insulation. Typical defects include:

- The short life of felt coverings: 10–15 years is not unusual, depending on workmanship.
- Rot to decks: chipboard makes a poor quality deck, as it is prone to disintegrate when damp. Marine plywood is better.
- Insufficient slope: a fall of at least 1:40 is required or water will not easily disperse.
- Inner dampness: flat roof decks are often uninsulated, so condensation from the rooms below can cause rot or black condensation mould on ceilings below. They are best insulated with sheets of polyurethene foam boarding placed above the deck (to create a 'warm' roof). Ceilings should be constructed with foil-backed plasterboard or with a polythene sheet applied on the room side of the insulation as a vapour-barrier.
- Symptoms on the surface – such as plant growth or the ponding of water into puddles – indicate that replacement is necessary.

**NB** Before carrying out work on roof structures, Building Control must be consulted.

## Defect: The roof sags

### SYMPTOMS

The roof slopes aren't level, but show signs of 'dishing' or curvature; the roof seems bowed under the weight of the slates or tiles, and on terraces the line of the party wall below the slates stands proud.

There are several possible causes:

### Cause  Overloading

Replacement concrete tiles are a very common cause of roof sagging because of their substantial weight.

### Cause  Weak roof timbers

**Roof timbers cut:** the result of bad DIY – such as a botched loft conversion or the installation of roof windows without the existing rafters being strengthened.

**Rafters:** undersized or spaced too far apart when house was originally built.

**Purlins:** deflection due to:

- Purlins being undersized, missing, or poorly joined – check for splits in the timber.
- Lack of struts or collars to provide support.
- Feet of struts being unsupported: they may rely on the internal load-bearing spine wall, which may have settled or been removed, causing the struts and purlins to follow and the roof to sag.
- Purlin ends resting in thin outside damp walls have rotted, or have slipped on broken corbels or brackets.

### Solution  *Strengthen roof timbers*

If movement is due to old historic settlement, it may have now stabilised and no repairs are required. Otherwise major remedial work may be needed:

- If the timbers are undersized, weakened, or overloaded then it may be necessary to beef them up with additional support, such as additional purlins and struts to improve support to the rafters.

- If the spine wall has dropped causing the roof struts to follow, a structural engineer can advise how loadings may be transferred to the main walls or the wall may be strengthened.
- Rotten purlin ends need to be cut out, replaced, or spliced. These then need to be supported on new metal hangers. Splits and poor jointing require the splicing of new timber alongside or bolting with metal plates.

### Cause  Roof spread

The rafters have pushed outwards, causing the top of the wall to bulge; the rafters may then sink, causing the roof to dip in the middle. This is common on rear lean-to roofs.

### Solution  The ceiling joists

acting as collars normally restrain the roof rafters from pushing out the walls. If they are not properly secured, they will need to be tied in to prevent further movement. In severe cases the tops of bowed walls will need to be rebuilt and tied in by the provision of collars or metal tie bars.

One method of preventing rafters from splaying out at the eaves is to fit timber bracing struts, 100 x 50mm (4 x 2in), which tie each rafter to the ceiling joist below. These are connected diagonally with bolts and toothed washers about 600mm (2ft) from the base of the

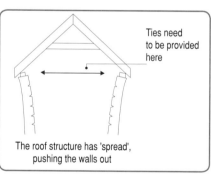

Ties need to be provided here

The roof structure has 'spread', pushing the walls out

rafter to about 1m (3.3ft) from the end of the ceiling joist. Further movement can also be restrained by bolting metal plates to the upper walls on either side, linked by a metal tie bar running through the building at ceiling joist level.

### Cause  Rotten or beetle-infested timbers

### Solution  *Rotten timbers must be treated or cut out and replaced once the cause of the problem has been identified*

It is comparatively rare to find beetle infestation in roof timbers to well ventilated lofts, although timbers around loft hatch openings are a favourite place. Even then, boreholes are rarely active (check for 'sawdust' around the holes) and may have been previously treated. See chapter 6. It's also unusual to find significant rot to roof structures, except where timber has been exposed to damp for long periods from leaks, such as the top wall plate in lean-to roofs due to defective flashings, or to timbers under valleys.

## Defect: Leaking roof

### SYMPTOMS

Water penetration; damp patches; wet rot to timbers where leaks have occurred over time. NB the source of a leak is not always directly above the damp patch on the ceiling – water may run down rafters and along ceilings. A simple way to locate the source of a leak is to look in the loft space with the light off (assuming the original roof has not been underfelted). Though there will normally be small, regular areas of light coming through the slates or tiles, you can easily pinpoint any large gaps – often at flashings (see next defect).

**Cause** **Slates or tiles broken, slipped, or missing as a result of damage (eg by storms or by people walking on the roof).**
See also 'defective flashings' on next page

**Cause** **Surfaces are flaking (delaminating) due either to age or poor quality of materials, allowing water to be absorbed and to freeze and expand**

**Solution** *If there are only a few slipped or missing slates or tiles, they can usually be refixed or replaced. If damage is extensive, or the coverings are obviously damaged, the whole roof may need re-covering. Creepers and climbing shrubs can invade eaves and dislodge slates or tiles, and must be regularly pruned or removed*
**Slates:** although slates are long-lasting, check for blistering, delamination, and softness on the undersides, which is where the erosion starts. Sulphates from acid rain pollution can react with carbonates in the slate, forming calcium sulphate that can eventually reduce slates to the consistency of cardboard. If they are easily scored with a knife, creating a powder, they are no longer of use. It is best to replace like with like: try not to mix artificial with natural slates on the same slope. Normally, however, slates are very durable, though the old galvanised fixing nails may corrode, causing them to slip. Refix using copper or aluminium alloy nails or, if just one or two are loose, refix using metal clips known as 'tingles' (see step-by-step

below). But if a roof has more than about half-a-dozen tingles, it probably needs stripping and re-covering. Heavy snowfalls can push tingles up, so they may occasionally need folding down again. Alternatively, small gaps between two slates can be covered by inserting a strip of lead sheet.
**Tiles:** can be prone to defects such as nibs breaking off, making them vulnerable to storm damage. To remove a broken tile, lift the two tiles above it with wooden wedges and use a bricklayer's trowel to lift the nib of the damaged tile off the batten and lift it out. The new tile can then be eased in and the nib hooked over the batten.
**Timbers:** battens or sarking boards are generally very resistant to decay but if affected by prolonged damp can suffer from wet rot, eventually sagging, and fail to hold further nails. Stripping and re-battening will be required. Verges commonly need pointing up where mortar has crumbled

**Cause** **Missing or loose ridge or hip tiles**
Neglected mortar pointing lets in damp, which can eventually affect the timber ridge plate and the tops of rafters. Loose ridge tiles are not uncommon and should be secured as they are a potentially lethal danger (as is any type of tile or masonry falling from height).

**Solution** *Re-bed loose ridge or hip tiles and point up*
Use a suitable mortar mix, such as 1:1:6 cement/lime/sharp sand. PVA bonding is sometimes used to improve adhesion. The mix should be fairly stiff. Treat any affected timbers. Hip tiles need a 'hip iron' at the base to prevent slippage.

**Cause** **Defective flashings and valleys**

**Solution** *Usually localised replacement in fresh leadwork will be needed.*
Leaks are sometimes due to loose flashings that require re-fitting and wedging into

mortar joints. Leaks at open valleys are sometimes due to loose mortar at the upstands/edges either side of the valley, and may just need pointing up. Adjoining timbers should first be checked for rot.

---

**Cause** **Defective flat-roof coverings**

**Solution** *Original metal-sheet roofs may require complete replacement with new sheets of lead if the problem is extensive. Modern felt roofs often have a very short lifespan and need replacing every 10–15 years.*

Metal-clad flat roofs can develop splits, wrinkles or holes in the surface. Corrosion over time can occur due to reaction with the wrong fixing nails, or moss growth and acid rain. Lead sheets may be inadequately secured, or with insufficient space for thermal expansion. Past attempts at sealing with bitumen paint may be evident.

Small defects can be repaired with soldered dots or patches of the same metal as a short-term measure. Sheets that have expanded should be shortened and new joints formed. Refix new sheets using copper fixing nails, but first check the condition of the deck and replace with marine plywood as necessary. Take the opportunity to insulate the deck prior to re-covering.

## Defect: General dampness in loft

**SYMPTOMS**
Damp ceilings; moist loft insulation and timbers

**Cause** **Condensation due to lack of sufficient ventilation and insulation**

Warm moist air from the house seeps into the loft, *eg* through uninsulated ceilings, recessed lights or the loft hatch. The warm air hits cold roof coverings and condenses back into water, which then drips down onto the ceiling. Damp insulation loses its effectiveness.

**Solution**

- Upgrade loft insulation over ceilings to at least 270mm depth of mineral wool.
- Fill any holes in ceilings and insulate the loft hatch.
- Replace recessed lights with sealed shower lights using LEDs and insulate above.
- Improve ventilation at eaves or fit vents to ridge or roof slopes, or an airbrick to side gable. Protect from bird ingress with mesh. Cut back any plant growth blocking vents.

## Defect: Water penetration without obvious signs of damaged roof coverings

**SYMPTOMS**
Leaks and damp; decayed timbers. Roof coverings may have been painted with bitumen compound.

**Cause** **Wind blowing rain under slates or tiles**

**Solution** A little occasional water penetration need not be an issue. Original roofs were built without underfelt, so small amounts of rain or snow entering the roof space in severe weather should disperse through evaporation if the space is well ventilated.

But if the problem is severe, complete re-roofing may be required – *ie* stripping down to the rafters, fitting new breather membrane underlay and battens, and re-cladding with slates or tiles. The type of roof covering should be appropriate for the degree of pitch, and must be laid to a sufficient lap, otherwise rainwater can track back underneath the slates or tiles, effectively running uphill. Slates can cope with shallower pitches than tiles but if wrongly laid re-cladding will be needed. Painting an old roof to stop it leaking is a desperate short-term measure that also renders the old slates unsalvageable, so re-cladding is the best option.

---

**Cause** **Defective pipes or tanks**

**Solution** *Check all pipework and tanks for leaks*

Pipe leaks are quite common in roofs where there is a lack of insulation, which allows pipes to freeze and burst in winter.

Drain down and repair. Lag pipework and insulate tanks.

---

**Cause** **Cracked mortar fillets or defective flashings at junctions**

A major cause of damp penetration. See chapter 3.

**Solution** *Replace with new lead flashings*
Mortar fillets at joints are particularly prone to cracking (due to differential movement) and may only last a few years, particularly

if a cement mortar mix was too strong. Original zinc or lead flashings may have split or corroded. This is a common cause of damp penetration, normally requiring replacement with new lead flashings. It is also advisable to rake out and re-point any eroded mortar joints to the brickwork near the flashing.

On a terrace or semi you may find problems where the roof slope of one property joins next door's roof over the party wall. Each house may by now have a different kind of roof covering, with inherent problems where they meet up at the junctions. Some localised stripping will be required, and the provision of strips of lead sheet (soakers) under the joints. Alternatively where party walls between houses were built up as short parapet walls, damp can leak down at defective flashings or via saturated brickwork. This may require a new flashing or lead-sheet cladding applied to protect the exposed masonry.

NB any proposed work on party walls means that legal notices under the terms of the Party Wall Act first need to be issued.

(particularly if rendered), and may only be of 115mm single thickness brickwork, or comprise rendering over timber lath. Light may be visible through some joints. Eventually this may cause rot in adjoining timbers such as purlin ends embedded in the masonry. NB This may be a clue that the whole side wall below is also of sub-standard 115mm masonry. See chapter 5.

(Solution) *Repoint or re-render*
Repoint brick or stonework with lime mortar. Make good cracked and blown render. Protect the masonry externally with new lime render or cladding (*eg* tiles). Treat any damp or decayed timbers. Dry-line internally with insulated plasterboard to upgrade insulation (with an airbrick for ventilation). Localised rebuilding may be required if severe bowing or structural movement has occurred.

 **Cause**  **Thin end gable walls**
Damp from driving rain can permeate cheaply constructed thin gable end walls to lofts and attics, *eg* through old eroded mortar joints or cracked render. In an end terrace house, for example, the gable wall may have been built using leftover old bricks

## Defect: Missing firebreak walls

### SYMPTOMS
You can walk straight through your roof space into the neighbouring lofts. This is a potential killer from fire spread, as well as a security threat. It is a common 'essential repair' retention item in mortgage lending.

**Cause**  **Cheap construction usually associated with poor support for the roof timbers**

(Solution) *Construct new firebreak wall to Building Regulations Standards*
Normally single thickness blockwork is adequate, or fire-resistant plasterboard fixed to a timber frame. Consult neighbours first –

they may be willing to contribute financially – and check with a chartered surveyor specialising in the Party Wall Act.

## Defect: Nests and vermin

### SYMPTOMS
Smells, droppings and straw in loft; damaged stored items; chewed electric cables and pipe lagging.

 **Cause**  **Easy access routes into the roof space**
via gaps at eaves, broken slates, open vents, plant growth etc. Small wasps' nests are particularly common.

(Solution) *Carefully remove nests, taking appropriate precautions (beware of wasps!). Use traps and bait or hire pest-control contractors.*

Clearly this situation is unhygienic and potentially dangerous (electric cables and even pipes can become damaged, or water tanks polluted), so future access should be barred by fitting wire mesh over any large holes to maintain good ventilation. Remember that bats are a protected species, so specialist advice should be sought.

## Inspecting your roof

Once every 12 months or so it's advisable to give roofs a quick visual once-over to spot any looming potential problems before disaster strikes. These are some of the common warning signs to look for:

■ Check for any slipped or missing slates or tiles. This could be a one-off defect, or part of a more extensive problem like 'nail sickness'. Despite being one of the most hard-wearing of all roof coverings, individual slates are quite fragile and can crack like crockery when trodden on. Installers of TV aerials, satellite dishes and solar panels clambering around in heavy boots on fragile roof coverings are a common cause of damage.

■ Check whether metal flashings are loose or cracked.

■ Where you have mortar fillets at joints they are likely to develop cracks, and are often best replaced with lead flashings.

■ Check mortar pointing at ridge tiles (and where relevant, hip tiles and verges). If it is badly eroded it will need repointing.

■ In terraces and semi-detached houses with continuous roofs (ie without a dividing parapet party wall), check the area above the party wall. Where next door's slates or tiles are relatively new, are they neatly integrated with those on your roof?

■ Where there's a parapet above the party wall, check the condition of flashings and coping stones. Damp can penetrate down parapet walls as there is no damp-proof course. Cladding the brickwork above in lead sheet is sometimes necessary.

■ In the loft, try switching off the light and look for any large 'chucks of daylight' light, ie big gaps in the roof.

■ Also in the loft, check for any damp staining that's still wet under weak points such as valley gutters, flashings and back gutters to chimneys.

London (butterfly) roof reclad – note gutter discharges to rear

Buildingadvice.co.uk

## Maintaining a slate roof

Maintaining a single-storey roof with a few slipped or broken slates should be a feasible DIY project. Obviously great care must be taken on roofs – a common cause of cracked slates is from people clambering on roofs without proper access equipment. Also, make sure no one is standing below!

### TOOLS REQUIRED

■ **Use scaffolding or work-towers for buildings of two storeys or higher. Ladders need a stand-off bracket to protect guttering**

■ **Roof ladder with a ridge hook (to go over the ridge) and wheels to move it up the slope of the roof**

■ **Slate ripper**

■ **Hammer**

■ **Ceramic tile cutter**

■ **Shears and pliers**

### MATERIALS

■ **Replacement slate(s)**

■ **Metal tingles (from builders' merchants) or a strip of lead, copper, or aluminium approx. 20–25mm (1in) wide and long enough to reach from the hole in the slate to its bottom plus 100mm (4in)**

■ **Galvanised clout nails 40mm (1.5in) long**

If new slates don't have holes, or second-hand slates have them in the wrong places, new fixing holes can be made (normally about half-way up near the sides). Use the old slate as a pattern, mark carefully, and drill using a No. 6 masonry bit. Work from the underside (the side without the bevelled edges).

A new product that is claimed to enable refixing without the need for tingles is the 'slate nib'. This is a V-shaped plastic hook that is pop-riveted to the back of a slate to create a protruding nib, as found on tiles. Having removed the broken slate, the replacement slate can be slid into place and the nib fastens over the batten. See www.victorian-house.co.uk.

**1** Working from the bottom upwards, loose slates can be refixed, or broken ones replaced by new slates of the same size and thickness.

**2** You need a tool called a 'ripper' that has two cutting blades to break defective slates free from their fixing nails.

**3** To remove a broken slate, slide the ripper under it and hook the end around each of the two fixing nails in turn and pull down sharply on the handle to cut them.

**4** The damaged slate is then free and can be lifted clear. Slates are nailed through twin holes in their top or middle. Unfortunately you can't replace a single slate by nailing it in the same place because the heads of the nails are covered by the slate above. Instead, the slate can be secured using lead or copper clips known as 'tingles'.

**5** Tingles can be purchased, or else made by cutting 20mm strips from lead sheets using shears. With a drill or a punch, create a nail hole about 25mm (1in) down from the top of the strip.

**6** Having removed the slate, nail the tingle to the exposed batten (which should just be visible under the vertical join between the two exposed slates) – this will later be covered by the new slate. The replacement slate needs to be matched to the existing ones not just in size, colour and texture but especially in thickness, as slates that are much thicker or thinner than their neighbours will not lie flat, and the wind can lift them.

**7** Slide the replacement slate into position under the two slates above with the bevelled edge uppermost, pushing it upwards until it is correctly aligned with the slates of the same row on either side.

**8** Once the new slate is in position, trim the tingle with shears so that about 30mm of it protrudes from beneath. Fold this upwards over the lower edge of the new slate, bend it double, and press it down flat to hold the slate securely in place.

**9** At the end of a row, slates may need to be cut to fit. To ensure an exact match, use the original as a template and mark the new slate to size by scoring a deep line with a ceramic tile cutter and a metal rule. Then use a bolster chisel and tap gently with a hammer along the line, or use an electric tile cutter.

# CHIMNEYS

The Victorians turned chimney pot design into an art form. Stacks sprouting battalions of outlandish and bizarrely shaped pots are a fabulous feature of many city skylines.

   But although the menace of 'pea-souper' smogs and under-age chimney sweeps are now a thing of the past, chimneys can conceal significant defects, such as crooked pots, dangerous leanings and mysterious leaks, that threaten the wellbeing of your home.

Welcome to one of the most potentially dangerous parts of the property – and yet one of the most neglected: the lonely world of the chimney stack.

Let's assume you've taken all sensible precautions and made it safely onto the roof and arrived next to a stack. Close up, it's the sheer size of these structures that is so striking. And the bigger they stand the heavier they fall, so it's rather worrying that many are left in such a poor state of repair. Hardly noticed from ground level, some stacks seem to be held together only by the TV aerial.

Exposed for well over a century to bitter winds, driving rain and frost, and under attack internally from chemical corrosion and intense heating and cooling, it's hardly surprising that some of the bricks and mortar joints may have become a little crumbly by now. You may even be able to prise the upper bricks off with your bare hands, the old mortar having eroded or lost its 'grip', reverting to a loose, sandy consistency.

From ground level, a decent pair of binoculars will be required for assessing all but the most obvious problems. And because scaffolding is a major part of the cost of repair work to chimneys, it may be cost-effective to have any roof works carried out at the same time.

# Chimney pots

Could you tell a 'Lancashire Bishop' from a 'Six Pocket Beehive'? Or maybe a 'Marcone Flange' or 'Cannon Head' from a 'Four Ring Roll'? Well, these are just a small selection of the dozens of historic pot types that grace our skylines.

Terracotta chimney pots were produced in many different sizes and patterns, glazed and unglazed. Some have been removed, but it is possible to buy replacements from salvage yards (they're commonly sold as garden ornaments). Elaborately moulded terracotta pots can be reproduced and made to order, at a price.

The addition of chimney pots helped reduce the problem of smoky fires by increasing the draught to the fire whilst helping

'Say 'ahhhh'; view down pot shows flue is not lined

New flaunching, but strong cement mix risks cracking

**Above:** Even in stone walls flues were commonly lined with brick

**Right:** Characteristic Victorian multiple flues

to resolve problems of down-draught. To protect flues with chimney pots from ingress of rain, a simple stainless-steel 'rain hat' with an integral bird guard should suffice. Purpose-made pots that incorporate a cowl are also available, but must be suitable for the type of fire or appliance.

When stacks were built, the pots would traditionally be placed on slivers of slate or tile laid across the corners of the stack, or the bottom rims wedged behind a course of projecting brickwork. To prevent gales from dislodging them they were bedded in thick layers of mortar flaunching to a depth of at least 150mm. But after many years of exposure, the flaunching can eventually crack and disintegrate. And although chimney pots are reassuringly heavy, there's a potential risk they can become dislodged by a severe gust of wind or even by the sweep's brushes.

# Masonry repairs

Even in areas where houses were customarily built of stone, brick was often used for chimneys and flues as it is better able to withstand intense heat (the clay having been fired at very high temperatures during the manufacturing process).

Although chimneys were sometimes given a render finish externally for improved weather protection, this isn't ideal because it requires more frequent maintenance and decoration, with considerable expense gaining safe access. Render also has a tendency to develop cracking due to thermal expansion in stacks over time, and being exposed to the full force of the weather means that any rain that penetrates will be at risk of freezing, loosening or blowing off the face. Fortunately, this is less of a problem with original lime renders as they were relatively flexible compared to modern cement-based materials.

Constant wetting can cause sulphates present in the brickwork, or in acrid chemicals from combustion gases, to attack the mortar in

Shrubs taking root in flaunching spell trouble (as does partially painted roof and mortar fillet!)

New lime mortar flaunching and period style pots

Repointed stack brickwork

Botched removal of pots leaves dangerous loose flaunching and risk of falling masonry – localised rebuilding required

Nice new flashing and apron

chimney walls. This tends to be more pronounced on the cold, windward side of the stack, which is wetter from internal condensation as well as from wind driven rain. Modern cement mortars can be susceptible to 'sulphate attack' – another good reason to use traditional lime.

But because of the extremes of exposure, a more durable hydraulic lime mortar mix is normally suitable, such as NHL 3.5 or NHL 5. But as is often the case with walls elsewhere on old buildings, sometimes past repointing in hard cement mortar has hastened the erosion of the surrounding masonry.

Where erosion or leaning is severe, rebuilding to at least the upper courses may be necessary. Alternatively, where stabilisation is required, a stainless steel tie rod and strap may be all that's needed. 'Stay bars' are a traditional way to secure tall or exposed chimneys and prevent the risk of rocking in high winds that could potentially lead to collapse. Any existing old stay bars should be checked for rust and security of the fixings. See page 55.

Another point to watch is the masonry just below the roofline, visible from within the loft. Because old stacks were built without the luxury of a Damp Proof Course (DPC) in the lower courses, in very severe weather damp can soak downwards, and can potentially be eroded by frost (or by salts crystallising as the water

evaporates). A similar problem can develop where party walls extend above roof level as short parapet walls. However, any such damage is only likely to require repair in the rare cases where erosion is severe. This involves opening up the roof around the stack so that any badly damaged bricks can be cut out and replaced.

## FLASHINGS

The joint where the base of the stack meets the roof is a major weak-point and a common cause of dampness in lofts and to the rooms below. Lead is the best form of weatherproofing, but Victorian houses often have flashings made from zinc sheeting or just cement mortar fillets, a cheaper and inferior form of jointing prone to cracking and leakage.

**Left:** The most durable flashings were of (more expensive) copper or lead
**Below:** Botched DIY tape job is a false economy

Ivy growth can erode mortar joints, harbour damp and block flues – before consuming the entire building!

Eroded mortar joints need repointing - very common

Vented pots help fires draw

Isolated thin flues are colder

# Flues

Flues became narrower during the 19th century, those in earlier houses typically measuring 230 x 350mm, but by 1850 the average size of a brick-built flue was 230 x 230mm (9in). To protect the inner masonry and prevent gases escaping through mortar joints there would be a layer of 'parging' – lime mortar traditionally mixed with cow dung or ox hair for reinforcement. Over time, the toxic products of combustion from coal fires (tar acids, ammonia, sulphates, and water vapour) will have combined to eat away at the parging and the mortar joints, causing expansion and cracks. Tarry deposits and salts could then be carried through the bricks and plaster to damage internal decorations – see chapter 11.

Chimneys built on outer walls are generally more exposed and colder than those built into party walls, and hence are more at risk. The flue may be only be separated from the cold outer wall by the width of one brick (115mm) so the hot gases from the fire will cool very rapidly, condensing into acidic water and corroding the masonry inside. Fortunately most chimney breasts were built into party walls and contained multiple flues which helped insulate them from the cold.

## THE ENEMY WITHIN

Victorian builders were sometimes tempted to make economies by using cheaper bricks laid by less skilled labour for parts of the building that were hidden or out of sight. So the quality of concealed flues wasn't always great, even when new. Add a hundred years or more of erosion from acid gases and old flues may have developed gaps that let fumes into the house. So there's a potential danger that a fabulous living-flame gas fire newly installed in an old fireplace could be a source of carbon-monoxide poisoning in the rooms above or next door. This is a deadly invisible and odourless gas, so if there's any doubt, its worth fitting carbon-monoxide detectors as a precaution. See chapter 11 for solutions to common flue problems.

Appliance served by lined flue with projecting terminal

Unlined flue

Flue gasses slowly eating away from inside can cause leaning

## Defect: A roof leak around the chimney

### SYMPTOMS
Damp patches on walls, ceilings and chimney breasts below.

**Cause** **Cracked or eroded mortar fillets or flashings**

These are a common weak point, especially the 'back gutters', which are the flashings at the back of chimneys on the upper roof slopes – the ones you can't see.

**Solution** *Replace with new lead flashings*
Ideally, all flashings should be of lead, but many Victorian flashings were of zinc, which may, by now, have suffered from corrosion by acid solutions resulting from air pollution or moss growth – typically more of a problem in industrial cities. Mortar fillets are a common cheap alternative, comprising a thick covering of cement or lime mortar spread over the joint. Differential expansion between the stack and the roof causes the mortar to crack and lose its bonding. Modern cement mortar is far more brittle and prone to cracking than traditional lime mortar. Rather than patch up defective mortar, it is always advisable to replace it with a lead flashing. See 'Renewing a flashing' project on page 57.

**Cause** **Metal flashing inadequately sealed to the brickwork**

**Solution** *Refix the flashing into existing joints with fresh mortar, or fix to a new position as follows:*

- Cut a new groove in the chimney about 150mm above the level of the roof covering, normally in a stepped pattern into mortar joints
- Turn the metal flashing at least 25mm into the groove.
- Fix it in place with metal wedges and seal with mortar.

**Cause** **There are insufficient soakers under the slates/tiles**
Soakers are special pieces of lead inserted under each slate at the sides of the stack (not necessary with interlocking tiles).

**Solution** *Provide new metal soakers to the side of the stack at every course of slates/tiles*

## Defect: Cracked or damaged brickwork or render

### SYMPTOMS
Visible cracks; severe weathering of masonry; damp to walls below.

**Cause** **Internal flue masonry suffering from expansion and cracking**

**Solution** *Install a flue liner*
As noted on the previous page, condensation of gases on cold surfaces inside the flue eventually causes chemical erosion of the masonry and mortar. Thermal expansion in the old flue can then cause cracks in the chimney at mortar joints. Installing a flue liner will prevent leaks and will improve thermal insulation. Selecting the right kind of liner is important and depends on the type of

Craig Hunter Allchimneys Ltd

fire or appliance used. Lining is a specialist job – see chapter 11.

In roof spaces, modern chimney breasts are rendered as a precaution to lessen the risk of fumes escaping. However, Victorian ones were seldom rendered, which is another good reason to have the flues lined.

---

**Cause** **External brick or stonework badly weathered**

**Solution** *Defective areas of frost damaged or spalled brick or stonework must be cut out and replaced*

If the upper bricks have come loose due to soft or eroded mortar, some localised partial rebuilding may be required. Make good old mortar joints as described below (under 'Leaning chimney') by raking out old joints and repointing. Where cracking or erosion is extreme, the stack will need to be taken down and rebuilt using sulphate resistant mortar and low sulphate bricks. Another cause of damage to chimney masonry is from poorly fitted TV aerials (or from attached power and telecoms cables), which may need to be professionally refitted. It's often possible to relocate TV aerials inside the loft.

**Cause** **Rendered surfaces cracked or loose, allowing damp penetration**

**Solution** *Patch defective areas of render*

Over time, small cracks may have allowed water to penetrate and freeze, loosening the render. Modern cement render is far more brittle and prone to cracking than the traditional lime variety. Loose, cracked, or hollow areas can be hacked off and made good with fresh render. See Repairing render step-by-step in chapter 5.

# **Defect:** Damaged pots and flaunching

## SYMPTOMS

Broken, missing, or crooked pots, problems with downdraught to fire; rain penetration; cracked and loose mortar at base of pots; some pieces may have fallen off.

**Cause** **Excessive weathering**

**Solution** *Replace broken pots. Secure unstable pots. Repair flaunching*

Up close, chimney pots are surprisingly large and heavy, so if you have to remove one you will need a helping hand. Look for frost damage in the form of flaking/delamination of the pots. The only practical solution for damaged pots is replacement, ideally with matching reclaimed or repro versions.

'Flaunching' is the large blob of mortar that holds pots in place, as well as providing a capping to protect the brickwork below. Cracks are commonplace, as the flaunching tends to decay more quickly than the pots themselves due to frost action.

But it is a fairly simple job to replace it. Take care not to damage the pots when cleaning off the old mortar. Also, cover the flue to prevent debris falling down and causing blockages.

If the flaunching is loose, it's best to remove it completely and replace it with a suitable mortar mix, carefully formed to slope outwards to disperse rainwater.

---

**Cause** **Eroded supporting masonry at the head of the stack**

**Solution** *Line the flue*

If the pots are sagging drunkenly into the chimney stack it may be due to a severe case of damp from condensing flue gases eroding the internal masonry, causing the brickwork to expand. The solution is to line the flue. The flashings are also likely to have suffered movement and may need attention. Redundant flues should be capped off and ventilated.

# Defect: Leaning chimney

## SYMPTOMS

Distinct leaning, often away from wind direction. Blown mortar joints.

### Cause

**Expansion of eroded mortar joints**

If a stack has a large exposed surface facing the prevailing cold wind, constant wetting can lead to expansion of mortar joints, causing the stack to lean away from the wind. When bricks get consistently very wet, any sulphates in the masonry or mortar can react, causing horizontal expansion cracks along mortar joints as a result of 'sulphate attack', jacking up on one side.

Fortunately, traditional lime mortar is not normally vulnerable to sulphate attack, but cement mortar joints can be at risk.

**Solution** *Rake out and repoint the mortar joints with fresh mortar*

A small degree of lean is quite common in old stacks. Many have leaned considerably more than the 'official' limit of 1mm in 100mm but in most cases are still stable. A structural engineer will need to confirm whether a lean is too extreme to be made safe. The remedy of last resort in severe cases is to take down and rebuild the stack. To comply with Building Regulations, a rebuilt section will need an internal diameter of 200mm, or a minimum of 185mm with a relined flue.

Old mortar joints may have become loose, crumbling or cracking with age, so that repointing is necessary:

■ Rake out to a depth of 20–35mm (0.8–1.4in), taking care not to damage the edges of the brick or stone.

■ Repoint with a sulphate resistant mortar. However, the strength of new mortar should be weaker than the old brick or stone so that the masonry can 'breathe' allowing moisture to evaporate via the mortar joints.

### Cause  Acid attack inside the flue

**Solution** *Install a flue liner*

Mortar joints may have also deteriorated internally. The stack may be colder on its windward side and have suffered more erosion there from condensing acidic gases, or expansion from sulphate attack may have caused uneven movement and leaning. See chapter 11.

### Cause  Structural alterations

**Solution** *Check support from chimney breasts below*

If a chimney breast has been removed but the remaining masonry above has not been properly supported, it can cause instability to the stack, which may then need to be taken down. See chapter 11.

# Defect: Damp penetration to chimney breasts

## SYMPTOMS

Damp and staining to walls and fireplaces below.

The main causes of damp in chimneys (other than the common problem of leaks at flashings) are ingress of rain from outside, and condensation from inside.

**Cause** **Condensation inside the stack**

**Solution** *Fit a suitable flue liner*
As described earlier, burning fuel produces water vapour, which turns to moisture when it hits cold walls up the flue, especially if the flue is very tall, very wide, or faces a cold outer wall. Fuels such as freshly cut timber are particularly wet, and give off a lot of water vapour when burning. Even if fireplaces are sealed up and you don't want to use them, there should be a through flow of air both top and bottom to prevent condensation leeching through the brickwork and causing damp, sooty stains. This may be visible on chimney breasts in bedrooms as well as in the roof space. See chapter 11.

**Cause** **Pots and flues exposed to rain**

**Solution** *Flues should be protected to exclude entry of rain by 'capping' the pot or fitting a cowl or hood*
Rainwater down an open chimney pot will soak into the internal chimney brickwork, as well as reaching further down the flue, where it mixes with old soot. The ensuing staining to

chimney breasts appears the same as that caused by condensation. Even where a flue has been lined rainwater that was previously soaked up by the old lime mortar parging may now dribble straight down the new flue liner, which acts like a downpipe, creating small puddles in the fireplace.

There is a whole range of caps, cowls and hoods available to protect flues that are in use, or to ventilate old redundant flues. Specialist advice may be useful here, as the wrong choice of cap can affect the way the fire draws, causing it to smoke excessively.

**Cause** **Thin and porous stack walls with no DPC or eroded mortar joints**

**Solution** *Ensure the pointing is sound*
Sometimes rain can get around an otherwise perfectly good flashing because there is no damp-proof course (DPC) above roof level in old stacks. Modern stacks have a DPC through the chimney at approximately 150mm above the roof and another one below the cap or the brickwork head. Repointing the brickwork joints can sometimes solve this problem. Otherwise, unless the old stack is being rebuilt, this can be difficult to solve, although fitting flashings that extend higher up and deeper into the brickwork is often an effective remedy.

# PROJECT: Renewing a flashing

New flashings should be formed using minimum Code 4 thickness lead.

Stacks that are located centrally over the ridge of the roof have flashings called 'aprons' to the front and back, and a 'saddle piece' in the middle. (see final photo below). Stacks located lower down a roof slope have the same apron flashing at the front but a special flashing called a 'back gutter' at the rear. Both have flashings and/or soakers at the sides. Flashings should overlap each other by at least 100mm.

The strips of lead flashing are attached to the stack by folding over their upper edges and tucking them into mortar joints in the brick or stonework and securing them with lead wedges and fresh mortar. Lead sheet is cut with special tools called tin snips and formed into shape with a bossing mallet.

**1** First fit the front apron flashing along the base of the stack brickwork by cutting a groove in a mortar joint in the chimney about 150mm above the level of the roof covering. Fix the apron at least 25mm into the groove and secure with metal wedges, then seal with mortar. It must overlap the surface of the tiles or slates below and be dressed around the sides of the stack by about 150mm.

**2** To the sides, separate lead soakers are inserted underneath each tile or slate. The soakers are folded up at the sides against the stack and then covered by a flashing cut into a stepped shape. For modern concrete interlocking tiles soakers aren't needed – instead they have a combined 'step and cover' flashing dressed over the top of the tiles.

**3** For 'back gutters' (at the upper edge of the stack) two pieces of lead are needed. One forms a gutter along the back of the stack and the other a cover strip flashing attached to the stack itself. The back gutter fits over a timber base, part of the roof construction. Recommended dimensions of a back gutter flashing are:

Upstand at rear of stack: **100mm**
Length: **width of stack** + at least **225mm** at each end
Sole of gutter: minimum **150mm**
Extension piece for roof slope: **225mm**

**4** For stacks located centrally on the ridge, the 'ridge saddle' is the last section to be installed. The saddle piece should be dressed over the ridge tile, ensuring a generous overlap with the side flashing. It should be cut with a stepped edge and anchored in two courses of the stack's brickwork.

# RAINWATER FITTINGS

These may not be the sexiest part of your house, but they're actually one of the most important. If the gutters and downpipes don't work properly, sooner or later water will invade the very heart of the property. Rainwater allowed to drip or leak down walls will lead to damp problems and, in time, rot. Worse still, consistently wet ground by the main walls can eventually undermine those shallow foundations and lead to possible structural movement.

**Left:** The rainwater system should be able to cope with sudden massive influxes

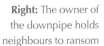

**Right:** The owner of the downpipe holds neighbours to ransom

Victorian houses can present a number of special challenges. Faults in gutters and downpipes are surprisingly common, causing rainwater to overflow and saturate solid main walls. The resulting dampness often appears at low level and is frequently diagnosed wrongly as rising damp.

In terraced or semi-detached properties the rainwater fittings act as one system for the entire building, so you and your neighbours are mutually dependent on each other. In most terraces much of the original cast-iron guttering will have been replaced – usually in all kinds of clashing styles and different materials in a perplexing variety of colours. As well as looking fairly gruesome this can cause connection problems at joints. Only where buildings are listed or located in conservation areas is it likely that everyone will have replaced the old cast-iron fittings in authentic matching materials.

You may also find that your guttering discharges via a neighbour's downpipe, which, on the plus side, saves you worrying about maintenance but means you are dependent on their DIY skills and enthusiasm for clearing blockages. So it's worth checking your legal rights, in case any emergency problems need to be swiftly rectified whilst they're away on a world cruise. And should the onus fall on you to provide an additional downpipe that benefits the whole terrace, there may well be a case for sharing the costs.

# Take a rain check

To judge whether your system is running smoothly, or is grimly clinging on by its last rickety fixings, it is worth taking a few minutes to stand back and contemplate your property. Starting at the top, visually follow the route the rainwater takes. You may find that all the water from the main roof discharges down a complex series of hoppers and downpipes, before ultimately cascading off an old lean-to roof, causing unhealthy ponding on the ground next to your walls.

The whole rainwater system should be able to cope with sudden, massive influxes of storm water without overflowing, so a simple practical test is to pour a few buckets of water into the gutters at the highest point (or carefully spray the lower roof slopes with a hose) and check for leaks, ponding, and overflows. Or invest in an umbrella and take a stroll outside on a rainy day.

# Falls

Nothing to do with Niagara – unless, of course, your stop-ends are missing. Guttering is supposed to slope or 'fall' slightly towards the downpipe to help the water flow away. The fall

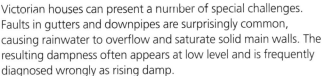

**Left:** Spot the join!

**Right:** Water feature – a missing stop end is easily fixed

should be between 10mm and 25mm per 3.5m run (up to 1in in 11ft). Too steep, and water will overflow at the downpipe outlet; too shallow, and a build-up of water and sediment may cause overflowing.

The manner in which the water leaves the edge of a roof is also crucial. Gutters should normally be positioned centrally under the roof edge, and no more than 50mm below it. But if falls are too steep, the lower end of the gutter may be too far below the roof edge, causing water to overshoot, and the gutter may even block the windows below from opening fully.

## DESIGNER GUTTERING

There is more to the world of guttering than you may think. To design a new system or to check your existing arrangements, you need to consider some key factors:

- First calculate the 'effective design area' (EDA) for each roof slope, which is roughly equivalent to its surface area in square metres. Where you have a lean-to roof against a wall, calculate the area of the wall above and add half this figure.
- To convert this to a figure showing how much water each gutter has to deal with – actual water run-off in litres per second – divide the EDA by the magic number 48.
- Work out the required size of the gutter and the number of downpipes needed.

Most roof slopes on terraced houses will have an EDA of less than 37m$^2$, which means they only need standard guttering with a 68mm wide downpipe. Standard 112mm 'half-round' gutters should be able to cope with almost 1 litre of water per second, assuming there's just one downpipe at the end. An approximate rule of thumb is at least 1 downpipe for every 10m of guttering, but the calculation is affected by any sharp bends in the guttering, which slow the rate of flow.

The optimum place for a downpipe is actually right in the middle of a gutter, as there it will double the amount of water the gutter can cope with. For even greater flows you will need to either use larger 'deepflow' type gutters or provide an additional downpipe.

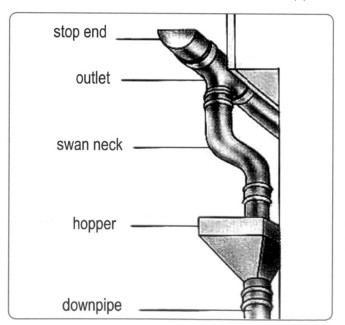

## BRACKETS AND FIXINGS

To avoid the nightmare scenario of heavy gutters toppling off and plunging earthwards, brackets and fixings need to withstand forces beyond the everyday – such as back-breaking loadings of snow and ice, and from ladders propped against guttering by decorators and window cleaners. Brackets were most commonly screwed or nailed to fascia boards or rafter feet, but sometimes driven directly into mortar joints in the walls.

Normally one bracket per metre run of guttering is necessary, with more closely spaced support where there are corners or junctions, such as to bays. Modern replacement brackets are available in traditional styles, but need to be made from galvanised or stainless steel for durability. Special 'rise & fall' brackets can be adjusted *in situ* with a threaded bar to the correct fall.

# Materials

Victorian gutters and downpipes were usually made of cast iron, which largely superseded the earlier Georgian lead-lined timber box gutters. However lead continued to be used for rainwater hoppers and downpipes in some more expensive houses. The hoppers might be cast or bossed (beaten) into shape, often with a decorative emblem like a rose, the year of construction or the owner's monogram.

### CAST IRON

Cast-iron rainwater fittings often still survive on Victorian houses and are very durable provided they are maintained. So if your house boasts original rainwater fittings, the best option is to overhaul the existing system.

Where replacement is necessary, or you want to replace inappropriate modern existing fittings or to extend the house, new cast-iron systems are readily available. There are a number of possible alternatives, each with their pros and cons. It's generally reckoned that the next best option to cast iron is cast aluminium, which looks pretty similar and is highly durable and long lasting – see below. But bear in mind that for Listed buildings, you need to apply for consent to replace rainwater fittings with anything other than like-for-like no matter how authentic a substitute material looks.

The main drawback with cast iron is its weight, plus the fact that it requires periodic decoration to prevent rust taking hold. (not forgetting the vulnerable back part by the wall). It's also dearer to buy than off-the-shelf plastic fittings from DIY stores,

Castirongutersupplies.co.uk

although if you factor in its long lifespan (subject to maintenance) in the long run it can cost about a third of the price of the plastic variety. Iron downpipes generally last longer than the gutters, so a mix of plastic guttering with original downpipes is quite common.

There are two main types of Victorian iron gutter: 'half-round' and 'ogee'. Ogee (and the similar 'moulded' style) has a wavy front and can normally be found screwed directly to the fascia rather than being supported on brackets like the more common half-round variety. Some old cast-iron gutters have wooden stop ends, which are frequently found to be in poor condition.

### WOOD

Some Victorian terraced houses (eg in Sheffield and parts of the Midlands) have moulded timber guttering known as 'spouting'. In good condition the material is robust and seldom becomes

damaged if a ladder is rested against it. But it needs periodic maintenance, and in dry weather generous coats of a bitumen-based paint should be applied. However, if the wood has started to rot, a skilled chippie should be able to replace defective sections with matching lengths made from well-seasoned timber. As part of the building's historic fabric, wooden guttering should be retained or reinstated wherever possible.

### PLASTIC

Many houses will now have replacement PVC guttering and downpipes of varying ages. These don't suffer from corrosion but can be prone to leaks at joints due to expansion and contraction,

*Sagging plastic gutters – often due to lack of expansion joints or insufficient clips*

Mayfield-roofing.co.uk

Alumasc

and older fittings tend to become brittle with age, degrading in sunlight. Resting a ladder against this type of guttering can damage it, so use a special ladder fitting called a 'stand off'.

Castironstyle.com

Unlike iron gutters, the PVC variety has a tendency to sag between brackets, over time, which affects the alignment causing discharge problems. The two main types are half-round and 'squareline', and they are available in various colours such as black, grey, brown and white. Imitation designs are now available in glass-reinforced polyester that replicate the texture and style of cast iron fittings and being self-coloured don't need painting.

## ALUMINIUM

Replacement aluminium gutters are available in traditional styles, and can provide an excellent low-maintenance modern alternative to cast iron. Crucially it's also considerably lighter, doesn't suffer from corrosion and can be customised on site. To save painting at height, heritage aluminium rainwater fittings can be ordered with a powder-coated finish.

But as with cast iron, it's not the cheapest option.

It is available either as durable cast aluminium or as the thinner 'extruded' type roll-formed on site using a mobile profiling machine in continuous seamless lengths with no integral joints. However the thicker cast type is a more authentic replacement for

*Authentic replacements for original cast iron in cast aluminium: 'moulded' (red), 'half round' (green and orange), and 'ogee' (blue).*

original cast-iron fittings and is more durable than the extruded type. Aluminium gutters are also available in replica ogee and moulded styles that can either be screwed directly to the fascia or rafter ends or supported with brackets.

## OTHER MATERIALS

At the end of the 19th century there was a brief vogue for copper guttering among Arts & Crafts architects, who liked its distinctive green patina, but because of its expense it is not found on mainstream housing. We've also not mentioned the joys of stainless steel or asbestos guttering as these are extremely rare on Victorian houses.

Period fittings ooze character – original copper gutters and downpipes (centre right) with distinctive green patina. But downpipes tend to outlive guttering, hence Victorian lead and cast iron downpipes sometimes serve modern plastic (upper left) or extruded aluminium (lower right).

## COMPARING MATERIALS

| | | Pros | Cons |
|---|---|---|---|
| **CAST IRON** | 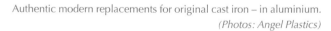 | ■ Traditional material that looks right<br>■ Long lasting and strong<br>■ Resistant to damage from ladders or snow loadings<br>■ Fairly consistent interchangeable sizes and bolted into place so parts easily replaced<br>■ New components come ready painted | ■ Expensive<br>■ Fairly brittle and if hit hard can crack or bits snap off<br>■ Prone to rust and needs periodic painting<br>■ Heavy and needs sound fixing |
| **CAST ALUMINIUM** | | ■ Very similar looking to original cast iron, so good substitute<br>■ Long lasting and strong<br>■ Resistant to damage<br>■ Light weight makes installation easier<br>■ Does not corrode, so is easy to maintain – no need for frequent decoration or can be left its natural state<br>■ New components come ready powder-coated | ■ Expensive<br>■ Not traditional and lacks surface texture of cast-iron (but thick paint can mimic)<br>■ Cheaper extruded aluminium formed in seamless lengths is relatively thin |
| **PLASTIC** | | ■ Cheap<br>■ Light weight makes installation easier<br>■ Some GRP types replicate the look of cast iron | ■ Non traditional<br>■ Looks cheap and is rarely acceptable on old buildings<br>■ Prone to warping and creaking, becoming brittle over time due to the effects of UV light. |

# Downpipes

Cast iron downpipes and hoppers were traditionally fixed with iron spikes driven into the wall through integral cast-iron 'ears'. They are often very close to the wall, so when replacing them, it's worth adding a spacer around the fixing bolt so they're positioned out from the wall by at least 25mm. This should allow for future redecoration. New fixings into masonry can be made using stainless steel screws or expansion bolts.

*Authentic modern replacements for original cast iron – in aluminium.*
*(Photos: Angel Plastics)*

## BLOCKAGES

Cast iron downpipe built into wall with consequent risk of corrosion and damp

Replacement cast iron downpipe screwed to wall via lugs

# Parapet and valley gutters

If you can't see your roof from the street, it may be hidden behind the front parapet wall, a feature of some early Victorian houses as described in chapter 2. Behind the parapet there is commonly a 'butterfly roof' with a valley gutter running down the middle. These 'M' shaped structures comprise two small lean-to roofs propped up against the party walls either side in a terrace. In between there is a lead-lined valley gutter, like a narrow flat roof, usually running front to rear. Valley gutters are also found on 'double-pile' roofs, a feature of some Victorian farmhouses. These effectively comprise two houses built back-to-back, each with its own separate gabled roof, forming a valley with a gutter running along the middle.

Unfortunately, being tucked away out of sight from street level these tend to be routinely ignored until a serious leak develops, often simply the result of blockages becoming clogged with silt, bird's nests, tennis balls – you name it, it'll be there. Overflowing water can eventually cause decay to the timber supporting beam below, which will then start to sag, causing water to 'pond' and making matters worse.

Hidden gutters sometimes suffer from design faults such as outlets being too narrow, so that with a build-up of melting snow, or during heavy rainstorms, water can't escape fast enough and overflows above the flashings, causing damage internally. More seriously, it is not unknown for the front parapet

Because it can be difficult to clear blocked downpipes without removing them, one preventative measure is to fit a leaf-guard (a wire cage) in the gutter on top of the downpipe outlet. Although it's sometimes pointed out that guards can themselves become blocked causing the gutter to overflow, the fact is, a blocked gutter is relatively easy to clear.

It's important to check whether there are sufficient numbers of downpipes to cope with a heavy downpour – for example to the front elevation of a typical terrace, you need at least one downpipe per three houses, which in practice, as noted earlier means at least about 1 downpipe every 10 metres.

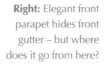

**Left:** Hidden from view – re-clad butterfly roof with central valley gutter

**Right:** Elegant front parapet hides front gutter – but where does it go from here?

walls to eventually fail, crashing to the ground with potentially fatal consequences – see chapter 2.

The lead linings themselves also sometimes fail. Lead sheet is very prone to thermal expansion and needs to be laid in lengths that are short enough to accommodate movement without splitting. The provision of an occasional shallow step where sheets overlap (known as a 'drip') is the conventional solution. Unfortunately, this is a key detail that the original builders didn't always get right. So at worst, the timber 'flat roof' substructure may need to be taken up and reconstructed with new steps formed approximately every 2.5 metres (for Code 7 lead). Replacing a lead lining needn't be too difficult, but the condition of the timber plate supporting it may not be evident until the covering is removed – and replacing this can seriously add to the cost.

Similar problems apply to properties with 'parapet gutters'

tucked away behind the front or rear parapet walls or balustrades at the top of the main walls. In order to discharge rainwater from boxed wooden lead-lined guttering running along the front of the building it was sometimes channelled at a 90° angle to 'secret' internal gutters running under attic floor voids. These would lead to a rear hopper and downpipe to the centre of the rear wall, or might be diverted to an internal downpipe in the centre of the building. Inevitably such designs have led to problems with undetected blockages and leaks in floors, and consequential damp and rot. In an emergency, a temporary repair with modern glass-fibre reinforcement may protect the property from leaks until a permanent repair or re-roofing can be undertaken. But temporary patching has a nasty tendency to be left for years. So if you come across old bitumen paint or roofing felt applied over old cracked lead, it should ring alarm bells.

# Underground dispersal

The final stage of your rainwater dispersal system is underground. The pipework is likely to be run in original vitreous clay, or may have been extended/replaced in modern plastic pipe, typically of 75mm or 100mm diameter.

Rainwater (known as 'surface water' or 'storm water') must normally be kept separate from foul drainage and bathroom or kitchen waste water, and often discharges into a soakaway or nearby stream or out into the street. For new pipework, it is not generally permitted to connect rainwater to the main system for fear of violent deluges of rain inundating the system during storm conditions causing tidal waves of raw sewage. Not pleasant. But historically in some large urban areas 'combined systems' were

operated, so it may in some cases still be permitted for rainwater to discharge into the same system that takes the sewage. But this must first be verified with Local Authority Building Control. One clue is where you can see that original bathroom waste pipes discharge into the same hopper or gulley that takes the rainwater from the gutter, rather than into a soil and vent pipe.

**Right**: Modern downpipes discharge direct to underground system

**Below**: New shoe and downpipe – with traditional 'ears' bracket – but not much use without a gulley. *(Photo: Angel Plastics)*

## GULLIES

At their base, downpipes normally discharge into a gulley, which is a water-sealed trap leading to the underground drainage system designed to keep smells at bay. The original gullies were normally made of salt-glazed clay, but if defective they can be replaced with new plastic ones.

'Ye olde overflowing water butt'

Original drainage channel in pavement for rainwater discharge to street

A common arrangement in urban terraces is for water to discharge via front garden pipes into the street (causing icy pavement in winter!)

Where downpipes terminate above the gulley, a 'shoe' is normally fitted to the bottom of the pipe to help slow and direct the flow. Being open rather than directly connected into the underground drainage system allows for easy inspection and maintenance, but it also means leaves and rubbish can accumulate causing blockages, so it's advisable to fit a grille covering over the gulley. A similar problem can occur where the 'U' bend traps can become silted up or clogged with leaves or rubble, eventually becoming obstructed. Modern 'back inlet gullies' are covered with a grid and can provide an additional drainage point for rainwater from surrounding hard surfaces (see chapter 13).

Where downpipes simply disappear straight into the ground without any sort of visible gulley, it might be worth gently excavating around them to see whether they're actually connected to an underground pipe and if so, in which direction it's heading.

## SOAKAWAYS

If surface water is allowed to collect near the walls, the ground will become marshy. In time this can rob the foundations of their support. So rainwater needs to be taken well away from walls and foundations. If connecting to the main system is prohibited, and there is no handy ditch or nearby stream then a simple solution is to construct a soakaway, assuming space and ground conditions allow.

A soakaway is the traditional 'hole in the ground' method of dispersing rainwater, which helps prevent overloading public sewers with water that does not need treatment. It takes the form of either a conventional pit filled with rubble, or a ready-made concrete chamber with holes in the walls, and must be designed to store the immediate water run-off from roofs and hard surfaces, and to then disperse this stored water into the surrounding soil.

If a soakaway is not built correctly there can be problems with overflowing or silting up. Worse, the property's foundations can be seriously affected if they are sited too close – they should be at least 5m from the house. The solution in both cases is to have them replaced. Maintenance can also be a problem – the main one being how to find it! Some form of access should be provided but often isn't.

Soakaways work best in non-clay, low-water-table areas, otherwise ground water can fill the soakaway instead of the other way around. First, you need to arrange for Building Control to check the soil and advise on the required depth and distance from the house, and to make periodic inspections during construction. This may involve digging a trial pit to the same depth as proposed, and filling it with water three times in succession to monitor the rate of seepage. The chosen site should also avoid any risk of waterlogging to downhill areas. In most cases where the soil drains well, and the soakaway is serving a roof area of less than 100m², you should be able to construct a traditional type of soakaway.

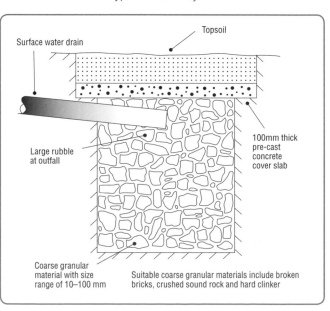

Surface water drain

Topsoil

Large rubble at outfall

100mm thick pre-cast concrete cover slab

Coarse granular material with size range of 10–100 mm

Suitable coarse granular materials include broken bricks, crushed sound rock and hard clinker

# PROJECT: Constructing a soakaway

**1** Excavate a hole of appropriate size for the total roof area and volume of water discharged (typically around 3m³) located at least 5m away from the foundations of the house and the boundaries, in accordance with advice from Building Control.

**2** Link the hole to the foot of the downpipe next to the house by digging a shallower trench and laying 100mm underground pipes with a gentle fall of about 1:100 towards the soakaway.

**3** Fill the hole for the soakaway up to the level of the trench with masonry, rock, or rubble, using large pieces around the sides to prevent them caving in. Leave large gaps between the pieces as, when complete, these will need to fill with water prior to its final dispersal into the surrounding ground through natural seepage.

**4** Place the end of the pipe so it discharges roughly into the centre of the soakaway. Then add more rubble to cover the pipe, except at the end, which must be left clear so that any debris carried in the water will fall down into the pit rather than accumulating at the point of discharge and blocking the pipe.

**5** Backfill the pipe trench with gravel around the pipe followed by a covering of topsoil.

**6** Place about 100mm of gravel around the top of the soakaway. Cover this with a heavy-gauge plastic sheet prior to covering with at least 100mm of concrete, ideally incorporating some form of inspection access. Finally, once the concrete has cured it can be prettified with topsoil for plants and grass.

## Defect: Leaking downpipes

### SYMPTOMS
White 'tide marks' on internal plasterwork; rusted pipes; wet, stained or slimy masonry walls with puddles to the ground below; apparent rising damp or penetrating damp in rooms; mould on walls, blown plaster, damp smells etc.

**Cause** **Downpipe blocked and overflowing at a joint or rusted through**

**Solution** *Overhaul or replace downpipe*
Before unblocking a downpipe, cover the gulley below the pipe to stop any debris entering the drainage system. Use a long pole or stiff wire to push down the pipe and dislodge the blockage.

**Cast-iron pipes:** a blockage that is allowed to build up will accelerate corrosion. Corrosion tends to be swifter where pipes are embedded in walls.

For maintenance of iron pipes see step-by-step.

**Plastic pipes:** in modern replacement systems, downpipes are sometimes connected directly to underground drains rather than discharging over gulley gratings, and access for clearing blockages can be restricted unless connected via a gulley with an integral 'rodding eye'.

**Downpipe split or cracked**

Solution  *Replace defective section*

**Cast-iron pipes:** if water in a blocked iron downpipe freezes it is likely to expand and split the pipe, saturating the wall to which it is fixed. Pipes that have rusted right through will discharge water straight onto the (porous brick) wall. Replacement cast-iron rainwater materials are available in imperial sizes – even replica downpipes complete with old-style fixing 'earlobes' – can be obtained from specialist suppliers.

Cause  **Downpipe hanging loose and leaking**

Solution  *Replace defective brackets or pipe*

**Cast-iron pipes:** the pipe is held by metal brackets or by integral metal 'earlobes' attached to the wall with special pipe nails driven into metal or wooden plugs. Sometimes there are spacers behind to prevent the pipe contacting the wall. See step-by-step.
**Plastic pipes:** unscrew and remove old defective brackets and replace with matching new ones. Brackets are required at every joint, and no more than 1.8m (6ft) apart.

**Plastic pipes:** a common problem with plastic downpipes is impact damage. Cracked and broken pipes should be replaced. A temporary remedy is to patch with bitumen-backed foil (sold in strips for flashings), but the surface must be dry and clean before applying.

# Defect: Gutters overflowing

## SYMPTOMS
Damp, stained, or slimy walls; splash marks to lower walls; internal damp smells; mould on interior plasterwork / blown plaster. There are several possible causes:

Cause  **Blocked gutters**

Solution  *Clear out gutters*
You may find small roof-gardens taking root and choking your gutters. To prevent leaves and silt getting into, and blocking, the downpipes, special wire or plastic 'balloons' can be fitted in gutter outlets (leaves etc. will then cause the gutter to overflow so you can clear it before it blocks the pipe). As a general maintenance point, you should clear out gutters, hoppers, downpipes, and gulleys each spring after the leaves have fallen in the autumn.

Cause  **Blocked hopper or downpipe**

Solution  *Clear out hopper or replace if corroded, split or leaking. Overhaul or replace downpipe (see page 71)*
A hopper is an open box on top of a downpipe, which collects water from pipes or guttering and channels it down the pipe

Hoppers are common on original Victorian rainwater systems, sometimes illegally adapted to also take waste pipes from bathrooms (which should be separately channelled to foul drainage system). They are a maintenance issue, being prone to blockage from leaves, debris, hair and so on, so need to be checked every few months.

## Cause — Defective underfelt

**Solution** *Check that the felt laps fully into gutters*

Victorian roofs were built without any form of underlay, which was only introduced on new mainstream housing from the 1950s. But many older roofs have now been reclad and have felt laid underneath the slates or tiles. The felt should project down into the guttering about 50mm (2in) to minimise the risk of rainwater running down the wall – don't be tempted to cut off any felt which projects in this way.

## Cause — Sagging or damaged gutters

**Solution** *Improve support and replace any defective lengths of guttering*

Cast-iron ogee gutters can be prone to splitting at the back where they are fixed to the fascia and rust has taken hold, causing dampness at the top of the walls and rotting the eaves timbers: replacement is normally the only option.

Plastic guttering can be prone to sagging or twisting due either to old age or lack of support from broken brackets or loose fixing screws. It is also easily damaged by having ladders leant against it. First remove the defective section of guttering, then secure any loose brackets before replacing the guttering. Usually one bracket per metre of guttering is necessary for good support, so you may need to provide some additional brackets. If there is no fascia to support them, the gutters may be held in place by 'rafter brackets' fixed to the rafter feet, sometimes hidden under the first few courses of slates or tiles. To save having to strip these to fit new brackets, an alternative solution is to use special brackets that screw onto rafter ends or directly into the wall. Another cause of damage to plastic gutters is where a boiler flue terminal has been positioned in very close proximity, causing the gutter to warp; a simple solution is to fit a protective metal deflector plate under the gutters.

## Cause — Not enough downpipes to discharge rainwater

**Solution** *Fit an additional downpipe*

If a well-maintained system overflows, the problem may be one of design. On older terraced properties there may be just one or two downpipes serving all the roofs of the whole terrace. As a rule of thumb, one downpipe should serve no more than three average-size terraced houses, as described earlier.

To provide an additional pipe you need to first consider the best position. Check the likely visual impact on the house, and plan a straight run that avoids windows and doors. Above all, think how it will discharge: will it be by connection to a soakaway or perhaps to the main surface water system in the street where permissible?

## Cause — Insufficient fall to gutter

**Solution** *Realign the gutter*

Some metal support brackets can be adjusted *in situ* to alter the fall of the guttering, but unfortunately most can't, and have to be fitted right first time. However, it may be possible to make small height adjustments by inserting strips of packing under a length of gutter where it sits on a clip.

If the brackets are fixed to a timber fascia board, the fascia may have rotted or may not be level, and will first need to be overhauled or replaced. Note the required fall described earlier (see page 59).

Ogee guttering differs from other types in that it is normally screwed directly to the fascia via integral screw holes. Rusted or loose screws need to be replaced.

## Cause — Corroded iron gutters

**Solution** *Replace or overhaul*

If the whole system has corroded, replacement with new cast-iron, aluminium or plastic fittings will be necessary. If it's only one section then it should be possible to obtain a matching replacement, but it's important to carefully check the shape and diameter. Cast-iron guttering is normally sold in 2- or 4-metre lengths, so it may have to be cut to size. See step-by-step.

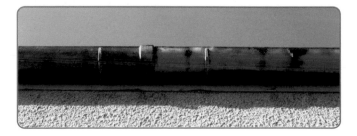

## Defect: Gutters leaking at joints

### SYMPTOMS

Localised internal damp patches to plasterwork; white 'tide marks' or green mould stains at joints; staining to walls with puddles below.

**Cause** **Defective gutter joints or fittings**

**Solution** *Repair or replace the defective gutter, connectors, or stop ends etc.*

**Plastic gutters:** listen carefully – you may be able to hear your PVC rainwater system creaking in the sun, because plastic is very prone to expansion in hot weather. Stresses occur at brackets, joints, and connectors, and in time the rubber seals can degrade or become silted and start to drip. Rubber seals can normally be replaced by undoing the union clip that surrounds the joint, cleaning the surfaces, fitting the new seal and then replacing the clip. Or the faulty joints can be replaced with new connectors, taking care to leave expansion gaps (of approximately 7mm) when

joining them. Alternatively, a flexible sealant may provide a temporary solution.

**Cast-iron gutters:** inadequate paint on cast-iron or galvanised-iron guttering leads to premature rusting. Unless the inside surfaces of metal gutters are fully protected, water that does not drain away will cause corrosion inside. Gutters are often tarted up outside with a quick lick of paint when a house is sold, so check the insides for rust. See step-by-step.

**Connecting old with new:** In terraced properties there can be particular problems when every house seems to have a different type of guttering: joining up your shiny new system with next door's decrepit old stuff may prove to be a bit tricky. Fortunately, adaptors for cast iron to PVC and from half-round to squareline PVC are available in standard sizes; otherwise it may be possible to fabricate a connection *in situ* with the help of a glass-fibre repair kit (although different expansion rates of different materials may eventually cause leakage). The only effective solution may be to replace the entire system – but talk to the neighbours first!

A satisfactory join between new guttering and old downpipes is normally fairly straightforward, with the new gutter outlet fitting inside the old pipe.

## Defect: Main walls damp at low level

### SYMPTOMS

As for rising damp – damp walls, blown plaster, damp smell, rot etc.

**Cause** **Leaks from gutters or downpipes**

**Solution** *See pages 59–61*

---

**Cause** **Downpipe discharging on ground next to house**

**Solution** *Connect downpipe to rainwater drainage system, or to street.* If the water from a downpipe is spraying everywhere except

where it should go – into the gulley. The pipe can be extended at its base with a fitting known as a 'shoe', so that it discharges accurately. Connecting pipes to a water butt is not necessarily a good solution as these very often overflow, causing ponding and damp problems.

The ideal solution is to connect to the existing underground rainwater system, but this may not be practical, depending on available space and proximity of the downpipe.

For many houses, front garden space was minimal or non-existent where fronting the street. So front downpipes were often connected to a piece of horizontal pipe running over the garden (or barely buried in the earth), through the front garden wall, and out under the pavement onto the street.

To the rear it may in some areas be possible to connect to the main system (subject to Building Control approval) or to run a new pipe away from the house to a ditch or a soakaway (see Constructing a soakaway project on page 67).

**Cause** **Blocked gulley**

**Solution** *Clear gulley and flush through*

'Ponding' of water around the gulley indicates that the water is not running away properly, which may cause damp in the walls. This is often due to a build-up of grease and solid matter in the trap, or a blockage caused by cement and debris washed down during building works.

A simple blockage may be solved by removing leaves and other detritus from the gulley grating, or can be cleared with a solution of caustic soda (take care!) or flushed through with a high pressure hose. A protective kerb surround should be placed around the gulley and a cover fitted to restrict ingress of leaves and debris.

Ensure that the water from the downpipe discharges without overshooting and splashing the wall. Modern gulleys have a direct connection to the pipe (with a 'back inlet' for clearing), or a shoe can be fitted as noted above. See also chapter 13.

# Overhauling cast-iron guttering and downpipes

Cast-iron guttering and downpipes enhance the appeal of a Victorian house, but they do require periodic maintenance. Small leaks and loose pipes can be rectified fairly easily.

However, major restoration work to old cast-iron systems tends to be a specialist job because of the need for working at height and manhandling cumbersome lengths of heavy guttering. Ideally, every component should be removed and dismantled before stripping and painting.

Cast-iron gutters were joined with caulking and putty that, over time, can become brittle and crack (as do modern neoprene gaskets). Where old joints are still sound they can just be painted over. But where rust has set in, joints need to be dismantled, stripped and painted before being re-joined using a suitable silicone mastic.

Any rusted areas need to be exposed by stripping back to bare metal. The presence of old lead paint means that chemical strippers are a safer option than power-sanding. Any sound paint is best left and lightly sanded by hand (taking safety precautions) to provide a key. Traditionally, it was common to coat the insides of gutters with bituminous paint, and any remaining bitumen will need to be removed otherwise it can trap damp. Rust can be chemically neutralised with special anti-corrosive gels or liquids. Bare metal can be primed with two coats zinc-based protective metal primer, which also provides chemical protection against future rust. This is followed by two further coats of MIO paint (micaceous iron oxide), which can also be used as a protective finish to the inside of gutters. Finally visible areas can be given a top decorative coat, either before or after refixing.

Weak spots at risk of corrosion can be found in any areas that avoided regular painting – notably to the back of downpipes and around gutter brackets. To check, feel behind downpipes for any roughness indicating rust. Hoppers should be checked inside for rust and small holes. If possible, they should be taken down so their backs can be prepared and protected by repainting.

One problem is where downpipes are partially built into walls or embedded in render. Because the surfaces in contact with walls are impossible to paint, eventually they will rust and cause damp problems. But removal is likely to cause damage. One option is to leave it in place but fit a new matching downpipe nearby as a substitute for rainwater dispersal.

When restoring an old cast-iron system, it should be possible to replace any individual components that are severely rusted with new matching replica parts. Because period cast-iron rainwater fittings were manufactured in standard sizes it's normally possible to source new or recycled parts. Where wooden stop ends were sometimes fitted, suitable replicas may need to be made up.

**1** This leaking gutter is located at the end of a run. It is secured by screws to the fascia rather than by brackets.

**2** Remove the bolt at the joint to the remaining guttering. Loosen this section of gutter and pull it away.

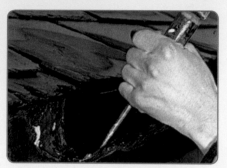

**3** Here, the leak is at the joint with the remaining section. Using a hammer and screwdriver, gently chip off traces of old sealing compound. Then scrape off old sealing compound from the joint of the guttering that has been removed. (Take care – this may contain lead!)

**4** Clean both ends of the joint thoroughly with a wire brush and clean the guttering. Take the opportunity to rub down and paint it inside and out with suitable rust-inhibiting metal paint – see below.

**5** Having applied new bitumen sealing compound, or a suitable mastic/putty, to the joint, replace the section of guttering and screw to the fascia. Insert a new bolt in the joint from above.

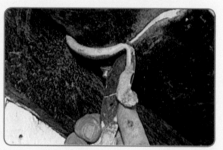

**6** Tighten the bolt with a screwdriver and spanner so that surplus compound squeezes out, and trim both sides.

# MAINTENANCE

**7** Clean out the insides of the guttering, ensuring that the outlets to downpipes are clear. Use a wire brush to remove all rust and flaking paint.

**8** Wash insides of guttering. Treat corroded areas with a rust remover. Leave to penetrate, then remove. Rinse and dry.

**9** Touch up any bare metal with metal primer (never leave bare metal exposed overnight). After 24 hours apply a coat of black bituminous paint inside guttering and allow to dry. After eight hours apply undercoat to outer areas, followed, when dry, with a topcoat. Use a cardboard strip to avoid smudging walls.

## TOOLS REQUIRED

- ■ **Ladder with a stand-off bracket (to protect guttering). For two-storeys or higher, use scaffolding or work tower**
- ■ **Screwdriver**
- ■ **Hammer**
- ■ **Adjustable spanner**
- ■ **Wire brush and sandpaper**
- ■ **Drill**

## MATERIALS

- ■ **Gutter joint sealing compound**
- ■ **Wooden or plastic wall-plugs**
- ■ **Pipe nails or screws**
- ■ **Metal paint and primer**

# DOWNPIPES

**1** This downpipe is fixed to the wall by pipe nails in wooden plugs, which have come loose. Remove nails by levering with a claw hammer (protect wall with a piece of wood).

**2** Loosen sealed joints (take care – these may contain lead). Then, having checked that the section above is fully supported, pull away the lower section.

**3** Remove old wall-plugs, then extend both holes using a 12mm masonry bit (holes approx 65mm deep).

**4** Hammer replacement plugs into both holes (cut round wooden plugs slightly larger than the holes). Take the opportunity to rub down and paint the back of the pipe with suitable metal paint.

**5** Replace the section of pipe so that the bracket holes are level with the plugs and hammer in two new stainless steel pipe-nails or secure with screws.

**6** To seal the joints, force some fresh mastic into the gap, wipe down and seal with bitumen paint.

## MAINTENANCE

**7** Thoroughly clean, and remove rust with a wire brush. Sandpaper rear of pipes.

**8** Ensure any areas of bare metal are painted with primer. Apply topcoats as described for gutters.

# THE WALLS

If our homes are our castles, it is surprising how vulnerable their solid walls are to attack. An unseen battle is constantly being waged with frost, damp and chemical erosion, as well as possible dangers from overloading or botched structural alterations. The results are plain to see – cracking, crumbling, bowing or leaning masonry, and hollow or cracked render. But if Victorian walls are known for one thing, it's their shallow foundations – with the potential for structural movement.

With few exceptions, Victorian and Edwardian houses were built with walls of solid brick or stone. Modern cavity walls were not widely adopted in most parts of the country until the 1930s.

# Solid brick walls

The front elevation of a typical brick house was built of the best available 'premier' bricks, often featuring fancy decorative designs such as string courses of dark engineering bricks or in contrasting coloured bands, perhaps topped with ornate cornices of specially moulded headers projecting at the eaves. These vibrant variations in multi-hued, patterned brickwork are one of the great joys of 19th-century architecture.

**Left:** Thicker than normal 13 inch solid wall
**Above:** Unusual solid 'header wall' with most bricks laid crossways

The Victorian builder really knew how to use the amazingly broad selection of bricks at his disposal, combining them in many decorative variations, usually laid in a Flemish or English bond. English bonds were more popular in the north and consisted of a layer of bricks laid lengthways (stretchers) alternating with a course of bricks laid crossways to show their heads (headers). A more economical version, known as English garden wall bond, used a course of headers for every 3 or 5 courses of stretchers. In parts of the north west, especially around Manchester, header bond was favoured, with each course consisting entirely of headers. But by far the most common style for main elevations was Flemish bond, with each individual stretcher punctuated with an adjoining header to create a strongly bonded wall in an elegant pattern that lent itself to 'polychrome' brickwork in contrasting colours.

Contrast this with modern cavity walls (two thin parallel walls with a gap in between) and their exclusive use of stretcher bond, with all the bricks laid lengthways. This can make matching a new extension difficult, not least because modern metric bricks are slightly shorter in length than the old imperials. If the wall is rendered or has some form of cladding on the outside, there is no easy way of determining the bonding other than checking any visible brickwork in the roof space.

The archetypal Victorian solid main wall is approximately 230mm (9in) thick, the length of one brick. On buildings of more than two storeys you may see thicker 327mm (13in) walls at ground level, which reduce by half a brick thickness per storey, the shelf thus created being used to support floor joists. But most houses had the same thickness all the way up, with the floor

**Left upper:** Flemish bond
**Left:** English 'garden wall' bond
**Above:** English bond
**Right:** Hybrid Flemish decorative bond

joists supported on protruding brick corbels or in pockets left in the brickwork. Consequentially joist ends embedded in damp walls for many years can potentially be at risk of decay.

It is not always appreciated that bricks of this era are actually quite porous and absorbent. If you take a bucket of water and place an ordinary Victorian brick in it, the amount of water it absorbs can be quite astonishing. This demonstrates the way old walls work – by temporarily absorbing rainwater until it can be released later in dry weather by evaporation. This is why modern paints, sealant fluids and cement-based mortars and renders are best avoided – since they inhibit natural evaporation, trapping moisture.

## Masonry killers

Brick and stone walls have some powerful enemies:

**FROST DAMAGE:** Unlike their modern counterparts today, Victorian bricks weren't graded for frost resistance and salt content. Consequently they can be vulnerable to frost attack when damp. Water expands by about 10% when frozen, so if bricks are unduly porous and wet, freezing will exert great pressure and cause them to 'burst' (delaminate). The exposed soft clay core will then quickly deteriorate. So keeping walls free of damp from adjoining sources of moisture, such as leaking gutters and high ground levels, is key to preventing saturation and frost damage. Walls that have been repointed in hard modern cement are especially at risk, as damp can no longer evaporate through the mortar joints, instead accumulating in the brick or stonework. Large shrubs can also harbour moisture and inhibit natural evaporation from walls. Innocent-looking ivy climbing up walls can be a particular concern since its roots erode soft lime mortar joints and allow damp to penetrate. If you can't live without shrub-adorned elevations, Virginia creeper is a better alternative that climbs by means of harmless suckers.

**SALT DAMAGE:** Frost often gets the blame for damage that's actually caused by salts. The ground on which your house is built contains natural chemical salts – sulphates, chlorides and nitrates – some of which are also present in building materials like

bricks (which is why modern bricks are classified by salt content). These salts readily dissolve in water so when walls become saturated, particularly near ground level, the salts travel in solution. The water will then evaporate out through the surface of the wall, leaving the salts behind in the form of crystals. The problem is, the crystallisation process creates pressure inside the pores of bricks, mortar or stone, which sometimes causes the surface to blow or spall – similarly to frost damage caused by ice crystals expanding. This doesn't just affect walls – salt contamination to damp plaster is well known (see chapter 6) and it also sometimes damages the undersides of slates and roof tiles (commonly seen at nibs).

**POLLUTION:** Some softer types of stone have poor weather resistance, particularly to acid pollution in the atmosphere. Coal was being burnt in prodigious quantities by the early Victorian period and coal smoke and sulphur dioxide from fires dissolves in rain as sulphurous acid. As well as causing erosion, smoke pollution rapidly changed the appearance of buildings. Whereas some sandstones adopted a dramatically blackened tone, Portland stone assumed a brilliant whiteness on faces exposed to the wind and rain whilst sheltered corners became filthy black. Brick was far more resilient to pollution. Today traffic fumes can cause similar problems of erosion.

**SULPHATE ATTACK:** Bricks that are continually saturated can potentially be at risk from sulphate attack. This can occur when sulphate salts in persistently damp masonry and mortar start to react and steadily expand, causing horizontal cracks along mortar joints. Potentially vulnerable areas include brickwork in exposed positions like chimney stacks or parapet walls, or under defective sills. Fortunately this is rarely a problem with lime mortar, so is more likely to occur where old walls have been repointed in modern cement mortar. Also, good building design should ensure that walls are protected from the persistent passage of rainwater, by the construction of overhangs like projecting roofs, sills on windows, and coping stones.

# Early cavity walls

Two single-thicknesses brick walls with a space in the middle were known to be far more proof against penetrating damp from driving rain than solid walls. A few Victorian and Edwardian homes pioneered this modern method of construction – often those sited in exposed coastal locations. Cavity construction is normally identifiable by its telltale bonding with all the bricks laid end to end with only their sides visible. However, some cavity walls were built in traditional Flemish bond using alternate 'split header' half bricks to resemble a solid wall, but should be at least 10–20mm thicker than a standard 230mm solid wall. It's worth noting that cavities of this age are likely to be relatively slim 'finger cavities' – which means they're unsuitable for injecting cavity wall insulation injected (the minimum acceptable width is normally 50mm).

The two leaves of a cavity wall are held together with wall ties at intervals. Some early walls were simply tied with the occasional brick laid across the very thin finger cavities, but most used metal ties fabricated from wrought or cast iron (in some industrial towns recycled iron hoops from cotton mills were used). Fortunately it was rare for cavities to be bound together with brittle slate ties, because over the years these are prone to snapping due to movement the walls, leaving the outer leaf dangerously unsecured. In a recent case the outer gable wall to a Victorian house in Southport was pulled completely away one stormy night by wind suction.

The required maintenance for a cavity wall is pretty much as for traditional brickwork. However, wall-tie failure is not unknown, particularly in coastal areas due to the corrosive effects of salty air. Failure is usually due to rust causing iron ties to expand, with resulting horizontal cracking at regular points in mortar joints corresponding to the position of the ties. But there's often more than one cause, such as where acidic 'black ash'

mortar was originally used on walls that are particularly exposed to the elements (usually at higher levels). Problems may also be exacerbated by recent repointing in hard, modern cement (rather than lime), trapping moisture in the wall.

In some early Victorian cavity walls the original iron 'fishtail' wall ties have rusted in the outer leaf, expanding from about 3mm to as much as 20mm. This has pushed up the surrounding brickwork, as the horizontal mortar joints are 'jacked up' causing the outer leaf to lean out and bulge. At worst, a few badly affected parts of the wall may need to be rebuilt, but normally the wall ties can be replaced and the wall repointed. Specialist repairs can be carried out using stainless-steel replacement wall ties (eg 'Dryfix helical'), which are drilled and fixed into place. But the existing old wall ties must then be 'neutralised' to prevent further damage. Where these have rusted badly, the ends can be broken off, otherwise they must either be removed completely or sleeved off. This requires a handful of bricks to be relaid in the immediate area around of each tie and repointed using a suitable lime mortar. Drill holes can be concealed by filling with resin coloured with brick dust.

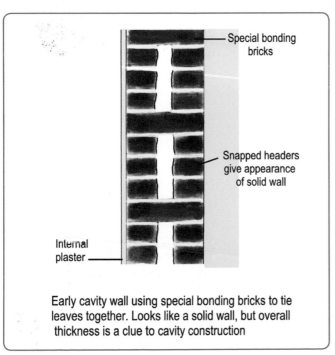

Early cavity wall using special bonding bricks to tie leaves together. Looks like a solid wall, but overall thickness is a clue to cavity construction

Special bonding bricks

Snapped headers give appearance of solid wall

Internal plaster

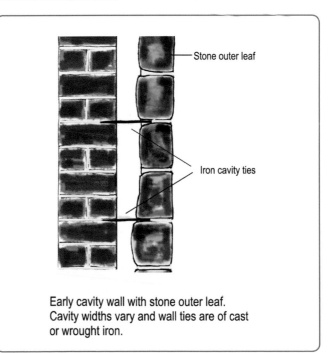

Stone outer leaf

Iron cavity ties

Early cavity wall with stone outer leaf. Cavity widths vary and wall ties are of cast or wrought iron.

# Jerry building

Victorian speculative builders suffered periodic financial pressures, which sometimes led to money-saving botches and shortcuts.

## Fake facades

To get the carcass of a terrace erected quickly and cheaply it might be hastily constructed using cheap single-thickness brickwork, comprising poor-quality underburned bricks. The facing elevations were later clad in good-quality brick by skilled bricklayers to give the appearance of expensive construction. Externally this might look like a traditionally strongly bonded wall with intermittent bricks (headers) running across both rows, tying them together. But this can be an illusion where 'snapped headers' (bricks split in half) were used. Sometimes the inner carcass would incorporate projecting 'bonding timbers' inserted at various heights in the walls in order to tie in the outer face, but being susceptible to rot, this eventually resulted in bulging front walls (a defect that sometimes also occurs where the party walls in a terrace were built in advance of the main front and rear walls but never properly tied in to them). Similarly, to create the impression of expensive solid fine ashlar stone construction at a fraction of the cost, cheaper inner walls of rubble or brick could be clad with very thin sheets of smooth stone. Or there may be no real stone at all – just painted stucco plaster craftily designed to imitate the real thing.

## Rat trap brickwork

To save money, unscrupulous builders also devised ingenious ways to economise on the numbers of bricks. For example, if what appears to be a standard 230mm (9in) thick wall has unusually tall bricks, it may actually be built in **'rat trap'** bond – a money-saving method used in some cheaper houses. Here although the bricks are laid in a Flemish bond pattern, they're actually placed on their sides, forming very thin

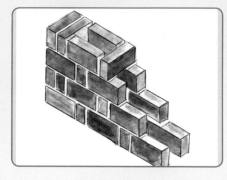

leafs with a series of small flue-like cavities within the wall (in an 'H' pattern viewed from the above).

## Thin walls

More commonly, many Victorian rear additions were walls built of cheap single-thickness brick. Also sometimes found in cheaper back-to-back housing, at only 115mm thick – half the width of a normal 9in solid wall – these are prone to damp and are potentially unstable if more than one storey high, which is why some mortgage lenders won't accept properties, even where only part of the building is single-thickness brick. However, it's not always obvious that walls are of sub-standard construction. To make it harder to spot, the wall may appear thicker (if it's been dry-lined with plasterboard), and the outer face is nearly always rendered, hiding the tell-tale stretcher bond brickwork. Even experienced surveyors can get caught out on this one.

If you have a rear addition of this type incorporating a kitchen or bathroom, they will be particularly prone to condensation and damp when warm steamy air hits the thin cold walls. All you can really do, other than complete rebuilding, is to line them internally with insulation and make sure the outer face is sound. If there's any sign of cracking or movement you will need to get a structural engineer's report, not least because this could raise significant concerns when you come to sell.

# Solid stone walls

In much of Scotland, Yorkshire, the West Country and the Cotswolds, local stone was the building material of choice.

There are 3 main forms of stone construction, from dearest to cheapest:-

## ASHLAR

Smooth ashlar cut from fine-grained prepared stone such as Bath and Portland stone, was reserved for the most expensive buildings. Precisely cut, squared blocks were laid in parallel courses and finished to a smooth 'dressed' surface, defined by joints often just a couple of millimetres thick.

## COURSED RUBBLE

This is roughly dressed quarried stone laid in courses of similar thicknesses. Rubble walls were often framed by smooth, squared

cornerstones or red brick quoins, with matching lintels and window surrounds. The quality of stonework can also be judged by the thinness of the joints.

## RANDOM RUBBLE

This is stone taken directly from the ground and laid in random sizes. Such rough-stone walls were commonly built to a thickness of half a metre or more, although they often have inner cavities between the inner and outer leaves, filled with a mix of lime, earth and small stones.

Despite its tough image, some types of stone have surprisingly poor weather resistance. Unfortunately, acid pollution was rife in the Victorian era from the innumerable coal fires belching vast quantities of smoke into the atmosphere (sulphur dioxide from fires dissolves in rain as sulphurous acid). This may explain why stone walls sometimes employed durable solid brickwork in vulnerable parts of the building, such as to stacks and flues, and below ground level. But the propensity for damage also depends on the quality of the original stone, hence cheaper buildings have generally tended to suffer the most.

So if you're wondering why your gargoyles aren't looking too perky, unfortunately it's just such ornately carved details as these that are made of the softest stone, which erodes the quickest. Spalling, staining and softening are common symptoms.

Limestone is generally the least durable stone and some varieties have suffered from erosion by acid rain caused by pollution such as traffic fumes. This can soften the detailing and blunt sharp edges, but it also has the effect of dissolving superficial grime, making surfaces appear clean. Sandstone is usually harder, though it tends to attract surface grime (for years, Glasgow was famous for its black stone buildings, now mostly cleaned, revealing a distinctive warm red colour). However, acid rain can ultimately dissolve the carbonate binding agent in sandstone. Exposed

**Above left:** Light surface erosion is part of an old building's character - not usually a problem.
**Above:** Erosion leaves pointing proud of limestone surface
**Below:** Glasgow's grey and red sandstone

sandstone sills and coping stones are particularly at risk. Granite, however, weathers well, and is extremely durable. It is common in Cornwall and north-east Scotland.

As with bricks, any wet porous stonework will be prone to frost blowing the face off, particularly where walls have been repointed with hard cement mortar instead of lime mortar. In coastal areas salt in the air or sea spray can penetrate the surface of the stone hastening erosion.

An important part of the solution is to ensure that the walls are well maintained, with effective lead aprons and flashings, watertight rainwater fittings, and window sills with fully throated drip grooves so that rain is thrown clear of the building. For large areas of stonework, regular washing with clean water can be a very effective form of maintenance (but avoid prolonged soaking in cold weather). Moss or lichen growth is not usually harmful, but ivy can damage old lime mortar joints and should be avoided.

**Left:** Sandstone attracts surface grime – Edinburgh tenements

**Right:** Moss or lichen growth is not normally harmful

# Repair or replace?

Problems tend to occur where walls are particularly exposed to the weather or where there's persistent dampness due to faulty sills, roofs and gutters etc. Once stonework becomes waterlogged over the course of many years freezing will cause expansion, and the natural layers that make up the stone will start to fracture. Where stone is exposed to pollution or extreme weather, the surfaces can eventually turn powdery as the natural glues that bind all the granules together start to fail.

The decision whether to repair or replace often depends on the location. At low level it may be better to simply leave an eroded face alone, but where stonework to parapet walls at roof level is damaged, it's probably best to replace it because it's at the mercy of the weather.

Repairing stonework is less straightforward than with brick. Where stones are cracked, make sure the cause is remedied otherwise it may simply reappear. Where you have a small number of individual stones that have eroded there are a number of options:

### ■ Do nothing

It's important not to overreact to signs of mild erosion. Naturally weathered stone needn't be a problem, and is simply a sign of age that adds character.

### ■ Check for 'water traps'

Ledges, sills and decorative string courses were originally built with a slight outward fall so they could 'throw off' water away from the house. But over time, slight settlement or erosion can cause subtle changes allowing water to 'pond' and penetrate through the walls. A discreet layer of lime mortar may be all that's needed to reinstate the necessary slope away from the building.

### ■ Enhanced protection

Limestone walls can be protected from erosion with a 'shelter coat' of limewash. Adding some powdered stone can help achieve a better match. Natural additives such as *casein* help bind and strengthen the mix. Limewash dries to a composition that's similar to limestone, and can last up to five years but is not suitable for some other types of stone

### ■ Cosmetic repairs

The stone face can be repaired using lime mortar that's specially tinted to blend in with the wall. Cosmetic repairs are often a better option than highly invasive cutting out and replacement. Localised surface decay can be filled using a carefully blended mix of lime mortar and stone dust or coloured sands or earth. Skilfully done repairs are indistinguishable from the surrounding stonework.

### ■ Fit a new face

The eroded area is chiselled back and a new matching stone face carefully pieced in. Some reconstituted stone products are made to match regional types of natural stone and can look surprisingly effective although they may weather differently over time.

### ■ Replace the whole stone

The damaged stone is completely removed so it can be replaced with a matching new piece, *eg* badly cracked stones can be cut out and replaced if structurally necessary.

Where stonework is so damaged that it needs to be cut out and replaced, either partially or fully, it raises the question of where to source suitable new materials. Where possible it's best to reuse existing stones. With luck you might have a redundant outbuilding of compatible construction. Finding matching new stone can be a tall order because it's unlikely the original quarry will still be in operation. The replacement stone should match the original as closely as possible, not just for aesthetic reasons but because different types of stone placed together can react chemically. For example limestone and sandstone shouldn't be used together because the acid limestone salts can set up a damaging process of erosion.

An area of new repair will inevitably stand out even if competently done, as it takes time to weather. Strange as it may sound, a useful 'ageing' technique is to thinly coat the new surface with a solution of water and cow dung. Brushed on and left to dry, this liquid manure will accelerate the growth of lichen or moulds.

Old eroded stonework replaced to lower floor wall only. Projecting new stonework could have formed a 'rainwater trap' on top of new upper course, so this is reasonably flush.

**Left:** Artificial stone cladding
**Above:** Value-enhanced restored property, left, in contrast with 'never paint again' spray-coated neighbour
**Below left:** Hard cement pointing causes serious erosion to stonework
**Below right:** Coated walls can't breathe; trapped damp freezes and blows face off brickwork

## Great mistakes

Guaranteed to knock thousands off the value of your house, artificial stone cladding is probably the worst ever DIY project. Made from coloured fuel ash, the tiles are glued on with mortar dabs. The trouble is they are often badly applied, leaving windows un-openable, airbricks covered, and DPCs bridged. Any cracks can trap damp, which soaks into the brickwork, being unable to evaporate externally; or can freeze, expand and blow off individual pieces.

Try hacking it off, but be warned: strong cement may leave soft brickwork damaged and pock-marked, requiring subsequent cutting out and replacement of damaged bricks, or covering in a fresh coat of render, or tile cladding. Similarly, spraying the front of an old house

with textured finishes such as 'Tyrolean' render can trap damp in old walls and should be avoided. It is non-reversible and can't be removed without damaging the brickwork.

# Crumbling and spalling

Brick and stone can become excessively weathered and increasingly porous with age, but some parts of the walls are more at risk than others. Decorative features such as projecting bands or string courses and cornices can allow water to settle and penetrate, so their top surfaces should slope down and away, and sills should have a drip groove underneath to disperse rainwater.

Victorian facing bricks have hard outer surfaces but soft interiors, although early handmade ones can be relatively porous (as can decorative reds or any under-burnt bricks). And as noted earlier, when bricks become saturated they become vulnerable to frost damage (as the water expands). This can result in crumbling,

Frost damage; persistently wet brickwork freezes and spalls; on right caused by water soaking into wall under sill.

Paint traps damp; frost blows face off

are extremely robust and were sometimes used for decorative work (as well as for DPCs).

Spalling is more likely where damp has penetrated the wall because the pointing has become severely eroded – or conversely damp has become trapped in the wall because the joints have been repointed with hard cement mortar (rather than traditional lime mortar). But spalling can also be caused by damp penetrating from inside, which is why you sometimes see bricks that have been eaten away following the line of a chimney flue up a wall.

Where a brick has crumbled badly it may be possible to cut it away and replace it, but where there are just a few spalled bricks set in old lime mortar, they can be carefully cut out, turned around, and reused.

If overall damage to the wall is extreme, it could be rendered or clad for protection, subject to aesthetic considerations.

dislodging the outer face, or the surface may start to fall away in thin layers exposing the soft interior, which has little resistance. Conversely, hard engineering bricks of a dark blue or black colour

# Repairing brickwork

The telltale sign of frost damage is 'spalling' – where the face of the brick has been blown off. Spalled brickwork looks like its been in a warzone.

There are three ways damaged brickwork can be repaired:

1  Individual decayed or cracked bricks can be cut out and turned round to expose the uneroded face. Because lime mortar is relatively soft, it can sometimes be raked out with a hacksaw blade. Angle grinders should not be used, but a series of carefully drilled small holes can help loosen the mortar around the brick, followed by a handsaw to cut around it. However, hammering with a bolster and mallet can sometimes weaken and shatter old bricks.

   The 'Allsaw' is designed to individually extract old handmade bricks, and claims to be able to discreetly cut into joints as thin as 3mm. It goes deeper than a grinder with less vibration to delicate structures. Once safely extracted, if the reverse face of the brick looks OK, it can simply be turned round, dusted off and refixed in place.

2  Where an old brick has become so eroded it's beyond reuse, the best option is to replace it with a matching brick (in imperial sizes). Replacement bricks need to match in terms of colour, size, porosity

and texture. Matching replacements may be found elsewhere around the house (eg old outbuildings), or reclaimed bricks may be available from local salvage yards. But you need to be cautious as some salvaged bricks may be under-fired and unsuitable for external work. For a larger number, a batch can be made specially by a local brickworks. But whatever the source, replacements need to be left to blend in naturally over time rather than trying to tone them down artificially. Once the new bricks are bedded in with matching lime mortar and fully weathered, they should be indistinguishable from the old wall.

3  Cosmetic surgery is an alternative way of repairing eroded bricks by building up the damaged face using tinted lime mortar blended with coloured pigments. This can be almost invisible to the eye if skilfully done but won't last as long as a replacement (perhaps 10–15 years). The first step is to brush away any loose surface material. Then mix a small amount of lime mortar to a stiff consistency with a matching tint. Just before it goes off, the mix is applied to the surface of the eroded brick. Finally, the carefully made-up face is finished with a special wooden tool. Once set, the mortar joints can be repointed around the 'brick'. The risk with this method is that the repair may weather differently to the surrounding brick over time. If only the surface is damaged, an alternative remedy is to cut a shallow recess about 30mm deep and fill it with a thin matching section of brick called a 'slip', fixed in place with mortar.' But these are hard to match and can potentially be prone to dropping off as old walls move.

# Pointing

The pointing is the visible edge of the mortar joints between the masonry. The walls of Victorian and Edwardian houses are commonly finished with discreet flush or very slightly recessed joints.

Thin joints were equated with top-quality construction and some extraordinarily fine workmanship can often be seen, far thinner than today's typical 10mm wide joints. The key thing about original pointing is that the mortar doesn't dominate the wall.

## TUCK POINTING

A cheaper alternative method was developed to give the appearance of fine-quality work applied to an entire elevation. Known as 'tuck pointing', this employed an ingenious deception. The brickwork would first be pointed perfectly flush using specially coloured mortar that matched the face of the bricks, making the mortar joints virtually invisible. Then a fine line just 2 or 3mm wide was cut along the hidden joint and filled with a mix of extremely white lime putty and marble dust or silver sand to give the appearance of a super narrow joint. Whether by design or error, tuck pointing sometimes ignored the real underlying mortar joints altogether with the fine line cut into the brickwork instead.

## REPOINTING

Old lime mortar joints are often condemned prematurely. Being a little soft or 'powdery' compared to hard modern cement pointing doesn't mean it's failed, it's just doing its job as the 'sacrificial element'. A fingernail can make an impression on lime mortar in sound old brickwork. But over the years, many perfectly healthy old walls with apparently 'soft' lime mortar have been damaged by needless repointing in much harder modern cement. The fact is, old masonry only needs repointing where the joints are so badly eroded that rainwater can settle in the grooves.

Super-fine tuck pointing – in contrast to mortar joints in side wall

Sound historic mortar should be left untouched.

Eventually, however, there comes a time when a certain amount of repointing is necessary. The art of repointing is that it should resemble the original as closely as possible. The emphasis is supposed to be on the brick or stonework and nothing mucks up the character

**1 and 2** Poor quality cement mortar pointing. **3, 4 and 5** Loosening cement with hammer and chisel. **6** But beware the risk of damage to masonry

# REPOINTING BRICKWORK

## Materials

Tub or bucket of lime mortar, hessian sheeting

## Tools

Raking out tool and hammer, fine water spray, pointing tool (or cutlery knife), small mortar hawk, stiff bristle brush

**1** Rake out at least 10mm or as required to get back to sound mortar.

**2** Clear away the dust with a dry paintbrush or bristle brush working from the top down.

**3** Wet the joint thoroughly before the work begins. If you apply new mortar to a dry joint it will not carbonate properly and will crumble and fall out.

**4** Place some mortar onto the hawk. Mortar should be of a smooth yet fairly stiff consistency.

**5** Before starting practise lifting a sliver of mortar off your hawk.

**6** Working from the top down, apply the mortar to a clean, damp joint. Mortar should be pressed in and left slightly proud of the brick face without spilling over. Avoid fiddling with it too much at this stage.

**7** Wait for the mortar to go 'green hard' – between a few hours and a couple of days, depending on the volume of mortar and the temperature. Prod it gently with a finger to see if it makes an impression – if not, the joint is ready to be finished. Any excess mortar can be scraped off.

**8** To finish off, once the mortar has begun to set, hit the joint with a bristle brush. This helps compact it slightly, forcing the mortar to the back of the joint, and closing any hairline shrinkage cracks. Finally, give the joint a fine mist spray and cover it with hessian. This slows the drying process and encourages optimum carbonation – particularly important on warm or windy days.

With luck hard cement will crack and drop off    Botched cement pointing storing up problems    Inappropriate protruding 'weatherstruck' pointing

of an old building like bad repointing. Often only small patches of walls require repointing because exposure to the elements varies, even within the same wall. South-facing walls at higher levels or in exposed locations will typically receive more of a battering from wind and rain. It's therefore highly unlikely that all the mortar will fail at the same time. So this is a job that's best done in small areas where needed, rather than blitzing entire elevations at a time.

It is a mistake to think that repointing with a strong, hard cement mortar will last forever; in fact it can actually accelerate the decay of old bricks and stones. The original philosophy was that the pointing should act as a conduit through which moisture could evaporate out of the walls. This meant that it was a 'sacrificial' material, because over 50 years or more the evaporation can cause gradual erosion; but it was better that the mortar should erode than the brick or stonework, as the pointing could be replaced more easily. Then along came the modern conviction that 'stronger is better'. It is not. It just upsets the balance. Problems can develop where repointing or repairs have been made using cement as it has the effect of trapping water in the wall. The mortar used for repointing should be slightly weaker than the surrounding masonry, and no stronger than the original.

Preparing a stone wall for repointing is essentially as described for brickwork. Then once the joint is clean and moist, the mortar

can be pushed to the back of the joint using a small pointing tool or cutlery knife, and left slightly proud of the face. Once the mortar has gone 'green hard', any excess can be scraped off to form a flush joint that's free from cracks. Finish with a stiff brush to give the joint a roughened open texture to help breathability. Cover with damp hessian so it doesn't dry out too quickly. In hot weather it can help to also give the joint a fine mist spray.

Repointing fine-jointed smooth stonework is very skilful and best left to the professionals. However, rubble stonework should be reasonably straightforward although the relatively large joints can be prone to shrinkage. So here the mortar for pointing should be mixed using coarser sand than for brickwork because the larger sand particles help minimise shrinkage.

## MORTAR MIX AND COLOUR

Victorian mortar was made from lime and sand mixed in a ratio of about 1:3, sometimes mixed with additives, such as ash where a darker finish was required. Although cement was used in some late Victorian and Edwardian buildings it was not as dense and impermeable as modern Portland cement, which did not generally start to replace lime until the inter-war period.

Repointing using lime mortar is a traditional skill that some builders may not be familiar with. Today Portland cement mortar (so called because of its grey Portland stone colour) has

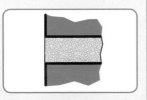

*(Continued on page 87)*

**Left:** Pointing tool used on irregular stonework
**Right:** Victorian black ash mortar

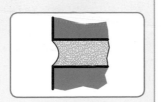

## Styles of finishing

### Flush

Flush finishing is achieved by drawing a strip of wood about 12mm wide, 6mm thick and 100mm long along the joints.

### Bucket handle

Formed by pulling a suitably shaped piece of metal (or a bucket handle!) along the joints. Also known as the 'hollow key' style.

# The importance of lime

Many old walls have been repointed or rendered with modern cement in place of the original lime-based mixes. It's now realised that because cement is so tough and waterproof it causes damage by blocking moisture from escaping, trapping damp in the walls.

Victorian bricks were fired at much lower kiln temperatures compared to those in modern brick-making and as a result they're relatively soft and permeable. Similarly, much old stonework is fairly porous. This wasn't originally a problem because the lime mortar used to build old walls tended to be slightly weaker than the masonry around it. This allowed the mortar joints to act as an escape channel, so any rain absorbed by the porous brickwork in wet weather could easily evaporate out later.

Moisture always takes the easiest route out of a wall, so if the mortar is too strong it will be forced to escape instead via the brick or stonework, which then becomes vulnerable to expansion from frost, and from crystallisation of salts, which can blow off the face of the brick or stone. But lime mortar is kind to masonry in another way. Over the years, many old walls have gradually distorted as old buildings shuffled about on their shallow foundations. Incredibly, the bricks haven't cracked or snapped thanks to the amazing flexibility of lime mortar than can accommodate, and even heal, small cracks.

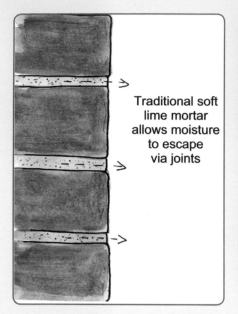

Traditional soft lime mortar allows moisture to escape via joints

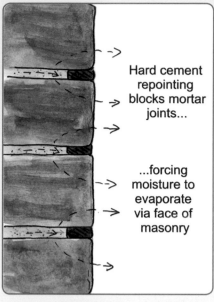

Hard cement repointing blocks mortar joints...

...forcing moisture to evaporate via face of masonry

## TYPES OF LIME

There are two basic types of lime: Non-hydraulic and Hydraulic:

### ■ LIME PUTTY ('Non-hydraulic' lime)

Lime Putty (calcium hydroxide) looks rather like thick white yoghurt, and is produced by burning pure limestone or chalk at 900°C to form 'quicklime', which is then slaked with water. N.B. this is a violent process as the lime cracks up and fizzes. Non-hydraulic limes set on exposure to $CO_2$ in the air, reverting to a form of calcium carbonate similar to the original limestone. Unlike cement, which sets by chemical reaction, lime putty needs exposure to air to cure and will

Creamy lime putty (right) Mixed with sand to make mortar (below)

not set in water. Hence lime putty can last indefinitely if stored in an air-tight container. Sometimes called 'Pure Lime', 'Fat Lime' or 'Slaked Lime', it is normally mixed with sand and other aggregates for added strength and bulk to make mortars, plasters and renders known as 'course stuff'. In its pure form, lime putty was used for very fine joints in stone masonry and gauged brickwork, and for finishing coats in plaster.

Ready-mixed lime putty is available in plastic tubs. This is the best way of buying it because the pre-mixed quality is more consistent. Some reckon that it improves with age, but containers must be kept airtight. It's also available as a dry white powder sold in paper sacks known as 'hydrated lime' or 'bagged lime', but this is not recommended because if the bags haven't been kept perfectly dry and airtight, the powder will have already reacted with the moisture in the air and reverted to useless chalk.

## ■ NATURAL HYDRAULIC LIME ('NHL')

Copyright: Studio Scotland Ltd 201

'NHL' is a tougher, more durable form of lime that sets comparatively quickly and with greater strength. It's produced from limestone containing natural impurities, such as clay and earth.

Unlike ordinary lime, NHL doesn't depend exclusively on drying by exposure to air. It sets largely by chemical reaction as soon as it comes into contact with moisture, when mixed with water, and is therefore more able to cure in wet applications (hence the name 'hydraulic'). This makes it more like cement.

Available in powder form, NHL lime mortars cannot be kept for long once mixed as they set even if kept airtight. Compared to lime putty, NHL has a relatively short shelf life and the bagged powder must be stored in dry conditions.

NHL is sold in different strengths, which reflect the amount of clay and impurities contained in the raw limestone. Stronger NHLs are more like modern cement, making them easier for mainstream builders to use.

There are three commonly available strengths:

■ **NHL 2** 'feebly hydraulic'
■ **NHL 3.5** 'moderately hydraulic'
■ **NHL 5** 'eminently hydraulic'

The numbers relate to the compressive strengths achieved after 28 days, NHL 5 being the strongest. For most pointing or rendering NHL 2 or 3.5 is suitable. But where you need a quicker-setting durable mix, such as for exposed work at high level to chimneys, stronger NHL 5 may be the best bet. Hydraulic lime mortar hardens slowly over time, and doubles in strength over the first 18 months or so.

| Choosing the right lime | | |
|---|---|---|
| *Application* | *Location* | *Lime strength* |
| Soft brick or stone | Exposed | NHL 2 or 3.5 |
| | Sheltered | Lime putty |
| Hard brick or stone | Exposed | NHL 3.5 or 5 |
| | Sheltered | NHL 2 or 3.5 |
| High level Roofs and chimneys etc. | *eg* ridge tiles, mortar fillets, flaunching on stacks etc. | NHL 5 |
| Below ground | *eg* footings, cellar pointing or rendering, retaining walls etc. | NHL 3.5 or 5 |

### SAND

Your choice of sand not only determines the colour of the mix but also helps achieve the right strength. When it comes to selecting the right sand, there is one major difference when mixing lime mortar compared to cement. You need to use well-graded *sharp sand* – rather than ordinary soft sand (aka building sand) that's used for mixing cement mortar. Builders who are not used to lime sometimes use the wrong sand, but soft building sand is not so good as a binder and is likely to lead to failure. Sharp sand comprises lots of tiny shards with angular edges and flat sides that interlock, leaving tiny voids between the sand particles, which give lime mortar its flexible quality.

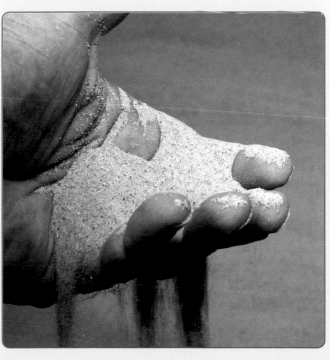

# Top tips – the art of rendering

■ **The background must be properly prepared**

Salts leeching from the masonry can cause loss of bond with new render (especially to chimneys and low-level walls). Brush them off with a bronze or copper wire brush (not steel as tiny bits can become embedded and rust). Any surface moss or lichen should be removed, *eg* with diluted bleach.

■ **Suction must be controlled**

High-suction backgrounds such as porous brick or stone must be dampened. On low-suction backgrounds (*eg* hard modern materials) or mixed backgrounds an initial 'pricking up coat' should be applied (aka 'slurry', 'bonding', 'scudding' or 'splatterdash' coat). Here a lime/sand **s**lurry is cast onto a damp surface using a harling trowel. The surface is then 'pricked up' by brush with a stippling action.

■ Apply each coat to a consistent thickness and key the surface before the next –

i.   Straightening ('base' or 'rendering') coat

ii.  Intermediate coat

iii. Flatwork – then rubbing firmly with float to take out any abrasions.

■ Allow adequate curing time between coats and protect new work. NHL render must be cured for around a week between coats without drying fully before the next.

■ Use physical guides (rules and screeds) to help form corners and angles, and to achieve a flat and straight finish.

All images courtesy of Studio Scotland Ltd

## MIXING MORTAR

For laying new brick or stonework, a weaker hydraulic lime such as NHL 2 or 3.5 is usually best, either mixed by hand or mechanically in a cement mixer. Alternatively, you can buy ready-mixed lime mortar, which saves all the hard work. For repointing, non-hydraulic putty mortars are used on older walls where the masonry is particularly soft. Mortars for repointing need to be relatively stiff and are best mixed by hand.

## LIME RENDERING

Lime renders and plasters perform best with two or three coats of similar thicknesses (although each coat can be a little thinner than the last). To help reduce shrinkage and cracking on drying, the mix should be as dry as possible yet still workable. Prior to the application of each coat, the surface should be sprayed down with clean water to prevent moisture from being sucked out too rapidly.

Lime render can take a relatively long time to set although for a stronger NHL mix the interval between coats can be as little as two or three days. However, it's essential to 'tend' it after application to prevent it drying out too quickly – or too slowly. In warm weather the finished render needs to be kept damp with a mist spray. To shield it against the drying effects of the sun and wind, it helps if new work is covered with moist hessian (also sometimes necessary for cement rendering but for a shorter time). Render is normally given a limewash finish to help protect it.

## LIMITATIONS

Lime mortars and renders carry on setting long after they've been applied. During this period they remain very vulnerable to frost, hence the use of lime externally tends to be restricted to the warmer months in late spring and summer. In fact it can take several years for lime mortar in an external wall to harden completely, and the material will always remain responsive to changes in temperature and humidity, allowing the wall to absorb and release moisture and accommodate a certain amount of movement without cracking.

| **HYDRAULIC LIME RENDER MIX** for external masonry walls | | |
|---|---|---|
| | **1 'Rendering coat** aka 'scratch coat' | **2 Floating Coat** aka 'straightening coat' |
| SAND | 2.5 parts sharp, well-graded sand | as per Scratch coat |
| LIME | 1 part NHL 2  or NHL 3.5 | as per Scratch coat |
| HAIR | Horse hair or goat hair 2kg per tonne of plaster | None |

| **ROUGHCAST  LIME RENDER  MIX** Roughcast is generally applied in two coats and for masonry walls is usually made with hydraulic lime NHL. The ingredients are mixed to a sloppy, porridge-like consistency and thrown onto the wall using a dashing trowel, with a motion similar to serving in tennis. | |
|---|---|
| SAND | 1.5 parts sharp sand (well graded) |
| LIME | 1 part NHL 2 or 3.5 |
| SHINGLE | 1 part washed pea shingle / pebbles |

replaced lime for new work as it is stronger and sets quicker, making it easier to use. However it is also very brittle, so on old buildings it is best to match the original lime. The use of hydraulic lime (NHL) is normally recommended (see page 86).

Mixing mortar of the right colour is key to a good job, and this is largely determined by the choice of sand. The sand used for repointing is sharp sand, which is quite coarse compared to softer 'builders sand' used to make cement mortar. A basic repair mix would be one part lime putty to three parts sand. Achieving the right consistency may require a mix of both sharp sand and soft sand of a suitable colour. There are also some time-honoured tricks of the trade. To help new mortar blend in, builders would sometimes add a little soot or strong tea to the mix.

But as long as the basic mix is right, lime mortar will naturally darken over time. Any wall that's freshly repointed in new lime mortar is likely to stand out slightly for the first year or so, while it weathers in to match.

Lime mortars and renders are very vulnerable to frost damage during hardening and should only be used during the drier months from May to September. Even then new work needs protecting ('tending') when drying out for at least a week by covering with damp hessian to prevent shrinkage.

Some builders add a dash of cement to the mix in a 1:2:9 cement/lime/sand mix, but conservationists regard cement as the devil's work to be avoided at all costs. It's also best to avoid the temptation to use 'white cement'. This is a product that builders sometimes throw into cement mixes to fake a lime appearance, as a substitute for lime. The problem is it looks too white and although not as strong as ordinary Portland cement, it still has cement's negative qualities.

### HOW TO UNDO OLD DAMAGE

There are some key differences between cement and lime pointing that can help tell them apart. Cement mortar very often has a drab grey (or dark greenish grey) colour, whereas lime tends to be a whitish sandy colour. However, the colour depends to a large extent on the choice of sand and strength of the mix. The real test is hardness. Try pushing the tip of a blade or screwdriver against a mortar joint – it usually sinks into lime mortar but not into cement. Or take a small lump of mortar out of the wall and try to crumble it between your fingers – lime will break up fairly easily but cement is far harder and remains in a rigid lump.

Cement is incredibly 'grippy' stuff that, once set, clings like a limpet. So where an old wall has previously been repointed in harmful modern cement it can sometimes prove difficult to remove without bits of the adjoining old brickwork also coming away, damaging the historic fabric. Thankfully, slack workmanship often comes to the rescue. Where the cement mortar was poorly applied it won't be adhering so firmly. It may have already come loose, cracked or dropped off and it may be possible to pick off small strips from the underlying lime. In contrast, soft lime mortar can't be picked out intact in pieces.

Problems tend to arise where the cement repointers did a really thorough job and first raked each joint out by 25mm or more. In such cases it may only be possible to remove the pointing with damage to surrounding soft masonry. A little careful preliminary drilling using a thin bit may help loosen hard cement.

Sometimes brickwork has been scarred where angle grinders

How to ruin a stone wall

used to rake out joints have skidded, and it may be possible to conceal any existing scars with skilled cosmetic surgery using brick dust mixed with lime.

# Detailing and finishes

Damp-proof courses (DPCs) were required by law on all houses built after 1875. Original DPCs commonly comprise two or three courses of slate bedded in lime mortar, dense engineering bricks or sometimes strips of lead or a layer of asphalt. Where damp has affected the lower walls it is very unlikely to be due to a defective DPC, so injecting a new chemical DPC should only be considered as a last resort. See chapter 6.

### PARAPET WALLS

Exposed to driving wind and rain at high level, parapet wall masonry is always at risk. It helps if the parapet was originally built with good-quality brick or stone and with decent copings on top (a good overhang with drip grooves to disperse water and a DPC underneath). It also helps if it is fully tied in to the roof structure behind. But there's always the additional risk of damp from flat roofs or gutters hidden behind parapets. The existence of unstable masonry at a great height has to be a serious concern and is best checked before disaster has a chance to strike. As with leaning chimney stacks, some rebuilding may be needed once the cause of the instability has been rectified.

## STUCCO

Stucco is smooth, external plasterwork that was a fashionable finish on early Victorian 'wedding cake' townhouses. It was usually incised with horizontal lines to imitate expensive finely cut stone but was generally applied over cheap brick or rubble stone, and may conceal a multitude of sins; hack off the imposing stucco finish on some grand London houses and you may well find cheap brick, tiles and rusting iron straps. The raw materials used to create stucco ranged from simple lime render (lime putty and sand reinforced with animal hair), to quicker-setting mixes with additives, or stronger hydraulic lime mixes. From the 1850s some builders were starting to use Portland cement but with lime retained in the mix to improve elasticity. Some more exotic stuccos known as 'mastics' were designed to repel water; these were based on linseed oil with ingredients such as limestone dust, silver sand, crushed pottery or glass and yellow lead monoxide (known as 'litharge' – a highly toxic powder added to speed up drying times). The topcoat was given a smooth, trowelled finish with grooves marked out to replicate ashlar stonework.

Stuccoing was a skilled job. It was typically applied in three coats to a total thickness of about 20mm, each progressively thinner and smoother. The perfect stucco finish would imitate real stone by being colour washed with limewater and copper sulphate or painted with oil-based gloss paint for extra realism. Minor ornaments would be pressed in or 'run' at the same time. For stucco cornices that projected further out, support was provided by bricks or thin roof tiles inserted into the wall. For decorative balcony balustrades or porches, hidden iron brackets and ties would be used. The late Victorians loved small pre-moulded features such as plaster heads over doors and windows. Ornate lintels and columns on bays and porches with 'carved' details such as leaves and fruit were also commonly made from artificial stone moulded from cement mixed with stone dust.

Stucco deteriorates rapidly once the

The late Victorians loved small pre-moulded features such as plaster heads over doors.

Stucco was commonly used for decorative features, such as flat window arches and surrounds to imitate expensive smooth stone.

Some outside corners have cornerstones of stone, stucco, terracotta, or coloured brick in a header & stretcher pattern. Known as quoins (from the French *'coin'* meaning 'angle')

Ornate lintels and columns on bays and porches with 'carved' leaves and fruit were made from moulded artificial stone

surface is broken. Problems occur from lack of maintenance allowing damp to penetrate, or when cracked areas are patched with a strong cement mix that then shrinks and cracks again. Splits are quite common around window openings but can be remedied by coating a new lime/sand render over a stainless-steel mesh that allows some internal movement. To be fully weather resistant it requires regular repainting with a suitably breathable paint such as limewash. Modern plastic masonry paints should be avoided as they prevent walls from breathing.

*Most stucco can be repaired with one of the following mixes:*

|  | **Base coat** | **Top coat** |
|---|---|---|
| Mix 1 | lime putty : sand 1 : 2.5 | lime putty : sand 1 : 3 |
| Mix 2 | cement : lime : sand 1 : 2 : 9 | cement : lime : sand 1 : 3 : 12 |

## ARTIFICIAL STONE

Manufactured building components became immensely popular in the Victorian period. Artificial 'Coade stone' was moulded into decorative forms that would have normally been carved from stone, such as lintels and columns sporting elaborate flower patterns and classical figureheads, as well as ornate friezes, statues and prancing lions etc. It is actually a type of stoneware made from baked clay that looks and feels exactly like worked stone. Other types of artificial stone were formulated from cement mixed with stone dust. Over the years Coade stone proved incredibly resilient to the assault of corrosive acids that have seriously eroded natural stone, surviving remarkably intact. Perhaps the best known example is the large lion on the south side of London's Westminster Bridge.

## ROUGHCAST RENDER

As the name implies, roughcast is a rough, lumpy render traditionally applied in exposed locations on account of its weather-resistant qualities. It became fashionable as a 'rustic' finish on some more expensive late Victorian and Edwardian houses,

often applied to gables and bays over a traditional timber lath backing. Sometimes renders were used as a handy way to disguise inferior brick or stonework.

Problems sometimes occur when small shrinkage or movement cracks in the render allow water to penetrate and soak into the wall, eventually causing the surface to deteriorate.

The rough texture is derived from shingle or pebbles mixed into the final lime render coat, which is thrown onto the wall (both masonry and timber lath) and given a painted finish. Because of its increased surface area, which maximises evaporation and protects against driving rain, roughcast provides a very effective way to protect walls.

Sometimes it's mistaken for pebbledash, a more recent variation similarly popular from Edwardian times to the 1930s. The main difference is that pebbledash is formed by throwing smaller chippings, or dry pea shingle, onto a wet final coat of smooth render.

'Old English' half timbering was a popular architectural treatment applied to many late Victorian and Edwardian

stronger and lighter than brick or stone it could be hollowed out and was used widely for decoration over doors and windows. Now usually thickly painted over, it can be hard to differentiate terracotta from stucco or stonework, although naked browny-red terracotta balustrades and grand porches can still be seen on some grander buildings.

Ornamental bricks, made from finer clay than common bricks, were used in chimney tops, decorative ridge tiles, and fireplaces. Moulded unglazed cornices and string courses, usually in red or buff colour, were especially popular in areas like Leicester and Liverpool.

Glazed terracotta tiles and bricks can be found adorning Edwardian facades and porches, and were common in areas like Cardiff and London – many pubs and public baths of the period retain their shiny impervious tiles in vivid greens, maroons, browns and turquoise.

## TILE-HUNG WALLS

Even when roofs were clad with slate, tile-hung gables and bays were popular on more expensive late Victorian and Edwardian houses, often laid in decorative patterns. As with roof construction, tiles were traditionally fixed from wooden battens, which being of good quality and well ventilated makes them generally resistant to decay. But because the tiles are hung vertically, each course of tiles needed to be securely nailed in place, although the overlap between courses could be less than on a roof, typically only about 38mm. But where tiles were applied to thin walls such as studwork, installing some form of insulation to reduce heat loss is a worthwhile improvement. Defects are similar to those found on roofs, such as the odd slipped tile due to corroded nails, and frost damage where tiles have become unduly porous. One particularly vulnerable area is beneath window sills, so a metal flashing or 'apron' was provided for weathertightness. More rarely, slates were hung in a similar manner to weatherproof walls in exposed locations.

gables. This 'fakework' comprised decorative timber strips bedded in mortar, and cracks caused by shrinkage of the timber can allow damp to penetrate. Once dried out, these should be filled with a breathable lime mortar mix prior to decoration. N.B. Some timbers to 'mock Tudor' gables of this era may be 'structural' – eg horizontal beams over main windows with their vertical supports either side.

Cracked or loose render must be hacked off and renewed once any trapped moisture has thoroughly dried out from underlying brickwork. Removing cement-based render from walls may not be so easy because it adheres so firmly to the adjoining surfaces that removal can damage the original fabric. Where render is poor or unsightly a good solution is to clad the wall with traditional tiling, which also improves insulation (ensuring that window and door sills still project adequately to disperse rainwater away from the walls).

## TERRACOTTA AND ORNAMENTAL BRICK

Terracotta ('baked earth') was made from fine sand and pulverised brick or burnt clay, usually left unglazed. Being

## LINTELS AND ARCHES

The time-honoured method of supporting masonry above openings in external walls was the arch. These were key architectural features, often comprising highly elaborate brickwork that formed a self-supporting head over doors and windows. Common types are the flat arch, the slightly curved camber arch over windows, and the semi-circular arch, common over front doors. Pointed Gothic, Venetian and Florentine arches are less common, perhaps because conservative speculative

builders didn't want their houses to look like churches.

More expensive properties would have super-fine joints to the arches above window and door openings made with pure lime putty. This 'rubbed and gauged' brickwork was achieved by rubbing relatively soft 'red rubber'

bricks to create a perfectly smooth surface allowing extremely fine joints, with each brick forming a close contact with its neighbour. Curved arches were formed with each brick rubbed into a slightly tapered shape, known as *voussoirs*. So thin were these mortar joints that the term penny joints was coined (being about the width of a chunky old penny). Today, the repair of such fine craftsmanship is a specialist job.

Behind the arch will be a hidden wooden lintel, only one brick away from the wind and rain. So if the mortar pointing is badly eroded, or has been repointed with hard cement instead of lime,

damp can penetrate or become trapped, potentially causing rot and sagging to the timber beam, and eventually allowing the brickwork above to sag. Dropped bricks in the arch may indicate decay in a timber lintel.

But skilled handmade brick arches became less common as manufacturing processes and transport improved from the mid century, making ready-cut lintels and pre-moulded arch bricks widely available. Stone lintels became very widely used as they cost substantially less than the alternative of building brick arches. Consequently the facades even of cheaper houses increasingly boasted lintels of solid stone, normally made from a single piece up to about 2.5m in length, about one brick thick, and three or four bricks high. Cheaper stone lintels and big projecting window sills were often painted from the time of construction as a protective measure to prevent erosion.

One type of lintel that can be prone to developing expensive defects are 'bressummers'. These huge timber beams were widely used to support the upper walls above flat-roofed 'balcony' bay windows – see chapters 2 and 7. Bressummers have an unenviable reputation because they were often overloaded when built and consequently could be prone to sagging – a common design fault. But serious deterioration can develop over time where damp from adjoining flat roofs and blocked gutters penetrates the wall, soaking the timber, causing rot and ultimately even structural collapse. It therefore pays to periodically check the walls over these kind of bays for fresh signs of movement, and to maintain adjoining roofs and gutters.

Skilled rubbed & gauged arches with super-fine joints

Stone lintel

Exposed timber lintel

Brick 'relieving arch' above lintels helps transfer loads away from windows

# Painting and cleaning

The Victorians often painted their front elevations at least in part. Decorative brickwork, stonework and stucco were routinely painted over with limewash or distemper. Even the brickwork itself was sometimes decorated with a light coat of red or purplish paint to help obtain a uniform appearance for slightly inferior bricks. However, this is not advisable today with modern paints because they can block breathability and trap damp. Sills, lintels and steps were often finished in oil paint. Pointing was sometimes highlighted in a contrasting colour like white, black, or red, using paint or coloured lime putty. See page 208 'Distemper and limewash'.

## CLEANING STONE AND BRICKWORK

Brickwork on old buildings may appear perfectly hard and resilient when in fact it's easily damaged. Bricks have an outer protective skin formed during the firing process, which can be

eroded by aggressive cleaning. And once exposed, the softer core will be highly porous and can allow water to penetrate through the brick. Similarly, soot staining accumulated over many years from smoke pollution formed a hard crust on the exterior surface of stonework. But this 'protective skin' can easily be stripped away, leaving the walls cleaner but more vulnerable. So abrasive treatments like sanding disks or sandblasting should be avoided.

Past attempts at removing grime using these sort of aggressive techniques have sometimes caused more problems than they've solved. Caustic chemical cleaners can leave permanent staining or roughen the surface of historic stonework, hastening its deterioration. Powerful high-pressure water jets in the wrong hands can erode pointing, loosen tiles and smash glass. Terracotta is especially vulnerable. And introducing large volumes of water to old porous walls can cause white, salty efflorescence marks and damp problems. Fortunately, today there are some gentle yet effective ways to clean both brick and stonework where there's a build up of soot and surface dirt.

Cleaning masonry needn't be a full-scale assault. The best approach is to start with some small-scale testing, and then carefully target the areas that need cleansing, avoiding sensitive areas. Simply brushing with a scrubbing brush and soapy water can often be surprisingly effective. Stone is best brush-cleaned in this way, using a fine spray of water so it doesn't become saturated. Brushes should be made of natural bristle or bronze wire – never steel which is too harsh and may cause rust stains. A plastic scraper can be useful to remove loose moss or lichen.

If brushing with water fails to remove ingrained dirt from the facings of old brickwork, a very dilute solution of hydrofluoric

acid is generally considered suitable on brick, sandstone and unpolished granite. The condition of the masonry needs to be checked before you start as any eroded bricks will be damaged further by the acid. The mortar pointing should also be in good condition. Protective goggles, masks and gloves need to be worn at all times. Dilute the acid, work in small areas at a time, and then wash it down thoroughly with a brush – trying not to soak the walls for longer than necessary. Work from the top down so that the areas you've already cleaned aren't then stained with dirt from above. Don't leave the chemical on the brickwork for too long, as making the bricks 'overly clean' may not look appropriate on a Victorian house. It is essential to carefully mask windows and openings, as the acid can etch the glass making it obscure.

Conservationists favour specialist systems using steam or superheated hot water in controlled amounts. These are the big brothers of the humble home wallpaper stripper. Operated by specialist contractors, such systems are designed to be gentle, although success depends on the skill of the operator, as jets of hot water are potentially capable of burrowing into mortar. Their use should be restricted to the warmer seasons to allow plenty of time for the fabric to dry out prior to winter, otherwise the walls could be at risk of frost damage. Note that Listed Buildings and those in Conservation Areas normally require consent for cleaning.

## PAINT REMOVAL FROM MASONRY WALLS

As a rule brick and stone should not be painted. Modern 'plastic' paints on old walls can trap moisture causing damp problems. So stripping is worthwhile. Rendered surfaces can be redecorated with a suitable breathable paint, or left naked. However, where walls are rendered with impermeable cement rather than traditional lime there's no point going to the trouble of stripping paint.

Where old paint is poorly adhering and flaking, it should be fairly simple to remove using a wallpaper scraper. But you need to be careful not to damage the surface of old bricks and lime renders, which are comparatively soft and are easily scored. Where masonry paint is more stubbornly attached to a wall surface it should respond to treatment with a wallpaper steamer. The most effective modern systems use superheated steam that can simply melt modern plastic paints. These are quite gentle and consume a minimum amount of water. The most challenging surfaces are roughcast or pebbledash renders which incorporate small pebbles, making the total removal of paint difficult to achieve.

**Left:** High temperature water jet system (120 deg C) suitable for the hardness and texture of the brick

**Below:** Transformed; wonderful 'polychromatic' brickwork released from behind a century of grime

Protinus.co.uk

Protinus.co.uk

# Structural movement

One of the key differences between period properties and their modern counterparts is the way they cope with movement. Modern houses are designed as rigid structures with deep concrete foundations normally extending at least a metre underground, penetrating below the relatively unstable upper layers of the soil that are prone to frost penetration and seasonal moisture changes. In contrast, old houses were built to accommodate a certain amount of movement, rather than trying to prevent it occurring. The use of lime was crucial in allowing gentle movement to occur over a wide area through mortar joints. In contrast, hard modern cement is rigid and inflexible, resulting in cracking.

## BOWING AND LEANING

A common design weak point is the side 'flank' wall (eg at the end of a terrace or semi). Because the ceiling and floor joists typically run from the front to the rear, parallel to the side wall, there may be very little holding this wall in place over its full height (apart from either end where it joins the main walls). In contrast, the front and rear elevations tend to be more robust as they're normally tied together by the joists. To make matters worse, in many houses the stairwell is next to the side wall, which as a consequence will be unsupported over two or more storeys where it abuts this large area of open space. Such a wall may develop a vertical lean, or sometimes a distinct outward bowing at first floor or roof level. This is why tie bars can often be seen on the end walls of terraces.

Front walls that support heavy gables or cornices etc. at roof level can also be prone to movement due to lack of restraint and poor design. Parapet walls on early Victorian 'butterfly' roof houses are particularly vulnerable. The main walls to the rear generally have an easier life, thanks to lateral support that's usually provided by an adjoining rear addition.

As we saw earlier, bulging can sometimes develop where the walls were originally built cheaply in two thin, separate rows with

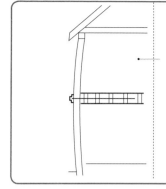

This side wall has bulged half-way up because the first floor joists are parallel to it and do not provide any support (or 'lateral restraint'). Ties can be built in at first floor level, fixed to the floor joists, if the wall is still reasonably stable.

only a few small bonding timbers tying them together. Traffic vibration or decay may have eventually caused the two rows to separate causing localised bulging. A similar problem due to 'out of sequence' construction can occur where the internal party walls (built by apprentice labour using cheap materials) were not properly tied into the main facing walls (built by experienced bricklayers) other than perhaps with a few bonding timbers. Again these may now have rotted, causing 'bellying' of front walls due to lack of restraint.

So if a solid brick wall shows signs of bulging, the first question to ask is whether the outer surface layer has parted company from the inner row. Alternatively the wall may have shifted as a whole, for example where joist ends have rotted away leaving a large expanse of wall unrestrained. Structural alterations are also a common cause of problems, such as where

*Right:* Bowed front to house in centre restrained with tie bar; house on left has rebuilt front elevation
*Below (3):* Signs of long established movement

**Above:** Newly installed tie bar
**Right and far right:** Distorted walls have now stabilised

concentrated loadings from new beams lead to bulging to the wall below.

Other than localised rebuilding of the wall, which may not be practical or economical, the solution tends to involve the insertion of tie bars or spreader plates in order to provide lateral restraint and tie in the affected parts of the walls. Discreet modern 'Bow ties' are frequently used to secure a bowing wall by anchoring it to newly installed floor joists. Or where a main facing wall has pulled

away leaving a gap at the abutment with an internal wall, grouted 'Cem ties' can be installed to reconnect them.

Where a wall is leaning outwards at its upper level, a common cause is 'roof spread'. This is where the rafter feet are pushing the top of a wall out, because the rafters aren't securely tied in by ceiling joists or collars. This can sometimes be spotted upstairs where gaps to window reveals are wider at the top than at the bottom.

Internally, the visible signs of bowing and leaning can be seen as vertical cracking to the plasterwork where the front wall meets the party walls; cracks between the ceiling and the front wall; and gaps between the floor and skirting.

However it's very common for old walls to show signs of past movement, so there's normally no need to panic. A wall that is out of plumb by up to 25mm, or bulges by up to 12mm over a single storey, is considered within acceptable tolerances and should not normally require structural repairs.

But a wall bulging out or leaning can cause further problems, because floor and ceiling joists are likely to come loose so they no longer meet the wall. One useful solution can be to install steel joist extenders that, as the name suggests, extend the joist so it can rejoin with the wall in its new position. Once reconnected to the masonry, this should restore lateral restraint and resist further movement. See page 174.

## Structural movement

Structural movement can manifest itself in different ways. Some common signs are:

■ New cracks appearing in the main walls or internal walls. New cracks can be distinguished because they're clean inside, whereas an old one reopening will contain dust or dirt. Stresses in the main walls are sometimes accompanied by fresh cracking to ceilings.

■ Any openings in a wall are structural weak points, hence cracking commonly runs diagonally from the corners of windows and doors. However, cracks to plasterwork above windows may be a sign of rot in a hidden timber lintel. And where windows and doors suddenly start to stick in winter this could simply be due to swelling from damp rather than structural movement.

■ Dropped arches: arches support the walls sitting on top of them by locking together under load and pushing out at their sides. Consequently they rely on a sideways compressive force to retain their shape. So where an old wall has moved around over the years, it can sometimes allow the arch to relax out of compression so it can drop or sag. With skill, dropped arches can sometimes be reset by inserting thin wedges. Or support can be provided from purpose-made stainless steel supports inserted underneath. Alternatively, arch lintels can be 'invisibly' reinstated by applying self-tapping 'Dry Fix' rods from below. These are inserted up through the lintel and into the masonry above, having first stabilised the brickwork over the lintel with horizontal steel reinforcing rods bedded in mortar joints. See page 100.

## FOOTINGS

The Victorian house is reputed to suffer from one very common defect – inadequate foundations. But millions of these properties have survived perfectly well until now, so do we really need to worry? Despite their lack of 'big boots', most old houses have managed to remain perfectly upright thanks to the walls being reasonably thick, with adequate restraint from floors, ceiling joists and internal walls.

Traditional foundations, known as 'footings' were built in a stepped configuration widening out at the base to spread the load, like sticking out feet. These pre-dated modern concrete strips, which superseded them from the 1930s. If the Victorian builders were competent they would start by digging a trench about six bricks deep (about 450mm) and at the base lay a footing of about three brick lengths (around 600 to 700mm). This would narrow with each layer of bricks stepped up to the wall at ground level. This was just about adequate on a reasonably firm subsoil, but being relatively shallow it was common for walls to settle in the early years after construction as the ground beneath compressed. Only where properties had cellars would the foundations have been much deeper. By the 1890s poured concrete was starting to be used in some foundation trenches but with stepped footings built on top.

Foundations need to be deep enough to avoid unstable topsoil that's prone to frost penetration causing expansion of the earth, and seasonal moisture changes, all of which cause the ground to move. But by today's standards many of those constructed by Victorian builders were not sufficiently deep. Fortunately, houses of this age were built to naturally

accommodate a certain amount of seasonal ground movement thanks to their use of flexible lime mortar, rather than rigid modern materials.

## SOIL TYPES AND SEASONAL MOVEMENT

The type of ground the house is built on is a major factor in determining its potential for movement. Chalk and rock are generally the firmest types of subsoil. Sand and gravel also have good load-bearing strength when well compacted. Worst of all is peat, which has very poor load-bearing capacity.

Shrinkable clay is one of the most common subsoils in Britain and is prone to seasonal changes, drying out in summer, shrinking and cracking in times of drought. In wetter months it is prone to swelling and heavy frosts can cause the ground to expand. All this is exacerbated by the effects of tree roots and changes to the underground water table.

Inevitably, houses with shallow foundations built on shrinkable clay (eg in London) will be more likely to experience seasonal movement. Hot dry summers can cause the clay to dry out and shrink, leaving many shallow foundations unsupported until the subsoil is later reconstituted as wetter conditions once again prevail.

But with our modern-day paranoia about even the tiniest of cracks, we sometimes forget that where a building has stood for well over a hundred years it will have been exposed to such movement many times before. By monitoring cracks over time, it can be established whether they simply open and close in tune with seasonal ground movement. As long as cracking gets no worse year-on-year, the chances are no further action will be needed. The advice from the insurance industry is that there's only likely to be a long-term problem if cracks do not eventually close, or if they continue to open beyond widths of around 5mm, in which case you should notify your insurance company.

## SUBSIDENCE

The mere word 'subsidence' is enough to strike terror into the hearts of all but the hardiest of homeowners. Whereas 'settlement' tends to imply an acceptable, limited movement of a new or altered structure as it settles down, 'subsidence' has an air of menace about it. The key difference is that subsidence refers to the ground beneath a footing or foundation giving way as a result of external conditions – nothing to do with the loadings placed on the ground by the building. If the ground under part of the footings shrinks (eg due to prolonged periods of dry weather exacerbated by trees extracting moisture), it can rob the wall of support, making it vulnerable to subsidence. This downward movement of the ground supporting a property is very different from a house that has settled over time as the ground is slowly compressed.

By way of contrast, all buildings settle in the years immediately following construction, or in response to major new structural changes such as extensions and loft conversions, as the ground adjusts to the new weight imposed upon it. Where the ground is clay, peat, or silt, settlement can continue for a number of years.

Subsidence tends to affect localised parts of a building after a change in the load-bearing capacity of the ground directly

# Tree-related subsidence

Most subsidence damage is caused by trees and large shrubs extracting moisture from shrinkable clay subsoils upsetting ground conditions. A mature deciduous tree can remove in excess of 50,000 litres per year. Some of the worst offenders are thirsty broadleaf species such as poplars, oaks, willows, ash, plane and sycamore trees, as well as fast-growing leylandii and eucalyptus. But you need to be wary of any species of tree, or even ornamental shrubs in very close proximity to buildings with shallow foundations. Roots can potentially extend more than one and a half times the mature height of the tree, so any structure within this 'influencing distance' may be affected.

Trees can also be an indirect cause of subsidence. For example where moisture-seeking roots enter underground drains causing leakage, the ground can eventually become soft and marshy. Despite these risks, there's a lot of paranoia about trees thanks largely to mortgage lenders overreacting.

Vegetation-related subsidence is cyclical – trees take moisture from the soil during the growing season from April to October. The damage usually reaches maximum severity by the time deciduous trees shed their leaves in the Autumn – which is when most insurance claims are submitted. But as long as there's a decent amount of rainfall over the winter the cracks may close as the ground recovers its normal volume.

## Insurance claims

Cover for subsidence, heave or landslip was introduced into domestic insurance policies in 1971 following pressure from mortgage lenders. Traditionally policyholders accepted small cracks as part of home-ownership.

In most 'subsidence' cases only cosmetic damage is caused, and the majority of claims are rejected as invalid, because the cause lies elsewhere. A large number of subsidence claims relate to differential movement, eg between conservatories and the original building – see below. Of the valid claims, around 70% are tree related and about 10–15% due to drains.

Peaks in claims correspond with years when there are long periods of drought, such as between 1989 and 1992, 1995 to 1997 and 2003. Most claims are in 'high plasticity' shrinkable subsoil areas such as London and the south-east. The average duration of a claim is 18 months, with a period of monitoring often required over 12 months.

## Solutions

There are two basic options: remove the offending tree or manage it. But if you simply cut down a large, thirsty tree, the ground may then react by swelling with moisture (which is no longer being absorbed by the tree). And if wet saturated clay soil freezes it expands even more, pushing the shallow foundations upwards. Known as 'heave', this is the opposite of subsidence, although it's a much less common cause of damage.

Tree management aims to reduce the amount of moisture uptake and the drying-out effects on the soil. One well known remedy is 'pollarding' – severe pruning. But not all trees will tolerate heavy crown reductions. For pruning to be effective in controlling water use, reductions of around 90% are required. And some species (eg willow, poplar, cherries) respond to pruning with vigorous new growth, which quickly restores the tree's moisture uptake. This is why specialist advice should be sought before deciding whether to prune, pollard, or remove a tree. Be aware also that cutting down a tree protected by a Tree Preservation Order or in a Conservation Area without consent can lead to prosecution.

Root barriers can provide an alternative method of tree management. This involves excavating a trench around 4m deep between the offending tree and the building and inserting special large sheets made of rigid plastic about 4mm thick. This should protect the whole building from the influence of the vegetation. This method can have considerable cost advantages (compared to the need for expensive underpinning), plus there's the amenity benefit of retaining trees. However, there can be practical limitations, for example barriers can't be inserted so close to the tree that the main roots are damaged, risking instability. It's also important that they are sufficiently robust and designed to accommodate any drainage pipes and underground services.

## Subsidence damage

Cracking in walls due to subsidence usually has the following distinctive features:

- ■ Cracks extend through the damp-proof course down to the foundations.
- ■ Vertical or diagonal cracks taper in width from top to bottom (*ie* are wider at the top or V-shaped).
- ■ Diagonal cracking is often symptomatic of localised subsidence, for example where the ground under a corner of the house has subsided due to a broken drain.
- ■ Cracking on a wall is visible from both inside and outside the property.
- ■ New distortion to the walls is likely to be reflected in doors and windows sticking severely.

Temporary support to base of main wall with acro props prior to partial rebuilding

beneath. Anything that disturbs the balance between the building and the ground can promote new movement. It is usually identifiable by corresponding tapered cracking to the wall. One common cause of such 'ground collapse' is where leaking drains wash away or soften the soil, causing the structure above to drop. The effects of tree roots, droughts and heavy frosts can also contribute to destabilising the ground. Old mine workings can be another cause, as can coastal erosion and any new nearby excavation work, such as an extension being built next door. As a direct result of such changes to ground conditions, the unsupported part of the wall immediately above can suddenly drop, separating itself from the rest of the structure.

### UNDERPINNING

Where part of a wall has cracked and sunk because the ground below has become unstable and can no longer support its weight, the standard response is to excavate down below the level of the footings to stable ground and pump the void full of concrete. Which can work fine for modern properties. But this is an enormously expensive remedy that can actually create future problems when applied to old buildings. The most likely outcome is that the problem will simply be moved along a bit, with new stresses set up between the part of the building that's now rigid and those that remain flexible continuing to move slightly as ground conditions change with the seasons. Where a building has been only partially underpinned (such as just one house in a terrace) this is likely to result in 'differential movement' causing fresh cracking at the junction of the repaired area.

Insurers and mortgage lenders have little understanding of old buildings, but this doesn't prevent them from sometimes stipulating inappropriate works. Yet houses that have been underpinned never seem to quite recover

from the stigma to the extent that insurers in future will often decline such properties!

Underpinning is highly invasive and disruptive and is too heavy handed a way of dealing with minor movement at the base of old walls. A better approach is to restore the support that's been lost rather than trying to bring part of the structure up to modern standards of rigidity. This can be done by 'underbuilding' the defective section of wall, laying down a few courses of brick or stone to re-establish contact with firm ground. The important point with structural repairs is that they should be in tune with the whole house and not just restricted to modernising one small part.

Fortunately, structural work is not normally necessary where movement has ceased and is unlikely to recur, or where the rate of movement is manageable and doesn't threaten the stability of the structure: just expect the need for occasional redecoration and adjustment to doors and windows. This was the traditional way of dealing with movement before insurance policies included subsidence as a 'claimable' defect. Sensible precautions taken in the first instance, such as pollarding (severe pruning) or removing nearby trees and repairing leaking drains, may solve the cause of the problem. See opposite page.

UNDERPINNED

New differential movement stresses

## DIFFERENTIAL MOVEMENT

As a rule, foundations should be the same depth for all walls taking similar loadings. So where you have a more modern extension built with deeper foundations, cracking can occur at or near the interface, due to different rates of settlement between the two structures. One way to accommodate this is to provide a flexible joint between the two parts of the building, allowing them to move harmlessly against each other without cracking.

Differential movement is, of course, nothing new. Victorian bay windows were often built with shallower foundations than those to the main house and may be prone to travelling at different rates over the course of the year. Similarly, in a typical terrace, each party wall will be supporting loadings from two houses, one on either side. But the end walls may only be required to take the load of one house. This can result in greater settlement to the party walls than to the end walls, with the houses at the end of the terrace adopting a weary lean towards their neighbours, their walls and floors having gradually settled inwards.

More dramatic consequences are sometimes seen in earlier Victorian houses built with basements beneath part of their ground floor. These effectively have two different foundations depths – the basement area extending relatively deep into the

ground, and much shallower foundations elsewhere. Over time, some properties have developed a 'skewed' appearance where the shallow half has settled and pulled away from the sturdier part over the deeper basement. In extreme cases this has caused underground waste pipes to fracture and leak; the resulting marshy ground has then robbed the shallow foundations of their remaining support, so that part of the house eventually sinks, splitting the side wall.

# Historic movement

Most old buildings show signs of having moved in the past. Leaning walls and cracks in masonry are often symptoms of old foundation settlement or other longstanding structural movement that has now stabilised. All kinds of bulges and leans may have existed for many years and the building has adjusted quite happily and found a comfortable 'position of repose'. Old cracks will often have been filled with mortar, and if the crack hasn't subsequently reopened it indicates that the movement has ceased.

Common areas where signs of historic movement can often be spotted are:

■ Distorted door openings where old doors have been trimmed into a tapered shape to fit.
■ Brickwork courses that rise and fall in old walls. The bricks were originally laid reasonably level, but as the ground has shifted slowly over time the walls have adjusted. Crooked walls are rarely a cause for concern.
■ Old spreader plates and tie rods fitted to provide lateral restraint and tie in bulging walls. Sometimes installed at ceiling joist level to prevent 'roof spread' where the rafters push out the walls.

■ Sloping floors where the resulting gap at the base of the skirting has been concealed by fitting beading or adjusting the skirting boards.
■ Parallel vertical cracks extending up the main walls. In houses where the window openings on each storey are positioned directly above each other – a common feature with early Victorian town houses – thin vertical cracking can appear in the wall directly above or below the window reveals. This is because loadings are concentrated via tall masonry 'columns' either side of the openings. Sometimes this is the result of damage from wartime bomb blasts.

## Bomb damage

Many urban Victorian houses were affected by bombing in the Second World War – both from air and ground pressure. Rendered end-terrace walls featuring external chimney breasts may indicate a former bomb site. Today many houses still sport discreet scars. Telltale signs include:

■ **Ground floors** humped up in the middle due to enormous air-blast waves entering via the air bricks. Whole floor structures were lifted up in the blast before crashing back down, but not quite settling back into place on the sleeper walls
■ **Vertical cracks** through brickwork between upper and lower windows, often two thin parallel vertical cracks. The brickwork 'panel' between the top and bottom windows was vulnerable to movement. Cracking typically extends from the bottom corners of the top window down to the top corners of the window directly below.
■ **Lath and plaster** ceilings damaged and hastily re-clad with fibreboard or asbestos cement sheets with battens across the surface.
■ **Windows** suffering distortion to frames and broken glass – often one or two panes would have been swiftly reglazed with cheap mottled, milky coloured semi-opaque glass.

## CRACKING

All buildings move and most crack. Very often the cracks are longstanding and of little significance. Sometimes cracking can be a danger sign, but more often it's simply telling you that the building is adjusting to changes in the weather and surrounding environment as the year progresses. As a rule of thumb, superficial cracks up to about 1mm wide are unlikely to be of any great concern. Minor cracking may simply reflect the fact that an old building is shuffling about on its relatively insubstantial foundations, as it always has. These tend to be fairly narrow and uniform in width, and can be filled during redecoration.

Diagnosing the causes of cracking is a specialist area and a structural engineer's advice should be sought. Investigation can initially involve digging small holes next to the walls to check foundation depths. Before recommending any action, an engineer will normally try to establish which part of the building has moved and which has remained static. This normally involves monitoring over time (eg by fixing calibrated glass 'telltales' over cracks) and recording any further movement to determine whether or not it's continuing. Even if it turns out to be 'progressive', it may still not be considered serious enough to warrant carrying out any major remedial work. Monitoring is commonly carried out over a 12-month period to check whether cracks are simply opening and closing with the seasons – which shouldn't normally be a concern.

But before employing expensive experts, it's often worth monitoring the cracking yourself. The simplest way to see if a crack is 'live' is to mark where it ends with a pencil and record the date. If the crack continues to grow, do the same again at regular intervals – rather like a height chart for growing children.

There are no foolproof rules for distinguishing the causes of movement, but when parts of a building sink or subside, any ensuing vertical cracking often appears wider at the top – a tapered 'V' shape. Conversely, vertical cracking caused by the ground swelling up – 'heave' – is generally narrower at the top.

When diagnosing cracking, points to note include its direction, its depth, and whether it tapers – ie is it wider at the top or the bottom. If a crack extends right through the wall to the inside, it's more likely to be serious. The Building Research Establishment (BRE) defines cracks up to 5mm wide as 'slight'. Cracks of less than 3mm (the width of a £1 coin) are less likely to be considered serious. In some cases the cause of settlement may lie below the highest point of a diagonal crack (eg a leaking drain). But these are only rules of thumb and should not be relied on exclusively – the cause and appropriate solution will depend on many factors.

One thing that is generally agreed is that cracks should be repaired to prevent water entering and causing further damage.

## Common causes of cracking

### ■ Shallow foundations

Problems with the foundations generally show themselves by fractures through the brickwork, often via window openings (the weakest point). Take a look at the bay windows in a row of terraced houses. Builders often skimped on foundations to subsidiary structures such as bays, porches and small rear additions. So these areas tend to move around more in tune with seasonal ground changes. As a result, bays etc. can be prone to settling at a slightly different angle to the main house with its deeper footings due to differential movement – see opposite page. In extreme cases, an attached structure with shallow foundations may have pulled away taking an adjoining part of the house with it. Matters are made worse where rigid modern cement mortars and gypsum plasters have been used as they don't have the flexibility of the original lime mortars that allowed a building to adjust without cracking.

### ■ Structural alterations

Problems can develop when old walls and floors are disturbed – such as when new openings are cut, a window enlarged, or internal walls removed, setting up new stresses. Structural alterations may have been carried out without taking account of the loads carried above, so all of the old load may now be thrown down through the remaining brickwork, causing parts of the foundations to adjust to the new unequal loadings and settle. A common area of cracking is over weak window lintels where the window has been replaced and the masonry above has been disturbed.

Sometimes mysterious signs of movement arise because of the activities of neighbours. In a row of terraces, each property is reliant on its neighbours for some degree of support, for example where floor joists are built into party walls. A terrace of houses is effectively one long cellular structure where the internal walls of individual rooms all help stabilise the complete row. It's not unknown for works such as illegal basement conversion excavations to undermine party walls, causing dangerous structural cracking in the adjoining houses.

Illegal basement conversion next door has caused serious structural movement

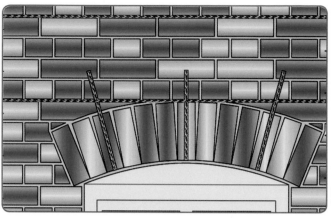

## REPAIRING CRACKS

External cracks should be pointed up with lime mortar to prevent the ingress of water, and smaller internal cracks can be filled with a flexible plaster filler. With wider cracks, it's worth recording the date they were filled in case they later open again after filling.

Where structural repair work turns out to be necessary, a masonry wall can often be stabilised by 'stitching' the crack. This involves the insertion of stainless steel 'helibars' bedded into horizontal mortar beds either side of the crack. They can also be inserted vertically from underneath. The bars are bonded in polyester resin and span across the crack, binding it together. Such repairs should be invisible once the wall has been pointed.

## Defect: Defective render

**SYMPTOMS**

Cracking or bulging of rendering.

**Cause** **Frost action on moisture trapped between the rendering and the wall**

**Solution** If the render is loose but the wall is intact, damage is often due to water penetrating through small cracks, then freezing and expanding. Defective areas of render should be hacked off. The wall surface then needs to be allowed time to dry out and any damaged masonry repaired. The render can then be replaced with a suitable new lime render. A breathable finish such as limewash can then be applied.

**Cause** **Differential movement to the wall**

**Solution** *New render should be carried across the joint on metal lathing to allow flexibility to accommodate future movement*
Cracking is commonly found where render runs across vertical joints, often between the main house and a later extension. Different foundation depths and loadings mean that these will move at a different rate, causing seasonal cracking. The render should be carried across the joint on stainless-steel lathing. If the cause is due to the wall itself being unsound, see the sections on 'Cracking to brick or stone walls' and 'Bulging brick or stone walls'.

**Cause** **Incorrect render mix**

**Solution** *Remove cement render and start again using traditional lime render*
A common problem is too strong a mix of cement render has been applied leading to shrinkage cracking and moisture penetration. If old lime render is patched with hard modern cement, the water, which cannot evaporate through the modern material, will instead be driven out through the old, causing it to fail at the edge of the repair. Hack it off and start again.

**Cause** **Physical damage**

**Solution** *Re-render wall*
Rendering a large area is a skilled job but patching is less difficult – see step-by-step. Alternatively, the wall could be re-clad with appropriate tiles to battens fixed over the old render.

## Defect: Bulging, bowing or leaning walls

### SYMPTOMS
Walls bow outwards; cracks internally to plasterwork.

**Cause** **The walls were not fully tied in when built**

**Solution** *Eliminate the cause of the bulging, then repair or rebuild the affected wall sections*
Inadequate restraint is common in side walls (*eg* to end terraces) and may require building-in metal ties from the outer wall to the floor structure. Remedial repairs may also be required where the outer skin of a thin 'snapped header' wall has come loose and is bowing.

**Cause** **Rotting timbers built into walls**

**Solution** Old bonding timbers may have rotted and expanded. Rotten timbers (*eg* floor joist ends) must be cut out, the cause of dampness remedied and ventilation improved. Defective lengths can be replaced with new pre-treated timber, protected with a DPC. This may require temporary structural support to floors.

**Cause** **Roof spread**

**Solution** *Provide additional restraint*
Rafters lean against the top of the main walls and can push them out if not restrained by collars or ceiling joists. The timbers may be poorly nailed together or may have rotted. Roof spread is more common in rear lean-to roofs (*eg* over kitchens), where the top of the wall can be seen to bow out and the roof above sags where the rafters have sunk. The solution is to provide additional restraint such as new ceiling joists or collars, and, in serious cases, to rebuild the upper wall.

**Cause** **Overloaded walls**

**Solution** *Increase support*
Walls built too thin, or alterations to a building can add to the loadings. Excessive loading of the structure (*eg* due to botched structural alterations) means the specific cause must be identified and additional support provided – a structural engineer will need to advise.

## Defect: Defective tiling

### SYMPTOMS
Broken or slipped tiles or slates; internal damp patches.
   Tile-hung gables or upper walls are fairly common in more expensive late-Victorian and Edwardian houses.

**Cause** **Decayed nails, battens or defective tiles or leadwork**

**Solution** *Replace slipped or missing tiles after checking condition of nails and battens*
Water penetration can occur due to poor sills and aprons/flashings above, or fixing nails may have rusted and timber battens may have rotted, or tiles themselves may be defective. See chapter 2.

Replace slipped or missing tiles and overhaul defective sills and aprons. In severe cases, strip and renew the tiles, fixing nails and battens.

# Defect: Cracking to brick or stone walls

## SYMPTOMS

External cracking in masonry, internal cracks in plaster; damp penetration.

Diagnosing the causes of cracking can be difficult; sometimes even the experts are flummoxed and need to monitor cracking for a year or more. Most houses show some signs of movement, the big question being, is it historic movement, or is it 'progressive' and likely to move still more?

There are two basic kinds of cracking found in masonry: those that run straight down through the bricks or stones, and stepped cracks that follow the mortar joints in a zigzag (either tapered or uniform in width). It is easier to make good stepped cracks, by pointing with lime mortar or using a clear flexible mastic, whereas vertical cracks normally require the damaged bricks or stones to be cut out and replaced.

Another remedial technique for areas of substantial cracking is to drill a series of holes along the crack and inject them with a special thixotropic resin grout. The face of the brick is then repaired with colour-matched mortar and should be undetectable. Alternatively, engineers sometimes recommend that cracks are stitched using stainless 'Helifix' small diameter rods cut into the masonry and bedded in epoxy mortar. However, before the masonry can be repaired, the cause of the movement must be dealt with. Common causes include:

### Cause   Expansion and contraction due to different materials or differing foundation depths

(Solution) *Create an expansion joint*
Thermal movement is likely to occur between different materials (*eg* old brick walls joined to modern blockwork walls), resulting in cracking. Differential movement is common where new extensions with deeper foundations abut an old house or when bays etc. were built with shallower footings than the main property. An expansion joint can be formed with a flexible mastic sealant to keep the structure weathertight.

### Cause   Corrosion of metal fixings

(Solution) *Replace with new stainless steel cramps*
Smooth-faced stone blocks were commonly held in position by iron cramps bedded into small pockets in the stone and 'caulked' with molten lead. Where the protective lead coating has cracked and allowed water to penetrate, the iron will rust and expand, blowing apart and cracking the stone it was meant to restrain. Cracked stone may have been repaired in the more recent past using a similar method employing iron cramps with unfortunate results, which is why today stainless-steel cramps are used instead. Repair work may require temporary structural support.

### Cause   Subsidence affecting the ground under shallow foundations

(Solution)

*Eliminate likely causes such as nearby trees and leaking drains. Localised structural repair and support necessary*
Cracking due to subsidence may in severe cases require the foundations to be partially rebuilt (or in extreme cases underpinned). Your insurance company may want to monitor the cracking to diagnose the risk of future movement. Subsidence or heave is more likely in older properties with shallow foundations on clay subsoil, on slopes, and with trees nearby. Drains should be checked for hidden leaks affecting the ground.

### Cause   Structural alterations, physical damage, or under-structured original materials

(Solution)   Movement due to poorly executed past structural alterations may require a structural engineer to design a new method of additional strengthening. All structural alterations should have been carried out with Building Regulations consent. Physical damage, such as from vehicle impact, normally only requires making good to the immediate area affected (unless there are structural implications). Original lintels that were too thin or are overloaded will need replacing.

## Defect: Flaking or eroded brick or stonework

### SYMPTOMS

Spalling of brick or stonework; blown or soft masonry; eroded mortar joints. Moisture can also leave salts appearing as 'blooms' on the surface, or cause pitting, powdering and flaking. More likely in very exposed areas like parapet walls at roof level. Common causes are:

**Cause** **Frost action or salt crystals forming in the brick or stonework**

**Solution** *Fix the source of damp and cut out and renew badly affected brick or stone*

Water that is trapped in porous masonry will expand on freezing, causing the masonry to disintegrate, or salts can crystalise damaging the surface of the wall.

Erosion at lower levels to walls adjoining roads is often due to spray from traffic. A relatively simple solution may be to lime-render and limewash the lower wall in the form of a traditional plinth. This should protect the surface from further erosion. Similarly, where defective window sills can't be fully repaired, rendering the wall below should help protect it.

Otherwise badly affected brick or stone can be cut out and renewed. At worst, walls may have to be partially rebuilt using matching brick or stone bedded in lime mortar.

**Cause** **Previous repointing with strong cement mortar**

**Solution** *Repoint porous mortar joints with suitable lime mortar mix to allow the wall to breathe so trapped damp can evaporate away*

Past repairs using cement, modern paints and damp sealants have often resulted in tapped damp hastening erosion. Some repairs using synthetic chemicals caused deterioration in the colour of the stone over time.

**Cause** **Excessive weathering, old age, or poor quality masonry**

**Solution** *Leave alone unless causing problems, in which case cut out and renew. Or consider cladding with a new tile-hung facade.*

See comments under 'Crumbling and spalling'.

**Right:** Natural signs of ageing – best left alone (but still requires maintenance, eg to ledges and guttering, to keep walls clear of rainwater)

# Repairing render

Where a patch of render needs to be replaced because it has come loose, the damaged areas need to be cut away and made good. Take back the edges until you reach sound material that's firmly adhered to the wall. If the patch is large, continue until you reach a natural break – such as a corner of the building, or adjoining half-timbering, or a score mark between 'blocks' of stucco. This makes it easier to disguise the new work as it's virtually impossible to make an invisible join without a bump or shadow where new render meets old. With patch repairs it's important that the depth of the base coats allow sufficient space for the final coat to be flush with the surrounding surface.

The appropriate mix will depend on the existing walls, so if it's original lime render, try to match it. But it's not just a visual match you need – texture, colour and thickness are also important, as are flexibility and porosity. Problems occur where old lime render is patched with hard, modern cement; because it's so rigid and inflexible it will be prone to cracking as the building moves. Any rain that gets into the cracks can then become trapped as it can't evaporate out again through the impervious cement. Instead it will be driven out at the edges causing the surrounding area to fail. Or frost can expand the trapped water loosening the surface. As we have seen, Victorian builders sometimes used a cement/lime/sand mix, which retained sufficient flexibility, so a similarly weak mix may be acceptable.

Ideally any existing modern hard cement rendering should be hacked off and replaced with a more traditional mix, but this is not always possible. So the next best option is to repair any cracking that could allow rain to penetrate and any hollow areas cut out and re-rendered.

Hairline cracks in lime render can often be 'healed' simply by applying a coat or two of limewash. Larger superficial cracks can be wetted and filled with a compatible mix.

Bigger patches need to be filled and built up in layers, traditionally applied with a wooden float. The wall should be lightly wetted prior to receiving the new render as a dry surface can suck too much water out of the mix, weakening it. The surface of each completed layer is roughened ('scratched') to provide an adequate key for the next, and each layer needs to be lightly wetted and requires 'tending' until set. This is because if render dries too quickly, it can shrink and crack. So to control the rate of drying, damp hessian sacking can be hung over the newly rendered wall.

## TOOLS REQUIRED

■ **Hammer and bolster/cold chisel**

■ **Wooden hawk/mixing board**

■ **Steel or wooden float**

■ **Paintbrush**

■ **Wooden batten**

■ **Scraper**

■ **Safety goggles**

## MATERIALS

■ **Sharp sand**

■ **Naturally Hydraulic Lime (moderate strength 3.5)**

■ **PVA bonding**

**1** Tap the surface with a fairly heavy object such as the handle of a hammer, and listen for any 'hollow' sounds where render has lost its key. It is quite common on old walls to find that part of the top coat of render has fallen away due to weathering.

**2** Starting at the centre, use a hammer and bolster to hack off the loose render until you reach a sound base. Cut away a patch to about 75mm (3in) beyond the damaged area. Loose render should come away easily.

**3** Clean off the masonry underneath by scraping with the bolster. Rake out brick joints and brush off any loose particles. Apply a PVA bonding agent to any smooth areas, like engineering bricks. Dampen rough areas with water.

**4** Now lay on the first coat (scratchcoat) with a suitably stiff mix – not too wet (try a mix of 1:1:5 cement/lime/sand). Place the mortar on a hawk held against the wall and, using a steel float, push the mortar onto the patch in an upward sweep. Build it up in layers until it is about 5mm below the surrounding surface. Force it well in and scrape off any excess.

**5** Leave it to 'go off' (set) slightly, then stipple the surface with a stiff brush to give the top coat something to adhere to. For larger areas, scratch the surface of the first coat with a combing tool whilst it is still wet. Leave to dry for at least 48 hours until hardened, or splits will occur.

**6** Once the base coat has dried, apply a top coat in the same way until it is slightly proud of the surrounding surface. Apply with upward sweeps of the trowel to help stop mortar falling off. This will be a slightly weaker mix – try 1:1:6 cement/lime/sand.

**7** Make diagonal or horizontal sweeps in a sawing motion with a straight wooden batten across the wet mortar to scrape off surplus and level it with the surrounding wall or 'sculpt' it to match the surrounding old render. Leave for an hour to dry, then dampen with a brush and water.

**8** Do the final smoothing with a wooden float. Roughcast render can be matched by adding small stones to the final mortar mix before it is applied.

**9** Finish by feathering the edges of the new patch into the surrounding area using a very slightly damp cloth or sponge. When dry, you can decorate to finish.

# DAMP, ROT AND WOODWORM

Many Victorian and Edwardian houses have been disfigured with unnecessarily injected damp-proof courses (DPCs) required as a condition of mortgages. So here we explore the national terror of damp and its unsavoury bedfellows, rot and woodworm. We look at the true causes and explain how you can tell whether your house is genuinely at risk.

The lower main walls are particularly vulnerable to damp. This house has good sub-floor ventilation and the ground level is not too high. But crumbling sills allow rain to soak down into the brickwork and rendered lower walls can trap damp

**Left:** Damp from roof, defective gutters and climbing shrubs spells trouble
**Above:** Damp carries white salts in solution which crystalise as water evaporates

It's the stuff of nightmares: rising damp, dry rot, toxic mould, death-watch beetle. These are words you simply do not want to see in your survey report. Along with the word 'subsidence', they have passed into modern property folklore.

The reason for so much consternation about damp is that it can eventually lead to timber decay and beetle infestation. But the fact is, most old houses at some stage experience a 'damp problem' and this is only potentially serious where structural timber floors and roofs etc. are exposed to prolonged dampness.

There are two basic sources of damp: water outside trying to get in, and water inside trying to get out. In fact, it's water from internal sources – such as leaks, condensation, and building works – that accounts for around 70% of the moisture in a typical home. So to keep your house free from damp there are essentially two things you need to do: prevent excessive amounts of water reaching the walls and getting in, and maximise evaporation to allow it to escape. The main route that moisture takes in order to evaporate out of solid masonry walls is via the mortar joints, rather than through the brick or stone. Over a long period of time the mortar that has been doing most of this 'breathing' can become a little worn and eroded, eventually needing repointing. But where this is done using hard impermeable cement, rather than traditional lime mortar, it can have the effect of trapping moisture inside the wall.

Dampness in old buildings is often the result of a combination of causes, such as earth banked up against walls, condensation running down steamy windows, or rain seeping through small cracks. But the presence of damp is

not always obvious: rot to structural floor timbers may be quietly taking place before any wet patches become visible, so a damp-meter can be a useful detection tool.

# Rising damp – a modern myth?

Musty smells. Clammy plaster. Rotten skirting. Spongy floors. That's the rising damp experience. According to conventional wisdom, moisture from the ground can rise about a metre up the walls if there's a problem with the damp-proof course (DPC). But DPCs only became widely installed in new houses late in the 19th century, so there may be no original DPC at all in some older properties. But this needn't be a problem if any damp in the masonry can freely evaporate in dry weather, and where floor timbers are sound. The fact is, true rising damp is very rare. There are many instances of specialist firms merrily injecting DPCs where it's simply not needed. Wetness in walls may look like rising damp, but it is often due to

**Below left:** Classic brown water stain from roof leak
**Below:** Damp patch to lower wall – but is it rising?

# Types of damp

The important thing is to identify the true cause and treat the source of the problem, not just the symptoms. Most damp at low level can be cured by common-sense improvements such as reducing ground levels.

There are five basic types of damp:

## 1. Condensation

When warm, moist air or steam hits a cold surface such as a wall or window it condenses back into water. Moisture from condensation running down cold internal wall surfaces and soaking into the masonry may appear as rising damp.

## 2. Penetrating damp

Water can soak through walls from the outside – due to defects like blocked gutters, cracked downpipes or defective sills – often combined with poor pointing. Where damp penetrates near the base of a wall (eg due to high ground levels, patios and rain splashing) it can be misdiagnosed as rising damp.

Damp is also encountered around old fireplaces – rain running down the flue and soaking into the soot and debris in the recess behind the fireplace. And climbing shrubs inhibit evaporation of moisture in walls.

## 3. Salt contamination

Where plasterwork has become damp, it often retains a residue of natural salts –

chlorides, nitrates and sulphates etc. – carried within the moisture. Some are 'hygroscopic' making a wall surface temporarily appear damp. Where the air in a room is humid, they can absorb moisture from the atmosphere.

## 4. Plumbing leaks

One of the most common sources of damp is defective plumbing and drainage. But it's not always as obvious as you might think. Old baths, showers and toilets can quietly go on leaking for many a long year, remaining completely undetected. Tiny leaks can develop unnoticed at the back of toilet pans, which over time cause rot to floorboards (exacerbated by steamy air condensing and dripping from cold ceramic cisterns etc.). Copper pipes can sometimes develop almost invisible pinhole corrosion leaks over time, and are particularly prone to erosion where buried unprotected in concrete floor screed, or if connected to pipes made from non-compatible metals, such as stainless steel.

Another common source of damp in properties of all ages is from mastic seals around baths and shower trays. Hairline cracks in grouting to wall tiles combined with power showers are a recipe for

trouble. Leaks from hidden pipework behind kitchen units are often only discovered when the units are replaced, by which time quite serious decay may have set in to timber floors. To prevent freezing and bursting, lag all water pipes run through cold areas, such as lofts, floor voids, larders, garages and cellars.

## 5. Rising Damp

True rising damp is extremely rare. Some experts (such as the RICS Conservation Faculty) believe that 'true rising damp is a myth and chemically injected damp-proof courses (DPC) are a complete waste of money'. There are two main sources of moisture to lower walls: excessively wet ground, *eg* from leaking drains or gutters, and high external ground levels where high paths of earth banked up against a wall can force moisture into the wall.

other causes such as condensation, plaster contaminated with salts, or water seeping back under window sills. So an injected DPC should really be a last resort. Just to make it more difficult to diagnose, there are a number of fiendish imposters. For example, warm steamy air inside the house will condense back to water when it hits a cold wall surface, running down the wall accumulating at the bottom, where it could then soak in and reappear in the form of 'rising damp'. A wall will attract even more condensation where it's already saturated – say from a leaking gutter. This is because a wet wall will more readily conduct cold from outside than a dry wall (known as 'cold bridging'). Moist indoor air can even penetrate conventional plasterboard, so even where walls are dry-lined, this may be taking place completely unseen.

You may be able to remedy the problem by first carrying out some common-sense improvements:

- Ensure that earth is not banked up against outside walls.
- Check that the external ground levels are at least 150mm (6in) below the DPC level, to reduce the risk of rain splashing off the ground and saturating the wall. Fit drainage channels to any paving near the house.
- Repair any leaking drains or rainwater fittings.
- Fit extractor fans inside the house to expel humid air and cut condensation.
- Install a gravel-filled shallow ditch known as a 'French drain' around the main walls. Damp in the wall can then evaporate by making contact with the ventilating air within the gravel instead of migrating upwards.
- If the walls are rendered externally the rendering should stop just above the DPC, curving out in a 'bellmouth drip' at the base so that rainwater clears the wall below.
- Remove old salt-contaminated plaster inside and let the wall dry before replastering.

**Right:** Looks like a nasty case of rising damp but…
**Below:** … is actually caused by banked up earth against the wall

Mike Parrett /Dampbuster.com

# French Drains

A 'French drain' is simply a shallow gravel-filled ditch excavated near the base of a wall. This helps disperse damp that would otherwise collect and soak into (and up) a wall, by allowing the

moisture to evaporate. In cases where the ground is seriously damp or waterlogged, a perforated pipe should be laid along the bottom of the trench before backfilling with loose gravel of about 20-40mm

in size. The groundwater then percolates down through the gravel and is carried away by the pipe to a soakaway, watercourse or surface water drainage system, like a mini land-drain.

Where the footings are very shallow, you may need to dig the trench about a metre away from the house. Where the garden slopes down towards the house, wet ground adjacent to the walls is almost inevitable. In storm conditions there's potential for deluges of surface water swamping the lower walls, and surging through air bricks. Fitting a simple drainage channel a metre or two away from the house can usefully intercept any mini tsunamis and divert them safely away.

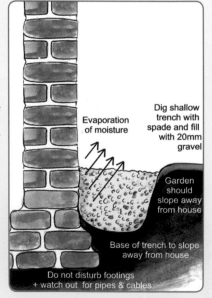

Evaporation of moisture

Dig shallow trench with spade and fill with 20mm gravel

Garden should slope away from house

Base of trench to slope away from house

Do not disturb footings + watch out for pipes & cables

## DAMP-PROOF COURSES

Damp-proof courses were first required by the Public Health Acts on houses built after 1875. An original DPC typically comprises two or three layers of slates bedded in mortar below the level of the ground floor joists. The DPC is usually evident as a wide mortar joint about two brick courses above the outside ground level, near the base of the main walls.

Slate tended to crack as the building settled so alternative materials with more flexibility were sometimes used, such as asphalt or tar mixed with sand, thick lead sheet, two or three courses of dense engineering bricks set in cement mortar, heavy paper coated in wax or pitch, or canvas soaked in bitumen. In the 20th century bitumen and felt became the norm.

Damp-proof courses are often hidden, typically where the walls are rendered down to ground level or if there is a plinth (a thick rendered base typically about 600mm/24in high). It is also common for the DPC to have been pointed over with mortar.

**Right:** Cement plinth hacked off shows how they can trap damp.

**Below:** Render should finish above DPC level

You can check if there is a DPC in the wall from the inside, by raising a floorboard and raking out a little of the mortar from the joint.

It is sometimes claimed that an old DPC may have worn out or broken as a result of physical movement in the walls. But a more plausible cause of low level damp is where something has physically linked the ground to the wall above, allowing damp can get past or 'bridge' an otherwise sound DPC – typically where earth in flower beds is piled up against the wall or paving stones have raised the ground level. In such cases, the ground should be lowered and paving should slope away to prevent rainwater 'ponding' against the walls.

Rendered plinths are sometimes claimed to be a cause of bridging. These were commonly applied to external walls as decoration to conceal ugly DPCs and to emulate more expensive projecting brickwork plinths at the base of walls. But a more likely cause is because they trap damp. Plinths made of impermeable cement (rather than traditional breathable lime render) are very prone to cracking, which can let water in but crucially won't allow the brickwork to dry out. So it's important they are well maintained. If you try to hack the cement render off completely it may look unsightly as it can take lumps of brick with it, but cutting out and removing the lower portion below DPC level can help moisture to escape. Or a cement plinth can be replaced with a matching new one made from lime render.

Where you've got solid floors internally there should be a connection between any Damp-proof membrane (DPM) in the floor and the DPC in the walls, to prevent any potential risk of bridging on the inside – see chapter 10.

## REMEDIES

The usual knee-jerk remedy to damp is to inject a new chemical DPC, but this is rarely effective. Water repellent silicone or resin is pressure-injected through closely spaced horizontal holes drilled into the mortar just below floor joist level. The silicone-based liquid soaks into the wall in a bid to prevent water rising through the pores in the brick or stonework (it doesn't work at all in flint or thick stone walls).

But regardless of how ineffective this 'remedy' often is,

mortgage lenders, seduced by the prospect of a 20-year guarantee, commonly insist that such works are carried out. The irony is that with many contractors the guarantees are worthless because of extensive exemption clauses and the fact that small firms tend to come and go. Plus the work is often very poorly executed with DPCs injected far too high.

If you do decide to have the walls treated, wait a while before celebrating the elimination of your damp problem. It can take months for the drying out of a saturated standard 230mm (9in) thick wall. And damp can be elusive. When a house is left empty for a while signs of damp may disappear, but when rooms are heated moisture can sometimes reappear on the surface of the plaster. It is also seasonal. It can disappear in the summer months only to reappear in the winter.

## PLASTERING

Even when you've fixed the cause of the damp, a strange thing can happen to the old plaster on the inside. You would imagine that after the damp in the wall had gone away and the plaster had all dried out, that would be the end of it. But no. The water that rose up the wall wasn't pure: it would have been contaminated with nitrate and chloride salts picked up from the earth or the masonry, and these hygroscopic salts will have found their way into the surface plaster. This is a different kind of damp altogether – each time the air in the house becomes a little humid the salts in the plaster absorb the moisture and liquefy, so patches of dampness on the plaster occasionally return. But when the atmosphere is dry, the damp mysteriously disappears again.

Salts can normally be brushed or vacuumed off, but don't try to wash them off as this forms a solution, sending them back into the wall. But when plaster has been heavily contaminated by salts that persistently reappear, the only remedy may be to hack off the affected area and replace with new lime plaster. But before replastering, the source of damp must be rectified and the masonry given plenty of time to dry.

Some experts say that most damp problems are actually a result of these salts having leached through from damp walls and taken up residence in the plaster, and that replastering to the right formulation is the key.

**Below:** DPC injection holes should be below floor level – here they're far too high, way above the original engineering-brick DPC. The injection holes should also be in the mortar course, not the bricks.

**Below:** Original hair lime plaster has lost its key to damp wall

# Obsolete damp treatments

There are several types of damp treatment that have been tried without much success over the years. Your lower walls may display one or more of the following past attempts at treating damp:

### Chemical injection DPCs

Banks sometimes make damp treatment a condition of a mortgage – in the belief that they're protecting their investment with a guarantee. But because all damp tends to get labelled as 'rising damp' it often results in totally unnecessary injection of chemicals and the scarring of old buildings. In fact this can actually make matters worse, interfering with natural breathability of old walls. The guarantees provided for this work are contingent on internal plaster stripped to 1m above floor and tanked in cement render – adding considerable disruption and expense. So if the DPC doesn't work, the render will hide damp from view so it appears to work (yet hidden floor joists may be silently rotting away).

The theory is that by pumping silicone-based fluid into a wall it will form a horizontal barrier to block any damp rising up. In practice what very often happens is that DPCs are injected at too high a level to protect floor timbers. To be effective, injection should be into mortar joints rather than the brickwork, otherwise damp can work its way past the injected bricks via the mortar. And if the walls aren't dry when injected, the chemicals may form an incomplete barrier. Doomed to failure, sometimes walls of hard, impermeable stonework such as flint are 'treated'. Even where done properly, there can be problems when the chemical fluids do not entirely fill up the pores – fluid takes the line of least resistance via the largest pores and cracks. And very often the drill holes are left unplugged, allowing rain to enter and settle. Even worse, as well as masking the problem, tanking in cement render prevents the damp inside the wall from drying out.

### Electro-osmosis

By applying a weak electric current to a copper wire fixed horizontally at low level around a wall it is claimed that water molecules are repelled as they try to rise up the wall (due to altered polarity). The copper wire is connected to metal electrodes inserted into holes drilled into the walls. In reality, such electro-osmotic systems have achieved inconsistent results and have suffered from corrosion.

### Ceramic vents (atmospheric siphons)

Porous ceramic tubes are inserted into holes drilled into the base of a wall. These are designed to draw moisture towards their surfaces by capillary action where it can then evaporate. This is the same principle as breathable lime mortar. But because evaporation is concentrated around one point (the tube) the pores on the surface eventually become blocked as mineral salts accumulate during evaporation. Worse, they are sometimes bedded in cement mortar, which actually blocks the path of water! Inserting ceramic vents into walls inevitably causes damage, although at least the tubes are flush with the surface. An aggressively advertised current version known as the 'Schrijver' method involves fitting several large projecting vents to a wall surface, which are both visually and physically intrusive, and therefore considered inappropriate for period buildings.

### DPC cut into wall

Retrospectively inserting a physical DPC involves cutting out a short section of wall and inserting strips of waterproof material, then repeating the process to the full length of the wall. Mechanical cutters such as chainsaws were sometimes used to cut open mortar joints. But such aggressive techniques can be highly damaging to traditional buildings with their irregular walls and uneven courses. Unsurprisingly, some properties suffered consequential structural problems.

### Damp sealants

It would be nice to believe that damp problems could simply be banished forever with a quick spray. Some 'miracle solution' products found on the shelves of DIY stores rely on containing damp behind a physical barrier, such as special thick paint. The trouble is, by definition they seal damp in, rather than deal with the cause of the problem. Damp sealants are also sold on their ability to stop water getting into walls. But this is like guarding the front door only to find that you're attacked from the rear. Old walls are rarely perfectly dry when damp sealants are applied, and moisture can enter by other means, not just from rain. A surprising amount of damp is due to indoor condensation, or soaking down from faulty gutters and sills etc., and 'damp-sealing' the surface will stop it escaping by evaporation – storing up future problems.

### 'Never paint again' renders

The advertisements that promise maintenance-free wall coverings guaranteed to last a lifetime are very appealing. They claim that by having a tough yet flexible coating applied to your walls you can rejoice in the knowledge that you'll never need to lift a paintbrush again (despite the fact that brick and stone doesn't need panting). But what the salesmen don't point out is that the new coating can actually have the effect of trapping moisture inside the walls. Also, the surface will only be as good as the condition of the wall behind it. Preparing and drying out the wall first is essential, and the temptation is to rush this. And of course older houses with relatively shallow foundations tend to naturally move slightly in tune with the seasonal changes to ground conditions, so if the paint or render cracks, rain will get in and stay trapped behind it. A more sensible variation on this theme, sometimes used in basements, is to apply a physical lining to a wall but leaving a well-ventilated air gap behind, thus isolating the damp from the new internal finish but, crucially, allowing the wall to breathe.

Lowered ground with shingle helps damp evaporate away

High ground levels cause damp walls

Eroded sills let water seep into wall

Hard cement pointing has led to eroded bricks

Paint traps damp in wall

Plenty of air vents keep sub-floor timbers problem-free

# SOLUTIONS - How best to deal with damp in old buildings

The key is to address the actual causes – often more than one:

1  Fix any obvious external sources of water, such as leaking gutters and drains, or defective sills.

2  Confront the enemy within – fix any obvious internal sources of water pipe leaks etc.

3  Prevent 'death by suffocation'. To allow damp to naturally escape by harmlessly evaporating away, carefully remove harmful modern materials eg cement render/mortar pointing, gypsum plasters and plastic paints that are trapping damp. Replace with traditional breathable materials.

4  Reduce high ground levels. Ensure that earth is not banked up against the walls. External ground levels should ideally be at least 200mm below floor level. Consider installing a 'French drain'.

5  Combat condensation. Dehumidifiers can provide a short-term solution, plus effective heating. The full solution is to ventilate, insulate and stop emitting moisture.

6  Allow time to dry out. Once each cause is eliminated, time is required to assess if it was successful – allow about a month to dry out for each inch thickness of masonry wall. Also, clear any external vegetation adorning the walls and keep the house heated.

## Defect: Suspected rising damp

### SYMPTOMS

Damp plaster at ground floor main walls, musty smells, rotten skirting or floors; white 'tide marks' on walls.

**Cause**  **Dampness near ground level that appears to be rising up the wall**

**Solution**  Dampness is usually restricted to within about a metre above the ground floor level by capillary action through porous masonry. However, what appears to be rising damp is commonly due to problems elsewhere – such as indoor condensation or water leaks from pipes, gutters, defective sills etc. Matters are made worse where damp is trapped in the walls by modern plastic paints or cement render or pointing.
NB damp to internal walls is often caused by leakage from nearby pipework run below or within the floor.

■  Reduce ground levels and clear away earth or debris. Ensure that the outside ground level is at least 150mm (6in) below

the DPC. Check that surface water in the garden runs away from the house.

■  Dig a shallow trench around damp walls to make a 'French drain', about 300mm deep x 300mm wide. The trench base should slope away from the house. Place a permeable liner at the base then fill with gravel (20–40mm size). Leave it loose – do not compress the gravel. This also helps prevent the risk of water entering the airbricks. Do not expose the base of footings. If the footings are really shallow, site the trench about 1m away instead of against the wall.

■  If the property already has a modern injected DPC, check the guarantee, as the work may be defective. Or the problem may not be rising damp at all.

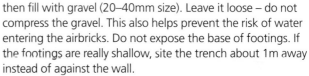

# Penetrating damp

Damp coming through the walls, rather than up them, can occur at any level but is more often identified higher up. Where the walls are exposed to wind and driving rain, or where the pointing is a bit iffy, damp can penetrate, leaving brown patchy stains on the plaster or wallpaper. If neglected, rot can take hold inside. In much of Britain winter weather usually comes from the south-west, so the effects of driving rain are most noticeable on the walls, doors and windows facing in that direction.

## DAMP PATCH PLACES

It is because old brick and stonework is porous that extreme driving rain can penetrate solid walls. And a surprisingly large amount of water can enter a building even through tiny cracks. In very exposed locations the outer face of the walls would traditionally often be rendered or clad with tiles for additional protection. This can also provide a very effective solution applied retrospectively. Tiles have the advantage of allowing any damp already in the wall to disperse, as they don't seal in moisture.

Brick and stonework is often coated with masonry paint to improve appearance, hide graffiti or blemishes, or reduce rain penetration. But such masonry was not really designed to be painted, and moisture in the walls may actually increase if modern vinyl paints seal it in. If rain has entered through joints

and cracks, the paint can restrict the drying out process. It is better to use breathable micro-porous paint, or traditional limewash. Mortars and plasters made from lime, rather than modern cement or gypsum, are permeable, allowing damp to escape by evaporation.

There are some specific areas where penetrating damp commonly enters the house:

- The sills of windows are usually made of stone or rendered brick. Over the years, with slight settlement, these tend to crack. The cracks allow dampness to penetrate the wall below, showing internally as damp patches or soft and loose plaster, perhaps leading to an attack of rot in the skirting. Sills must project sufficiently away from the wall and have a drip-groove or 'throating' underneath, otherwise rain will run back under the sill and soak into the wall.
- Windows and doors: rain quickly runs down and seeks out the numerous joints between the frames and the walls, or the glass and the putty.
- Overflowing gutters or downpipes that are too short are another common cause of damp through the walls, as are leaking overflow pipes from WCs and watertanks.
- Decorative string courses or any protruding ledges on walls present an invitation to water. Unless sloped away from walls, rain will pond and soak into them.
- Roofs and stacks are another major area of damp penetration, commonly at flashings and down chimney breasts, not forgetting damp coming through from inside the flues – see chapter 3.

Other common sources of dampness include condensation, faulty plumbing, flooding, damp rising through floors, adjoining garden walls and via creeping plants on walls. As noted above, even once the sources of dampness have been eradicated, it can take a while for the fabric to dry out. And dampness drying out can look identical to a continuing problem, especially if plasters and finishes have absorbed salts. Like an unexploded 'damp-bomb', they can come back to haunt you with their strange ability to 'self-dampen' long after the source of the problem has been rectified, unless correctly replastered.

## BELOW GROUND – BASEMENTS AND CELLARS

'Dank and festering' describes the usual condition of cellars and basements. Particularly vulnerable are the cave-like vaults similar to small railway arches that were sometimes built below the pavement for coal storage in some urban areas.

Basements are generally only found in early Victorian town houses, evolving into semi-basements by mid century. However small cellars used for coal or storage were incorporated in many urban villas throughout the century.

If you're the proud owner of a musty old cellar, the best advice is to leave well alone, as the headroom is usually very limited, and excavation is both ruinously expensive and fraught with structural risks. The good news is cellars provide useful access to check the condition of floor joists built into periodically damp walls, so it's important that these areas are well ventilated to prevent rot in joist ends.

Fairly obviously, one of the main challenges faced by Victorian builders when constructing basements for habitable use was to make them watertight. The most effective solution was the commonly adopted 'semi-basement' design. Here a trench at least a metre wide would be excavated around the walls, so they were effectively at ground level and could be built conventionally. This, however, necessitated the construction of a retaining wall to form the side of the trench to the garden at a higher level (see chapter 14) and providing effective surface water drainage to the trench.

But when it came to constructing conventional underground basements, Victorian builders often struggled to provide effective damp proofing. This might be because servants and staff who occupied basement rooms were expected to take dampness in their stride without complaint!

Basements were usually lined with an exterior coating of asphalt (or sometimes slate) before backfilling with gravel in an attempt to waterproof the side in contact with the surrounding ground. However, the floors may only have been given a thin covering of stone slabs or an ineffective layer of weak concrete. But even these rudimentary measures were generally omitted in uninhabited cellars and pavement vaults. If you're lucky, the cellar may have since been professionally lined internally to keep the surrounding moisture from the ground at bay.

The causes of damp in basements include water pressure from waterlogged ground as well as penetrating damp and condensation. The possible solutions begin with excavating externally and laying porous ground pipes to divert ground water and reduce lateral pressure on the walls. And it may be necessary to incorporate an internal drain system to collect any damp that does get through and disperse it via a sump – see chapter 15. It may be better to console yourself with the thought that a damp basement is said to be better for storing wine!

## Defect: Damp patches on main walls

### SYMPTOMS
Damp smells, damp and stained plaster often at higher levels.

**Cause** **Leaking gutters or downpipes**

**Solution** *Overhaul or replace as necessary. See chapter 4*

**Cause** **Roof leaks**
Defective flashings or stack brickwork often showing as damp and stains around chimney breasts and upper walls.

**Solution** *Repair stack/roof defects. See chapters 2 and 3*

**Cause** **Mortar joints badly eroded, or repointed in hard cement mortar. Cracked render or masonry**

**Solution** *Overhaul brick or stonework*
Overhaul brick or stonework and pointing to mortar joints. Replace modern cement mortar or render with suitable sand/lime mix. Make good defective render. Fill any gaps to walls, *eg*

around pipes. Check any tile cladding for gaps.
_____

**Cause**
**Penetrating damp**
(often mistaken for 'rising damp' where found near base of walls)

**Solution**
*Eliminate source of damp and allow to dry out (reduce ground levels, install French drain, repair gutters etc.) – see page 115*
Dampness penetrating through walls at ground level is rarely true rising damp. It can sometimes penetrate from adjoining garden walls and structures such as outbuildings which are best demolished. Alternatively install a vertical strip of plastic DPC to form a waterproof barrier between your house and any adjoining structures that might remain damp.

# Condensation

Americans call it 'toxic mould'. Steamed up windows, damp walls, and moist ceilings will eventually attract black speckled fungal growth. Even clothes and carpets can acquire the unpleasant smell of damp when there is excessive indoor humidity. Patches of mould growth resulting from moist air condensing commonly occur where there is little air movement, *eg* behind bulky furniture and unventilated cupboards. In extreme cases the same black mould can even grow in the lungs, exacerbating bronchial or asthmatic conditions.

When steam or warm, humid air in your house hits a cold surface it will cool and condense back to water. And the source of this water vapour? You and the family, just doing normal everyday things. Some of the biggest living emitters of moisture are actually dogs and children. This is how many litres of water vapour are typically produced in a day:

| | |
|---|---|
| A bath or shower | 1 |
| Tumble drier | 4 |
| Cooking | 2 |
| Two adults breathing and sleeping | 2 |

Add to this pets, washing machines, kettles, gas heaters and houseplants and you'll have well over 10 litres (up to 3 gallons) of water vapour a day, and it has to go somewhere.

The solution is to do exactly what you do in the car when it steams up: wind down the windows and turn up the blower, *ie* improve the ventilation. If you also improve the insulation and reduce emissions of vapour in the first place, the problem should be solved. See chapter 15.

## VENTILATION

The Victorians used heavy curtains, screens and hangings to try and reduce cold draughts, but foreign visitors were forever complaining about the draughtiness of English rooms. Open fireplaces ensured a steady flow of air through the house, which helped disperse damp air before it could condense into water. The 1875 Health Acts required additional vents in walls to ventilate rooms from noxious fumes produced by gas and oil lamps. And if that wasn't enough, draughts from floorboards and sash windows ensured fairly consistent movement of air into the home.

Today the modern obsession with sealing up old houses to eliminate all possible sources of draughts can result in condensation unless controllable ventilation is provided to compensate. It can also cause problems to hidden areas such as lofts and underfloor voids, which need a good flow of air to keep them dry.

Today's lifestyles generate large amounts of water vapour, so it makes sense to tackle the source of the problem by fitting extractor fans (ideally with humidistats and heat recovery) in kitchens and bathrooms to expel moist, humid air before it has a chance to cause trouble.

## WEAK POINTS

Condensation weak points are found anywhere with a cold surface that humid air can reach. Insulation was not incorporated in the construction of Victorian houses, but by now most lofts will have been at least partially insulated. But that still leaves a number of cold, uninsulated areas such as dormer windows (the ceilings and side cheeks often have no insulation), flat roofs, thin rear addition walls to kitchens and bathrooms, and uninsulated pipes. See chapter 15.

## Defect: Mould staining

### SYMPTOMS

Recurrent black mould growth; general smell of dampness in the house; peeling paint and wallpaper.

**Cause**   **Condensation of humid indoor air. Poor ventilation and insulation**

**Solution**   This is a common problem, particularly evident to the ceilings, walls, and windows in bathrooms and kitchens (where most steam is produced). The steamy water vapour hits a cold surface and condenses back in to water. Mould staining occurs particularly often in thin-walled rear additions. The solution is:

- Insulate cold surfaces like walls and bedroom ceilings.
- Get moist air out of the house: open windows and close

bathroom doors.
- Improve ventilation. Fit extractor fans with a built-in humidistat to kitchens and bathrooms. Open up fireplaces. Fit windows with trickle vents.
- Reduce emissions of humid, moist air at source by ensuring tumble driers are ventilated to outside, and if possible minimise indoor clothes drying. And it can help to put the dog in the garden for a while and cut down on boiling food!
- Use a heavy-duty dehumidifier.
- To get rid of the mould, clean off with water and a suitable fungicide. Finish with a coat of mould-resistant paint.

# Injecting a damp-proof course

If your house has a row of little holes in the main walls near ground level, it is probably where a chemical damp-proof course (DPC) has been injected in recent years – perhaps unnecessarily. You may notice that the injection holes are actually above the level of the floor joists – in which case they will not be doing the intended job of protecting floor timbers from damp and rot.

The usual reason for attempting to 'retro-fit' a DPC is because someone has spotted damp to the lower walls and assumed that the cause is due to the original DPC no longer working. In actual fact this is normally the least likely reason! True 'rising damp' is extremely rare. It's reckoned that as many as nine out of every ten cases are misdiagnosed. So injecting a new one should be a last resort.

If all the other solutions have been tried, the best advice is to seek the advice of the Conservation Officer at your Local Authority. If there is no other alternative (eg where the source of damp can't be entirely eradicated, such as from an adjoining property), injecting a DPC is a job that is within many people's DIY capabilities. You should be able to do a better job yourself than many 'timber and damp' contractors, at considerably less cost, although of course you won't get a paper 'guarantee'.

Modern best practice is to work from inside the house (to avoid disfiguring external walls), and to use special damp-proofing cream applied with an applicator gun. Damp-proofing creams are a recent development that make the job easier. They are safer (being non-caustic and non-flammable) and are OK for tricky areas like party walls. Once installed, the cream diffuses into the wall before curing and forming a water-repellent resin that forms a barrier to damp. They only require a hand applicator gun instead of conventional heavy-duty electric pumps, hoses and drums.

**1** First lift the floor boards adjacent to the wall to be treated and check the position of floor joists. The DPC should be injected below the level of the joists. Make sure ground levels are at least 150mm below the new DPC position.

Carefully lever off adjacent skirting boards and remove any damp plaster. The void under the floor should allow access to the lower wall.

**2** Using a 13mm masonry bit drill holes at intervals of about 100mm along a mortar course, with the holes angled downwards slightly. Holes should be about 150mm deep, not right through the wall (typically 230mm thick). For solid floors this should be as close to the floor as possible.

**3** Insert DPC cream cartridge into an applicator gun and slowly inject the cream into the first hole, allowing for it to expand slightly.

**4** Repeat until all the holes are filled and the wall saturated. Leave the wall to dry for a month or more and remove any salts before re-plastering.

# Fungal decay

Most people have heard horror stories about 'dry rot', the dreaded fungus that over the years has gained a fearsome reputation in the popular imagination. Like something out of a Hollywood 'B' movie we imagine it relentlessly spreading, reputedly capable of consuming entire buildings – at least if urban folklore is to be believed.

'Wet rot' is the other well known, if slightly less exotic, type of timber decay that also occasionally pops up in surveys to scare the pants off house buyers. But fungal decay comes in a variety of shapes, sizes and colours and can sometimes be a little difficult to distinguish. So where rot is afflicting a property, the first task is to identify what type you're dealing with. However, there's one thing all types of fungi have in common – they thrive in moist conditions. So for wood to be attractive for rot, its moisture content has to rise above 20%.

# Wet rot

As the name suggests, this type of fungus has a preference for wetter conditions. External joinery is commonly affected where water accumulates and can't drain away freely. So window frames are especially vulnerable around the lower sill area. But it also attacks window and door frames on the inside, as condensation drips down the glass. Wet rot thrives best in damp, dark, and

*Sills and lower frames to poorly maintained windows can suffer water ingress and wet rot*

**Above left:** Timber embedded in damp wall has suffered beetle attack and rot

**Above:** Localised wet rot to floorboard

**Right:** Wet rot has affected suspended timber ground floor with blocked air bricks

poorly ventilated places: timber lintels and the ends of floor joists embedded in damp walls are areas typically at risk if moisture isn't free to evaporate out. Rotten wood has a soft, spongy feel, caving in easily when prodded with a screwdriver. But it's not always obvious, particularly since painted timber can sometimes appear perfectly sound whilst decaying away from the inside.

## ERADICATION

It can be hard to tell wet rot and dry rot apart in the early stages. But all rots thrive in damp conditions, so by eliminating the source of the damp, they die.

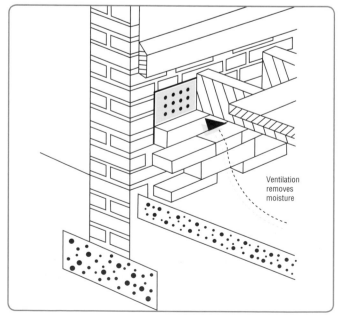

Ventilation removes moisture

## How durable is wood?

If you peel back the outer layer of bark on a tree, the wood you've exposed is known as sapwood. This is the living part of the tree with relatively little durability, so it's far more vulnerable to beetle and fungal attack than 'heartwood' - the dead inner part of the tree which is particularly resilient to decay. When timber is used for building, it's common for a certain amount of sapwood to be left on the corners of the timbers, and these can potentially be attractive to woodbeetle. Heartwood has a natural resilience to decay provided it's kept reasonably dry.

Dry rot fungus

Mike Parrett /Dampbuster.com

Poor ventilation is often a big part of the problem as it prevents damp from evaporating harmlessly away. Most ground floors are made of timber boards over joists, which are potentially vulnerable to damp and attacks of rot. Airbricks are essential to ventilate and dry out moisture under floors, but they're often blocked. To see if they're working, hold a smoke generator or a smoke match (or a lit ciggy) against the grate – the smoke should be drawn towards the opening (depending on the wind direction). If the airbricks are blocked, rake them out and clear them. You should have two or three airbricks on opposite walls, depending on the size of the house, for a decent through-flow of air.

Minor damage (eg to window frames) can sometimes be repaired by drying the affected wood and then using a wood hardener, a resin that soaks in. More serious damage requires the wood to be replaced; damaged joists, for instance, can be cut back to good wood, treated with preservative, and have an extension bolted on. Any wood in contact with a wall can be protected by placing a strip of plastic DPC underneath.

## Brown and White Rot

Although rot spotters everywhere are familiar with the terms 'wet rot' and 'dry rot', experts today consider them a little simplistic. Fungal decay to timber can more accurately be categorised as brown and white.

Included in the brown camp are all types of dry rot and its close relative 'cellar fungus'. In the white camp are most types of 'wet rot', but not all. The key difference is in their respective methods of attack that destroy different parts of the timber. Brown rots work by removing cellulose from the wood, leaving behind a brown matrix of lignin (lignin is a chemical found plant cell walls that makes them rigid and woody). White rots also consume the lignin, leaving behind blanched looking decayed wood with an anaemic tinge. Despite this, brown is actually far more destructive than white.

# Dry rot

Take a deep breath of fresh air – and the chances are you will have just inhaled some spores of dry rot fungus. These are present in the atmosphere of most towns and cities, and a Victorian house with lots of nice moist timber can provide a welcome invitation to the fungal spores to feed and breed. Damp, badly ventilated spaces under timber floors and in cellars are most suitable. Otherwise old window and door frames, timber panelling, or beams in contact with damp walls will do nicely. Given time, the fungus will consume its wooden host, all the while emitting a decaying dank aroma.

When the airborne spores settle on damp timber, they send out thin grey web-like root strands that spread across the surface of the wood. The fungus feeds by sucking the moisture from the wood, making the timber dry, brittle and structurally useless. These fungal root strands (hyphae) quickly multiply in moist conditions and rapidly become matted, completely engulfing the wood. They then adopt the appearance of cotton wool, becoming grey with lilac tinges and yellow patches, until finally morphing into a mature, pancake-shaped, rust-coloured fruiting body, which emits millions of spores that float off into the air.

Cuboidal cracking, usually indicative of dry rot

Dry rot (or to give its proper Latin name '*Serpula lacrymans*') is often misdiagnosed because its appearance is constantly changing. It can sometimes look similar to other types of rot, depending on what lifecycle stage it's at and where it's growing. But whether correctly identified or not, because of its fearsome reputation people are sometimes panicked into taking drastic action at the slightest whiff.

'Cellar fungus' thrives in the same sort of conditions and is frequently mistaken for dry rot because it displays very similar symptoms with an almost identical fruiting body. But unlike its more famous cousin its *mycelium* strands are blackish brown rather than grey-white.

## ERADICATION

Damp wood + poor ventilation = rot. Some claim to be able to detect dry rot by smell (it is variously described as smelling like urine or mushrooms), and it is true that detector dogs have been successfully used to locate it.

The best way to kill all types of rot, like any living organism, is simply to cut off their water supply. So all that may be required to deal with an outbreak is to simply isolate the timber from damp surroundings and boost ventilation to reduce moisture.

However, profit-driven timber treatment firms have an incentive to hype up the danger to justify expensive works. Perversely, the water-based chemicals they often use have the effect of introducing large amounts of moisture into the building, the very thing the rot needs to thrive! As well as blanket spraying with chemicals, for several decades the standard approach for treating suspected outbreaks of dry rot has involved aggressive cutting out of timber to a metre or more beyond the last known area of outbreak and then burning it, as well as hacking off all the plasterwork in the vicinity. Timber ground floors are the area most commonly targeted for such works.

The important point is that, without damp, the spores will not

germinate and the timber will be safe. Dry timber will not rot even if the spores are sitting on its surface. So to successfully treat rot the source of water must first be found and dealt with. Then the area affected must be allowed time to dry out and effective permanent ventilation installed, for example via new airbricks inserted in the walls. The timber must be well ventilated and able to breathe on all sides. A dehumidifier can help speed up the initial drying out process by extracting excess moisture from the building.

Most outbreaks are simply due to a leak that's gone unnoticed for months or years. One common source of problems internally is from persistent leaks to shower trays and bath seals.

Where rot has attacked structural timbers like lintels then physical repairs may be needed. But before condemning an affected piece of timber, you may find that the core is still sound and it is salvageable. With the aid of micro-drills (also used for beetle – see below) it's surprising just how sound old pieces of timber can prove to be despite outward appearances. Where badly rotten timbers are not salvageable, they will need to be cut out and replaced with new pre-treated timber. Where joists are in contact with damp masonry walls they can be protected by inserting strips of plastic damp proof membrane.

One of the things that tends to alarm people about dry rot is its reputed mobility. In the past it was believed that, left to its own devices, the foul fungus would spread relentlessly through plasterwork and solid walls in search of nourishment, 'wetting up' fresh timber causing new outbreaks until whole buildings were consumed. But the fact is, once the source of the outbreak has been controlled and allowed to dry out these strands should naturally die off. So there should be no need to strip away historic plasterwork. Even so, it's a wise precaution to check for secondary outbreaks. If in extreme cases it's considered necessary for timber treatment to be applied as a precautionary measure, targeted use of a boron-based fungicide is probably the best option. But normally all that's required to kill the beast is a dry environment and good cross ventilation.

**Right:** Bad news: blocked air vent and high ground levels

**Below:** DPC strip on pier protects timber floor plate

# Beetle infestation

'Woodworm' is the generic name for tiny wood-boring insects that like to nibble away at the wooden components of old houses. Furniture beetle is by far the most common type of woodworm; longhorn beetle and woodboring weevil are less common – but climate change may yet see exotic continental termites dining out on your skirting boards.

Death-watch beetles prefer old hardwood, more often found on period buildings like churches; you may hear their trademark 'tapping' noise (like a fingernail tapped on wood a few times in quick succession).

As they munch their way through the nutritious cellulose, their boreholes have the potential to eventually destabilise wood by hollowing out its insides. But the fact is, most old buildings have a few boreholes as evidence of past beetle attack. As with damp and rot, the mere suspicion that beetle once trod the boards in an old house, can provoke hysterical overreaction and a rash of inappropriate works. It's estimated that as many as 50,000 unnecessary treatments are carried out each year in Britain.

There are good reasons not to panic if you encounter boreholes. For a start, the infestation may well be of some considerable age and no longer active. It may also have been repeatedly sprayed every time the house has been sold or remortgaged, at the insistence of the banks. But the main reason for guarded optimism stems from understanding the way beetles interact with timber.

Wood beetles are particularly fond of surface sapwood since it is much easier to consume than the tougher heartwood. This explains why beetle infestation rarely extends much deeper than

the outermost few millimetres, and hence is unlikely to affect the structural integrity of the timber.

Prodding with penknife can quickly determine the condition of the heartwood timber below surface. If you can sense firm resistance, the timber may be perfectly sound with only the outer layers close to the surface affected.

Some common places where you might find woodworm include:

- Floorboards, particularly where infested antique furniture might have stood, or around WCs (the grubs like the proteins in urine-soaked floorboards).
- Under the stairs, or in old cupboards, especially in plywood panelling.
- Behind skirting boards.
- The roof space, often around loft hatches (where it is warmer – they like to be cosy).

But consistently moist timbers anywhere in the house can potentially be at risk. Look for the signs of telltale 'flight-holes', about 1mm in diameter (or 3mm for death-watch). The big question is, is it an active colony? You may be looking at 20-year-old holes whose occupants have long since departed.

### COMMON FURNITURE BEETLE

Furniture beetles are matt brown and about 3mm long from head to tail. As the larvae consume the nutritious cellulose in a piece of wood, they form tell-tale exit holes of about 1 or 2mm in diameter.

Furniture beetle attacks the outer sapwood (of both softwood and hardwood) if the environment is sufficiently damp with a moisture content above about 15%. This means that poorer quality timber containing a lot of sapwood is most at risk. Softwood floorboards are potentially vulnerable due to their relatively thinness. On rare occasions this can result in the odd isolated board becoming weakened with a risk of collapse when walked on. However, where boards have been stripped, you can often see the remnants of little bore tunnels, looking a bit like Arabic script. But this evidence of beetle activity is normally confined to an area close to the surface.

Decay detection drilling using the Sibert microdrill

TRADA Technology Ltd

## MONITORING BEETLE ACTIVITY

Each year, the larvae buried in the wood start to hatch. The insects grow as they munch their way to the surface, finally emerging as adult beetles from early spring to summer, creating new exit holes. Active infestation is revealed by piles of very fine timber dust, known as frass, left behind near the flight holes. Existing holes are often reused to deposit eggs, starting the lifecycle all over again.

To judge whether infestation is still active, take a close look at the exit holes. If they're clean and light it's indicative of recent activity. As you might expect, old, inactive holes tend to be dark and dirty. The simplest method of testing is to glue tissue paper over the exit holes using wallpaper paste or alternatively wax polish timber surfaces that are potentially at risk. Then observe whether the tissue or wax is chewed through as the insects emerge from around April/May onwards, either from new flight holes or via reused old ones. Emerging wood beetles instinctively head towards the light, which explains why you sometimes find dead beetles around window sills or on floors in front of windows. Based on this principle, special 'light traps' can be used in dark spaces like lofts, designed to attract beetles and trap them on sheets of sticky paper.

Where a length of load-bearing timber shows signs of beetle attack, clearly the big question is whether its strength has been compromised, or whether the damage is merely superficial and limited to the outer surface area. To assess the condition of the

core, non-destructive testing can be carried out by carefully drilling into the timber using a micro-drill with a long, fine bit. Where the core is sound it will offer resistance making penetration difficult, whereas a rotten core will allow it to easily pass through.

To undertake professional testing, a specialist firm will need to be employed. Although this will involve paying a fee, it's important to bear in mind that in most cases the cost will work out substantially less than a 'free survey' from a sales-driven

firm that recommends a lot of unnecessary and expensive replacement work.

## OVERKILL

Past approaches to treating suspected beetle infestation primarily involved indiscriminate attempts at mass poisoning. Spraying chemicals that are now considered too toxic for safe use, resulted in unpleasant side effects to human health as well proving largely ineffective. Pressure impregnating timbers with chemicals has implications in terms of pollution and toxicity to wildlife and humans, and some Housing Associations specifically ban this method. Bats, a protected species, can be harmed by sprays.

In spite of such 'chemical warfare' techniques proving ineffective, mortgage lenders still routinely insist on instructing 'Timber and Damp Contractors' intent on selling such obsolete treatments. But blanket spraying of chemicals rarely reaches the infestation, and simply results in toxins being distributed throughout the building. In any case, chemicals are difficult to apply beyond the outer surface of the timber, only penetrating a couple of millimetres.

Insecticides also have the undesirable side-effect of killing off 'friendly' natural predators of wood beetles and lavae, like spiders and shiny blue-black house-beetles. Another problem is that the most serious outbreaks tend to affect damp structural timbers such as joist ends that are embedded in walls, where access is almost impossible.

## TREATMENT

Today, property professionals have adopted a more intelligent and relaxed attitude to dealing with suspected beetle infestation in old buildings. The focus is now very much on correct identification of the problem in the first place. As with fungi, wood beetle will only cause serious damage where there is damp. So the key to eradicating beetle is to cut off their fuel supply and reduce the moisture content of affected timbers. Once the dampness has been resolved and timber allowed to dry out, the beetle colony should diminish over time.

As wood dries out it becomes less attractive to beetle and activity decreases. Once moisture content reaches below 12% they struggle to survive and their numbers decline naturally. Heated buildings are very hostile environments for them.

To fix the root cause of the damp means repairing leaks

and removing any modern sealants, paints and renders that are preventing the damp from naturally evaporating out of walls and floors. Applying gentle background heating to affected areas will help to reduce moisture content in the surrounding timbers.

If you have a serious and active outbreak, then in addition to the above works, a targeted application of insecticides to the affected areas can sometimes be justified. Until a permanently drier environment has been achieved, a brush treatment can be effective if applied sparingly and locally to only the worst signs of live infestation. Where chemicals are used, special gels and pastes containing *permethrin* are most effective. By spreading toxic pastes directly onto affected timbers, the theory is that when the beetles emerge they eat through the paste which poisons them. But as noted earlier, this can also kill natural predators that otherwise could have done the job for you.

## The four stages of timber treatment

To avoid unnecessary treatment with chemical preservatives, there are four key tasks to carry out:
1. Check whether the outbreak is live or historic. Most bore holes are simply inactive remnants of old infestations that have been dead for years.
2. Check whether damp conditions still exist to fuel continued activity.
3. If an outbreak is current, identify the source of the moisture and fix it.
4. Dry out and ventilate the affected area and repair any damaged timbers.

## Defect: Decaying timber

**SYMPTOMS**
Wet wood is spongy and appears cracked and wrinkled along the grain.

**Cause**
**Timber becoming consistently wet over time**

**Solution** *Eliminate the cause of the damp*
Timber can become wet over time, often near or below ground level, due to damp walls or plumbing and drainage leaks etc.

Having eliminated the cause of the damp, cut out and replace the affected areas. Treat adjacent wood with rot fluid. Wet rot will only persist while timber is wet, hence the importance of good ventilation. New timbers should be pre-treated and protected from damp, *eg* on a strip of plastic DPC, or supported in joist hangers rather than in walls.

**SYMPTOMS**
Small cube-shaped cracking pattern; fungus growth; mushroom smell; grey strands extending out; wood is a light brown colour and feels dry and brittle.

**Cause**
**Timber infected by fungal spores**

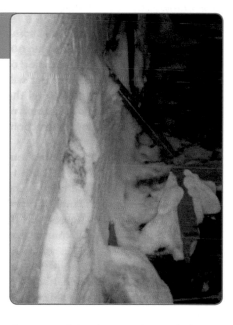

**Solution** *Eliminate the cause of the damp and treat timber*
Timber becoming moist (but not wet) in a cool, poorly ventilated area allows the spores of the dry rot fungus to thrive.

Eliminate the cause of the damp and improve ventilation. Badly affected timber cannot be treated and must be cut off and burned.

## Defect: Wood beetle infested timbers

**SYMPTOMS**
Fine wood-dust around flight holes

**Cause** **Insect larvae in wood**

**Solution** Insects fly in and lay eggs in wood; larvae hatch inside wood and eventually bore their way out. Old furniture stored in lofts can contain beetle, which then spreads.

Rectify causes of damp by controlled drying out of timbers and improved ventilation. In extreme cases carry out localised spray treatment to affected timber.

**CHAPTER 7** Victorian House Manual

# WINDOWS, DOORS & EXTERNAL JOINERY

Without doubt, the windows, doors and external joinery are some of the most important architectural features on Victorian and Edwardian houses. Yet more money is wasted fitting inappropriate replacements than on just about anything else. And making the wrong choice can mean, in effect, paying good money to knock thousands off the value of your home.

Nothing you can do to a period house seems to cause more outrage than messing with the doors and windows. Everyone loves the look of the originals and estate agents all agree that period features should be retained. But living with draughty old windows that rattle, and rotten doors that leak copiously, isn't everybody's idea of fun. So the idea of making your home well insulated, quiet, and secure by means of modern replacements can be a very seductive proposition. It's just a pity they so often ruin the character of an old building – which is why if your house is Listed or in a Conservation Area you will need to talk to the Planners first. The trick is to try and achieve the best of both worlds.

In order of desirability, the ideal solutions are:

**1** Overhaul and upgrade the original units: specialist firms can do this, or there are DIY kits available.

**2** Where this is not possible, replace the original units with matching new timber equivalents.

**3** Replace with matching new units in a different material.

The quality of naturally seasoned Victorian and Edwardian softwood was far superior to much of the fast-grown, kiln-dried variety used in a lot of modern housing. Whereas today's fascia boards will often have started decaying by their mid teens, many timbers on 19th-century properties are still going strong. Problems are mostly associated with excessive weathering due to lack of protection, as all external timbers require periodic decoration to survive.

If you're aiming for authenticity, popular colours used on Victorian windows and joinery included Brunswick green, maroon and reddish brown, as well as woodgraining to imitate expensive oak or mahogany.

Classic Edwardian projecting oriel window

# Windows

The eyes and soul of a typical Victorian home were made from naturally seasoned timber, predominantly taking the form of sliding sashes. These were produced in a wide variety of styles,

along with softwood or cast-iron casements, the latter manufactured in a range of quite elaborate designs often featuring traditional leaded lights.

Well over a century later, there's a pretty good chance the originals have been replaced with (often inferior quality) double-glazed units. These might comprise sashes or hinged casements made from timber, aluminium, plastic or steel, being either single, double, or secondary glazed. Other variations include grotty 1950s louvres with small horizontal glass slats (which are infamously 'burglar-friendly') and modern tilt-and-turn windows.

## BAY WINDOWS

The bay window was the Victorian must-have fashion accessory – with a bit of ornate detailing they could make the frontage of an otherwise fairly plain house look sophisticated and expensive. Projecting bays catch sunlight, attract fresh air, and allow a good view of the street, so even many cheaper dwellings could boast at the very least a single-storey bay window. In fact, they were so popular that it was considered highly desirable to have another one at the back, to let light in to the rear parlour, or the side wall of the kitchen. A stylish variant, common in Edwardian houses, was the small, projecting upstairs 'oriel' window.

There were two main styles of bay: polygonal ('splayed') or square. These would be either single storey or full height, with a choice of hipped, gabled, flat, or lean-to roofs. Flat roofs were usually finished in sheet metal, some surrounded by a medieval-looking parapet. The detailing on bays often reflected local skills and traditions: small wooden bays in Birmingham, carved ornate woodwork in the North-East, stonework in Scotland, brick and artificial stone in London. Some multi-storey bays are of lightweight timber-frame construction with a rendered finish, which allows a larger glazed area than the more usual heavy stone or brickwork panelled structures with their bigger supporting piers.

Elaborately decorated bays are one of the great glories of the

*1: Intricate leaded iron casements to crenellated 'battlement' single storey bay 2: Two storey late Victorian bay with classical columns and fashionable small panes to upper sashes. 3: Early lightweight wooden bay. 4: Gabled Edwardian square bay. 5: Slim and elegant c 1885 – but prone to bowing. 6: Leaning bays.*

Victorian house, but can sometimes be a source of concern technically.

Builders were sometimes tempted to skimp on foundations to bays and, to save money, some were built only of weedy single-width brickwork. In a row of terraced houses they may all have settled at a slightly alarming angle due to differential movement between themselves and the main building. Along with porches, bays are one of the most common areas in which to find cracking and settlement. The thinking was that because the bay was often a fairly lightweight structure compared to the main house, the foundations didn't need to be so deep. But we now know that to be clear of seasonal movement in the topsoil even dwarf garden boundary walls generally need foundations to a depth of half a metre or more. And where a lack of foundation depth is combined with subsoils of shrinkable clay, sloping sites or large trees growing nearby, there's a good chance that a shallow bay will go travelling independently from the main house, particularly after a long, dry summer (see chapter 5).

When the windows and glazing in bays are renewed, it is important not to remove any timbers supporting the roof or floor structure above. The structural columns or piers on bays were sometimes undersized and are frequently seen to have bowed out. In severe cases rebuilding may be necessary, but over the years most will have settled down to a stable position – until contractors arrive armed with crowbars and sledgehammers to rip out the windows.

## GLAZING

Traditionally windows were divided into small panes with timber or metal glazing bars, or by means of strips of lead to form 'leaded lights'. But in 1838 a cheaper method of producing larger sheets was developed – known as sheet glass. This had most of the virtues of plate glass but cooled much faster and needed less polishing. Sheet glass became common in mainstream housing from the mid-Victorian period, changing window fashions as old multi-paned glazing became obsolete; the fashion now was for large expanses of glass. But things came full circle later as Queen Anne and Arts and Crafts styles, which flourished during the 1890s, reacted against the mainstream by glazing the upper parts of windows with small coloured panes, which were also popular in fanlights above front doors.

Today sheet glass is available in thicknesses between 3mm and 6mm. Glass used to be sold by weight per square foot, commonly either 15, 21, or 26oz. Original rolled plate glass was very thin by modern standards, with more durable 21oz sheet glass only appearing towards the end of the 19th century, but this was still less than 3mm thick. After manufacture it was sorted into different qualities ('bests', 'seconds', etc.), but much fourth-quality glass was used in mainstream housing.

You can usually tell original glass by moving your head from side to side to see the distortions, as it had more imperfections than the ubiquitous modern float glass, and produces interesting

1 and 4: Coloured glass fanlights c. 1900. 2: Timber 'crow's nest' late Victorian bays in Exeter. 3: Multi-paned glazing became fashionable around the turn of the century. 5: Elegant metal casements.

effects as the light passes through. Preserve it where possible, as replacement with thicker modern materials can lose some of the charm of imperfect 'wavy' original glass. And because it's heavier, changing it can alter the balance of a sash window.

## GLAZING REPAIRS

Glass in Victorian windows was bedded in the rebate of the frame with linseed oil putty and secured with small metal pins called sprigs. A common problem is where cracked or loose putty is letting in water, or the odd bit has dropped off. The ideal solution is to remove cracked putty and replace it, but unfortunately this can be almost impossible to do without breaking the pane. Although heat guns can soften old putty, their use is not advisable because the glass is similarly heated, which can cause it to crack. In most cases, however, it should be sufficient to periodically fill cracks in putty by warming up some new linseed oil putty and rubbing in with your finger. Where sections have come loose, gently remove them and simply re-putty the gaps.

Where complete removal is required, a specialist joiner can use a 'putty lamp' that applies focused infrared heat, softening the putty prior to removal. The infrared rays pass through the glass without heating it so the pane is not damaged.

## SASH WINDOWS

Any window you want – so long as it's a sash. That was the general rule in Victorian times, with the exception of small casements to some WCs and sculleries, and in many rural properties and later Arts-and-Crafts-style houses. The timber sash window (or 'double hung vertically sliding box sash window', to give it its full name) is synonymous with the Victorian house, despite its Georgian and Dutch origins.

The sash is a clever piece of engineering. It allows a room to be ventilated very effectively. The raised bottom half of the sash allows cool, fresh air to enter, while the opened top half lets the warmer, stale air escape. In kitchens, the windows were often set as high as

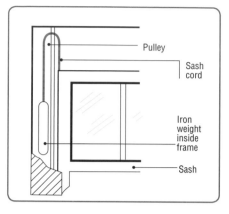

possible for the most efficient ventilation, as the big open ranges for cooking generated huge amounts of heat.

Sash windows work like this: To prevent the two sashes from sliding down (with gravity) when you want them open, the windows are counterbalanced by weights. Flax cords (or chains) are nailed to a groove in the side of each sash, looping up and over pulleys in the upper frame and down to heavy iron weights concealed in boxes each side of the frame. The timber sides of the boxes are removable to provide access to the weights. You can spot later models by their thicker frames with small projecting 'horns' at the lower ends of the upper sash frames and to the tops of the lower sashes. This stronger structure was necessary to support the increased weight of larger, heavier sheets of glass. To make it easier to manoeuvre them up and down a pair of handles were sometimes fixed to the bottom rail of each sash.

If you can't open the small sash windows either side of a main one, it may be because they're fakes, never meant to open, to save on space and cash – they just look like they should.

Internal security shutters were a popular addition. When not in use they folded back into neat compartments next to the window or slid down like sashes below, but they were out of fashion by the end of the 19th century. However, these timber shutters can provide additional sound and heat insulation, as well as improving security, so they are well worth renovating if you have them.

Modern replacement timber or UPVC sash windows tend to employ friction fittings between the sash and the frame, or use springs, dispensing with all the bulky gubbins of weights, pulleys, and cords.

**Below:** Non opening decorative sashes
**Centre and right:** Modern uPVC sliding sashes – but how long will they last?

## CASEMENT WINDOWS

The casement is the simplest form of window. These 'modern' outward-opening windows are hinged from the frame at the side or top, usually with an iron 'hook and eye' stay at the bottom to allow graduated opening, and a

'cockspur' fastener at the side. But because the Victorian window of choice was predominantly the timber sliding sash, in the majority of urban houses casements are only normally found in small windows to the rear. Houses in rural areas were more likely to defy the vagaries of fashion and remain loyal to tradition with casements throughout.

Timber windows of all kinds are potentially vulnerable to rot from damp – either externally from lack of maintenance (painting) or internally from condensation. Typical problems with casements include windows that are hard to close, cracked glass, and worn fittings, but they are usually quite easy to refurbish.

Rarely, you may encounter original iron-casement windows, often in Gothic style, with thin iron glazing bars. Where these have suffered from rust, the expansion of the corroded metal can sometimes cause cracking to individual panes. This is likely to require removal of the window for dismantling and rebuilding in a workshop. However, when the rust is removed there should usually be sufficient sound metal underneath. Use a rust inhibitor and metal primer before painting with a topcoat.

### ORIGINAL AND GENUINE?

In many houses today the original windows will have already been replaced. But by now, these may be getting rather shabby and be in need of replacement themselves, ideally with new high-performance replicas of the originals.

To discover what kind of windows your home was built with take a look at the back, where some of the old windows may have survived in the less

**Right:** Many replacements look wildly inappropriate

Replica uPVC bay sash windows

important rooms. Also, some neighbouring houses may still have their original windows, which can be copied. Undoing disfiguration by replacing inappropriate modern windows with sympathetic 'original replicas' is likely to pay off handsomely in terms of 'kerb appeal', ultimately adding to the property's value. This can be more of a challenge in properties where the size of the original opening has been altered to take some garish 1970s monstrosity because you need to reinstate the opening with matching brickwork.

### DOUBLE GLAZING

Although we don't share the Victorian love of fresh air, controlled ventilation (*eg* from trickle vents) is now recognised as important in the prevention of condensation problems. However, a certain amount of condensation on single-glazed windows in cold weather is hard to avoid. The obvious compromise is to retain the

New replica originals custom made with slim high E' double glazing

'misted' after only a few years, defective seals having allowed in moisture that looks like condensation but is actually inside the cavity. The bad news is that there's nothing you can do except replace the units – and when one unit goes, the others are likely to follow. Hence the importance of warranties.

## SECONDARY GLAZING

Sometimes the best solution is to retain the existing windows and fit custom-made secondary glazing tailored to the individual opening. Secondary glazing is an independent system of windows fitted on the inside of the existing window that can dramatically boost thermal insulation, seal draughts and reduce noise. This way you can get the best of both worlds – improved thermal insulation whilst retaining the special 'sparkle' of original period windows – plus it usually costs less than complete replacement. This can be a particularly good option for Listed buildings and those in Conservation Areas.

Individual units can be designed to discreetly match and be aligned with the original frames, making them virtually invisible from outside. And thanks to the deeper air gap between the main windows and the inner secondary glazing, sound insulation performance is actually superior to that of double glazing. You can choose from a range of designs with neatly hinged openings, sliding units, or lift-out magnetic panels that can be removed in summer.

Best of both worlds – originals with high performance inner secondary sealed units

original sashes or casements and simply replace old window glass with new thermally efficient sealed units. Specially developed thin double-glazed units have been developed to fit the thinner rebates in old buildings

To comply with Building Regulations, replacement windows now have to be double-glazed, even though this actually ranks fairly low in the league-table of cost-effective energy conservation measures. Double-glazing consists of two vacuum-sealed panes of glass with a gap typically of at least 10mm between them. The panes are sealed at the edges into a single unit that is fitted into the window frame. But it's surprisingly common to find that modern sealed units

In many old houses frames will be distorted; but timber windows can be trimmed to fit

**Above:** A challenge – I like it !

High-performance secondary glazing can actually comprise double-glazed sealed units, which match the thermal performance of replacements, whilst of course retaining the originals. It's also worth installing thick curtains and making use of any internal shutters – methods the Victorians found so effective for draught-proofing.

**Right and below:** Three types of Victorian sash fastener – screw, lever and cam (clockwise from right)

## RATTLES, DRAUGHTS AND COMMON DEFECTS

The problem with original windows is that they wear out with age. The sashes can become loose and rattle. The timber beading that holds the sashes in place can work loose or split. Dropped arches, decayed lintels and movement to the walls of the house over time can cause frames to distort and windows to stick. Many uneven openings still bear the scars of wartime bomb damage, but the windows have been planed and adjusted to accommodate the movement.

Sash windows are fairly straightforward to overhaul and can nearly always be repaired. To do this properly, it's best to remove the sashes from their frames, which can be done from inside. This involves levering off the nailed beading that retains them, which needs to be undertaken with care to avoid splitting the beading. In Scotland the lower sashes were sensibly designed to open into the room for cleaning and ease of maintenance.

One common problem with sliding sashes is that the cords become clogged with paint or become frayed and break. They also have an endearing tendency to stick or become jammed, also due to a build up of paint or a lack of lubrication. To ensure smooth running, the channels in the frame on either side in which they slide should be lubricated with linseed oil or wax. Also the pulley wheels over which the cords loop on the upper sides require a periodic drop of oil – and while you're at it, check that they haven't seized.

Although the original high-quality Victorian timber is remarkably resilient to decay, it's important that rain is free to disperse rather than be allowed to 'pond'. So check that each part of the external window overhangs the part beneath it, and that all horizontal surfaces slope slightly towards the front, with no ledges that can harbour water. If moisture is allowed to penetrate joinery over a period of time, rot is likely to develop; this can sometimes occur unnoticed to enclosed spaces such as the box frames either side of sash windows that house the weights.

Sashwindowrenovation.co.uk

Sashwindowrenovation.co.uk

Well-maintained timber sashes should not rattle or admit draughts, and it is fairly straightforward to upgrade your existing windows with draught-stripping and DIY window kits. Annoying rattles can be solved by fitting replacement beading with a draught-proof strip together with cam-style brass catches ('fitch fasteners') or screw fasteners that pull the sashes together when closed. Loose glass is often due to defective joints between the glazing bars and the top and bottom rails.

## TIMBER REPAIRS

A day or two spent rubbing down and painting every four or five years can work wonders. Any surface cracks in putty, or gaps that have opened up against the glass, should be filled and painted before the frost gets a chance to blow it loose or decay sets in. As noted earlier, horizontal surfaces should slope slightly towards the front so rainwater can disperse. Sills and lower joints tend to be the most vulnerable areas, and flaking paint may conceal early signs of decay, particularly to bottom rails and lower sashes. Small areas of rot can be cut out and filled with putty or filler, which can often be applied without having to remove the windows.

B. Mulford

*As good as new; localised decay cut out and new timber strip expertly spliced in*

## REPLACEMENT WINDOWS

The difficulty in making sash windows draughtproof or fitting them with double-glazed sealed units has encouraged many owners to rip them out and replace them with modern casements, many of which look wildly inappropriate. But today many new casements are designed in styles that approximate to the original architecture, imitating the look of sash windows.

Building Regulations (Part F) now require new windows to meet certain thermal standards, so replacements normally need to incorporate double glazing. But if your house is Listed or located in one of the UK's 10,000 plus Conservation Areas, the requirements of Building Control may be overridden by those of the Conservation Officer. Strange as it may sound, when it comes to rectifying past wrongs, such as ripping out gruesome 1970s horrors to install appropriate handmade replicas, with Listed buildings you still need to obtain consent. So it's always best to check first.

Many ready-manufactured replacement windows come only in standard sizes, which may not match the old openings. And as older buildings have often settled over time, the openings may have become a little irregular. The important thing is to at least get the width right; the masonry underneath can usually be built up and the sill raised. Commonly available replacement options are:

- Timber casement windows: if chosen carefully these can look very similar to original sashes. Because old buildings tend to move with the seasons, timber is a good choice, being relatively conducive to 'easing and adjusting' – *ie* can be planed to fit. But poorly chosen timber replacements have the double drawback of not only often looking out of place but also of needing periodic maintenance.

- UPVC double-glazed (plastic) windows: much derided, but they are easy to live with and provide good insulation. The trouble is unless carefully designed, they can ruin the look of an old building, and many are poorly manufactured or carelessly fitted.

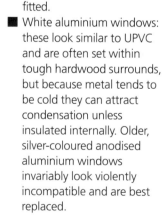

- White aluminium windows: these look similar to UPVC and are often set within tough hardwood surrounds, but because metal tends to be cold they can attract condensation unless insulated internally. Older, silver-coloured anodised aluminium windows invariably look violently incompatible and are best replaced.

If you've read chapter 6 you'll know the importance of minimising condensation. It is often after the installation of unvented sealed UPVC double glazing that serious condensation problems arise elsewhere in the house. Trickle vents (which can be fitted retrospectively) are designed to improve ventilation.

**Above:** House on right has widened window openings and ugly replacements, with loss of period character and value

**Right:** UPVC replacement window makes no attempt to match recessed originals (which are salvageable)

**Below:** Rectifying past wrongs; new replica timber box sashes to upper floor replacing ugly 1970s aluminium horrors

**Above:** Stripping away plaster reveals timber lintel nestling behind facing brick arch

**Right:** Exposed timber lintel formerly concealed by small bay roof; note cracking above

Other problems with replacement windows include dropped lintels placing stresses on frames that in turn cause casements to jam. All frames should be sealed at the joints to the surrounding masonry walls to prevent ingress of moisture, using a suitable mastic sealant for gaps up to 10mm (0.4in).

## LINTELS

Timber was the traditional material for lintels spanning openings above doors and windows. Even where the outer face of the wall has a brick arch or stone lintel, there is usually a secondary beam sitting behind it made of wood. Internally, timber lintels provided a useful surface for fixing laths for plasterwork, and therefore tended to be plastered over rather than being left exposed. Where problems do develop over time, they tend to manifest themselves in the form of cracking to the external render or to internal plasterwork. But even where a lintel has started to rot, it may not automatically need replacing. The important thing is to expose it

and allow it to dry out. This often requires the removal of cement renders or impervious modern paints that are trapping damp.

## REVEALS

Traditionally windows were always rebated – usually set back from the wall about 100mm (4in) – to comply with legislation and to provide protection from driving rain. Today some replacement windows ignore this and are fitted flush with the wall face, which can look a bit odd.

A common maintenance issue is at the junctions of window or door frames with the surrounding masonry reveals. These are normally pointed with a thick strip of mortar, but quite large gaps are often evident due to erosion or movement in the house as a whole over time – it just happens to show here most obviously. In particular, roof spread (where the roof pushes the upper walls outward) can often be seen at top-floor window reveals as a gap in the mortar, widest at the top.

The solution for gaps over 10mm is to point up with a 1:2:9 cement/lime/sand mix. Smaller gaps can be sealed with mastic. For large irregular gaps, injecting expanding foam filler first can provide a useful base for the mortar. It's quite common for reveals to be lime rendered and many have suffered from cracking or come loose. This allows damp to penetrate and the affected area must be hacked off and re-rendered.

Upper wall pushed out by rafters, with tapered gap to window reveals

Cracked sill will channel rainwater to (poorly repointed) brickwork below

## SILLS

Because Victorian windows are set back within their openings, the small sills at the base of windows (often of solid timber) depend on the large projecting main stone sills beneath to disperse rainwater. Better-quality houses employed stone for window and door sills (eg Derbyshire grit, York or Bath limestone). Sometimes cheaper slate or tiles were used, or in poorer houses just rendered brick. Some modern replacement 'stone' sills are actually cheap-looking concrete. The most important thing is to check there is a clear groove underneath projecting sills (under the front edge set back about half a centimetre). This allows water to drip harmlessly off the sill instead of tracking back underneath to the wall, where it can penetrate and cause rot. If it is clogged with paint, the blocked groove can be carefully cleared with a chisel.

Rainwater should run swiftly off the tops of sills, otherwise even stone and concrete will eventually crack from frost action and wooden sills will rot. They should be sloped away from the window, more steeply if made of timber, tile, or brick. If they are flat, or small puddles form, wooden sills can be planed and stone sills resurfaced with mortar, the old surface having first been rubbed down and primed with a PVA adhesive.

## ROOF WINDOWS

You may come across original small cast-iron skylights (rooflights) built into the roof slopes of attic rooms. These windows were

either non-opening or hinged and, like any single-glazed window, are prone to condensation running down the glass. Somewhat ingeniously, the moisture was often prevented from dripping into the room by collecting internally at the sill in a small lead channel and being conveyed away via holes or grooves to the roof outside. Nonetheless, rust and decay may by now have taken their toll.

New dormers and rooflights should not jar with the existing architecture and planning consent will be needed, at least for the front elevation. Replicas of traditional rooflights are available, or double-glazed skylights, supplied complete with flashings, may be suitable. Any new openings in a roof must not weaken the roof structure. The adjoining rafters need to be doubled up for strength and, as for all structural work, Building Control must be notified in advance. The aim should be to use the smallest number and size of windows as possible, and to replicate the proportions, glazing bars and profiles of the originals.

## PROJECT: Reglazing a window

Replacing a damaged pane should be fairly simple (unless windows are leaded or have stained glass). Always wear thick gloves and eye protectors, and proceed as follows:

1 Very carefully chip out the old glass with a chisel, keeping your legs well clear of any falling shards. Old hard putty can be softened with paint stripper then chipped out. Clear and rub down the old frame rebate and, using pliers, remove the small metal fixing pins known as 'sprigs' that retain the glass.

2 Measure the opening, and get a new pane of glass cut to size (horticultural glass has imperfections like old glass). Remember to deduct about 1.5mm at each end from the frame size to allow sufficient clearance when fitting. Modern 3mm (0.125in) glass is suitable for sizes up to 600 x 400mm; otherwise 4 or 5mm thicknesses will be required.

3 Clean and prime the rebate before puttying in a new pane, to prevent oils in the putty from being adsorbed and drying out too quickly. Roll some fresh linseed oil putty into a line between your hands and press into the rebate in the frame. Carefully place the new glass into the puttied frame. To secure the pane, gently tap in some new sprigs, protecting the glass from the hammer with a sheet of thick card or Perspex leaving a small gap between the sprigs and the glass to prevent fracturing.

4 Make another line of putty to press over the edges of the glass, using a putty knife dipped in water to smooth into a bevel shape and cut off any waste – note that it requires plenty of patience to achieve a neat finish! Wait at least two weeks for the putty to harden before painting.

## Defect: Rot to timbers or gaps at corner joints to sashes or casements

### SYMPTOMS
Soft decayed wood, flaking paint, sticking windows.

**Cause** **Water penetration causing deterioration of timber**

**Solution** To test for rot, check to see if the wood is soft and spongy and can be dug away with a screwdriver. Sills and lower frames are especially vulnerable. Only in very severe cases is complete replacement usually necessary. Normally if parts of the timber are in bad condition, it should be possible to cut away the damaged area, treat the remaining sound wood with preservative fluid, and graft new sections of replacement timber into place. Localised damage from rot can be dealt with as follows:

■ Cut out areas of rotten wood and remove any loose material.
■ Using a two-part wood repair system, first use the brush-applied chemical hardener to strengthen the wood by reinforcing any decayed timber.

■ Then apply the filler paste that sets to a solid material that can be sanded and planed. Preservative tablets can be inserted in holes drilled in the frame, which release a preservative into the timber.
■ Finally apply a primer and undercoat followed by two topcoats of exterior paint.

**Cause** **Corner joints worked loose**

**Solution** Flat stainless/galvanised steel L-shaped brackets can bridge the damaged area after cutting out any decayed timber. The brackets can be countersunk and the joint repaired with epoxy filler. Filler repairs should be painted as they degrade in sunlight. Alternatively, secure with three glued dowel pegs: two through the side of the stiles and one from the front. Cracks and gaps at joints can be treated by cleaning off flaking paint, cutting out any rotten wood, and then treating, priming and filling. Cracked or loose glazing putty must be renewed. Finally, paint with undercoat and one or two topcoats.

## Defect: Sashes are stuck or stiff, or won't stay open

### SYMPTOMS
Sashes are heavy and prone to jamming, or slide down uncontrollably.

**Cause** **Broken sash cord, or pulley not running freely**

**Solution** Renew sash cords and lubricate pulleys. See step-by-step overleaf

**Cause** **Modern replacement glass is too heavy for the counterbalance weights**

**Solution** Adjust counterbalance by adding lead weights

**Cause** **Timber beading on frames that holds the sashes in alignment is loose or broken, or the windows have been painted closed**

**Solution** Repair or renew beading
Prise off beading, dismantle and rub down. Plane off any distorted wood, lubricate sides with candlewax, and oil the pulleys. Strip any paint from the sash channels in the frame and protect them with linseed oil. Fit handles to the bottom rails of sashes. Re-fix beading with discreet brass screws.

# Overhauling a sash window

It is normally better to overhaul the original sash windows than to replace them. Most problems are simply caused by a build-up of paint, box frames distorting, or broken sash cords. These can be rectified and windows decorated when dismantled.

**1** Before removing the beading, score through the paint with a knife to prevent layers of paint from breaking off.

**2** Prise off beading to both sides of frame using a chisel. Start half-way up, working up and down until you can 'spring' the beading out (it is wedged in by mitred joints top and bottom) – try not to split it.

**3** To prevent the sash weights falling inside the box frame, cut the sash cords (if still intact) and lower the weights to the box bottom, then pin their cords temporarily to the frame.

**4** Lift out the bottom sash, then with a chisel prise out the vertical 'parting' bead located centrally on the frame sides that separates the sashes (they may break up) and rake out the groove. Cut the cords to the top sash, lower the weights, and lift out the top sash.

All photos courtesy **DIYsashwindow.com**

**5** The weights are hidden behind thick wooden 'pocket covers' in the side tracks ('pulley stiles'). Insert a flat screwdriver blade into the 'hidden' horizontal joint at the bottom of each cover and lever it off (some have screws at the bottom).

**6** Having first removed the temporary pins holding their cords, lift out the weights, noting where each weight belongs. Remove old cords and nails from sides of sashes.

**7** Smooth running is often prevented by old paint on the side tracks. Remove excess paint with a scraper until flat and smooth. Repair any rotten timber and fill any holes to the box frame. Check pulleys run freely – oil or replace.

**8** Remove any excess paint from the frame sides and sand the edges until they run smoothly in the frame. Repair any open joints and fill holes. Rub down to prepare for decoration.

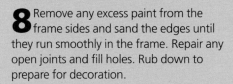

**9** This is a good time to fit draught seals (optional), but grooves may need to be cut for the seals with a router, which is not necessarily a DIY job (see points 10 and 13). If fitting seals or new glazing, weigh the sash before the alterations.

**10** Now the weights can be replaced. If you've added heavier glass or draught seals, the balance may have been altered, so you need to weigh each frame again and then both its weights. Small increases in frame weight can be corrected by adding lead pieces to the iron weights.

**11** To fit new sash cords, first measure from the top of the pulley down to the window sill and multiply by two. Then cut four lengths of cord this long. Starting with the top sash, tie a heavy nail to a long piece of string as a 'pathfinder' and feed it over the pulley so that it drops down the frame. Tie the other end to the new sash cord so it follows down, and pull the cord back out at the lower pocket. Attach to the weight with a knot.

**12** At the top end, pull the cord to lift the weight up as high as possible. Temporarily restrain it by pinning the cord just below the pulley so you can then attach the end of the cord to the sash without struggling.

**13** Resting the top sash on the bottom of the frame, nail about 350mm (14in) of cord in the side grooves with tacks, leaving the upper 50mm (2in) un-nailed. Trim excess cord at both ends. Remove temporary pins and check the sash slides to the top of the frame and is held closed by the weights. Now stand the lower sash in its correct closed position. Any gaps at top and bottom can be sealed by planing the bottom sash or packing the top sash with a slim timber wedge.

**14** Refit or replace the central parting bead. This can be done as two pieces each side. With the top sash in place, fit upper beading into the grooves. Pin and glue in place. Then lower the top sash and fix the lower beading. Check that the sash runs smoothly.

**15** Apply a line of mastic at the beading joint and smooth off with a wet brush.

**16** Re-hang the bottom sash after attaching the new cords. Check that both sashes are aligned at their horizontal centre rails.

**17** Close both sashes and refix the beading on both sides (start by bending it into place). Nail or screw loosely at intervals of 300mm (12in), check sashes run smoothly, adjust and then fully secure. Apply a small amount of mastic. Refit or replace catches.

All photos courtesy **DIYsashwindow.com**

Classic Victorian terrace paired hallways

Original Victorian porch doors of superb quality…

…open to vestibule porch

# Doors

You can date your terraced house by the position of the front door. Earlier terraces repeated the pattern of one front door and window followed by the neighbour's front door and window, and so on down the street. In later models, doors and windows were in pairs: your door was next to your neighbour's, followed by a window next to a window, and so on. The reason for this change can be found at the back – they were cheaper to build with the rear additions of two houses joined together under one roof. Externally this created the classic 'side return' layout with its enforced intimacy due to side-opening back doors facing those of the neighbours across a small yard.

In all but the cheapest Victorian houses the entrance would be ornamented to be as imposing as possible, with the front door normally recessed within an open 'vestibule'. Only very rarely would outer doors be fitted, although many have since been added. Smaller cottages might have a simple storm porch, rarely without some embellishment in the form of carved timberwork. The temptation to glaze-in an open vestibule is best avoided as it can spoil the appearance of the house.

Rear return, with new period style French windows

Victorian cottage porch was originally open

Classic stepped main entrance, with separate steps to semi-basement below

timber, and the large upper panels were often glazed with decorative acid-etched or sandblasted glass, perhaps with a coloured glass fanlight above the door frame. From the 1870s front doors became more elaborate, some with their panels arranged in three columns or with coloured leaded fan lights, with six-panel designs regaining popularity from the end of the century.

At the base of the door a stone threshold of hard York or Portland stone was usually provided, except in cheaper houses, where timber would suffice. Houses with basements or semi-basements required short flights of stone entrance steps enclosed with side parapet masonry walls. The treads and landings were made of solid stone slabs rebated into each other and fixed with iron dowels, but the risers were sometimes only of thin concrete mixed with stone dust.

For security reasons the front door was always the heaviest. Good-quality front doors would typically be 50mm thick and at least 825mm wide with moulded panels. Frames would typically be of 112 x 75mm spruce with moulded stops and upper glass lights. Today, front doors are generally thinner, typically 44mm thick, and panels may only be 6mm thick, which as well as being a potential security risk can also be prone to heat loss and condensation.

Front doors denoted status – the more expensive the house, the more elaborate the mouldings. In grander homes a projecting front porch with its own roof would dominate the front elevation, perhaps featuring elaborate Gothic columns and exotic moulded or carved stone components and fine timber detailing. External doors were invariably painted, unless they were made of expensive hardwood, in which case they would be varnished. Popular colours were chocolate brown, olive green, deep red, dark blue and later in the century black or creamy white. Early Victorian front doors generally have six or sometimes eight panels, arranged in two columns. Classic Victorian four-panel front doors predominated from around 1850.

The lower panels were of thick solid

To the rear of the house, the back door from the

kitchen would have solid timber panels, or often with the upper pair glazed with cheaper rolled glass. Houses in rural areas might have a traditional 'cottage style' ledged and braced door, made from vertical boards nailed to Z-shaped timber bracing strips on the inside. These would be painted the same colour as other external woodwork.

You might also find narrow double 'French window' doors opening from the back parlour or dining room into the garden. Victorian draught-proofing largely comprised of thick, lined curtains fitted behind the front door, which can cut down heat loss surprisingly effectively.

alternatively the house number might be painted on the door or frame. More expensive houses had elaborate gilded and shadowed numbers hand-painted on the interior face of the transom light over the front door, or etched into the glass.

Designs for doorknobs, knockers and locks echoed classical themes, such as shields, lions' heads, Medusas, dolphins and urns. Wrought or cast iron would either be given a lacquered 'japanned' finish, stove enamelled in 'Berlin black' or painted in fake bronze or the same colour as the door.

Today, stripping multiple layers of accumulated paint can reveal wonderful original detailing on these small antiques that are an important part of restoring authentic charm to a period property. Original fittings tend to have a reassuringly solid feel, and are not quite as elaborate as much of the modern imitation repro stuff.

## LOCKS

Modern cylinder locks like the 'Yale' became widely used from the mid-Victorian period along with traditional mortise locks. Where original locks have survived, they're an important part of history of house and are best retained, particularly where surface mounted. Old locks can usually be brought back to life with a little oiling and adjustment. New keys can be cut to replace long mislaid originals.

However, such antique security measures are unlikely to offer the level of security required today by insurance companies. So they may need to be augmented with discreet new locks and bolts. But such works need to be undertaken with care because historic doors and windows can be damaged by trying to fit new security devices.

It makes sense to upgrade all door locks where possible, using new ones that won't damage the fabric – a five- or seven-lever mortise deadlock is recommended. The best solution is to buy locks that fit existing mortise openings without having to cut or drill (see page 145).

## COMMON DOOR DEFECTS

Being constantly exposed to the weather, an external door may become damaged beyond repair if neglected over a period of time. It is quite time consuming and costly to fit a replacement door, so it is worth repairing original ones where possible. Joinery near ground level is always at risk, particularly the lower door frames, which can become saturated with moisture from the ground. Weatherboards at the bottom of the door are designed to protect the lower door from rain, as well as scuffs and kicks.

## IRONMONGERY

Victorian door furniture was mostly mass produced in cast iron – polished solid brass was reserved for the better off.

The introduction of postal services from the 1840s required the provision of letter plates; although relatively small these are well worth preserving. Some quite elaborate examples were produced combined with pull-bar handles or door knockers,

often with the word 'letters' cast into the flap as an aid to identifying their purpose!

Postal deliveries also required that houses were clearly numbered for the first time. Oval plaques of glazed ceramic or enamelled iron still survive on some properties;

Late Victorian 'Arts & Crafts' house shows signs of movement, but original door has adjusted to accommodate change

They should be periodically checked and may need to be repainted, refixed or replaced.

Small areas of rot in an original door may be solved by cutting out affected areas and using special filler, as described earlier for windows. Larger damaged sections can be replaced with well-seasoned wood to match the original. However, if the existing door is only a cheap modern one, replacement is likely to be preferable.

Choosing a replacement door for older houses isn't always easy. Sizes are often non-standard, frames may be distorted, and period styles can't always be matched, although specialist firms will make replicas – at a price. Doors manufactured in UPVC come as a complete unit including the frame and are claimed to be 'maintenance free'. But because old houses tend to shuffle about on their shallow foundations in tune with seasonal ground changes, UPVC can pose a problem – it can't be 'eased and adjusted' like timber doors and windows. But above all they are visually inappropriate for a period house. Timber is still a very popular material, but avoid the temptation of buying a significantly over-large door in the hope of cutting it to fit, as

anything more than minor trimming can structurally weaken them by slicing through joints.

If you're planning on forming a new opening, it is important to ensure there is an effective DPC below the sill, aligned with the main DPC in the walls to reduce the risk of damp.

Victorian doors are generally heavier than modern ones and need to be hung with three 100mm (4in) brass butt hinges. Hinges should be fitted about 150mm (6in) from the top and 200mm (8in) from the bottom, and always fitted to the door first, not the frame. If original doors or windows need to be stripped, it is best to use a paint stripper, rather than sending them away for 'hot dipping' in tanks of caustic solution which can dissolve old animal glues making joints widen and warp.

Common door problems include:

- ■ Warping: It is not normally cost-effective to try to fix a badly warped door, although a good carpenter can usually adjust or repair light warping.
- ■ Split panels: Repair should be fairly straightforward. The mouldings that retain the panel can usually be prised off and the panel replaced with plywood.
- ■ Loose joints: If the top or bottom rails have come loose at the join with the stiles, the door will have to be dismantled and the joints glued and clamped.
- ■ Rotten bottom rail: The rail can be replaced by removing the door from its hinges and sawing through each joint between the rail and the vertical stiles to the sides. The plywood panels can be slid out and replaced if necessary. Depending on the size of the door, a new rail can be made from three pieces of planed 100 x 50mm treated softwood, cut to shape and positioned between the two stiles to form a new bottom rail approximately 300mm high. It is secured by drilling through the stiles from the outer sides and driving hardwood dowelling pegs into the new timber, having first applied exterior PVA glue, then clamping the structure until dry. Finally a new weatherboard and water bar should be fitted and the door re-hung.
- ■ Sticking doors: If the door sticks in wet weather, it is likely the door has distorted. If it sticks in dry weather, this usually indicates that the frame has changed shape. Frames should have a DPC around them to prevent damp from the adjoining masonry causing swelling and decay. Loadings over frames need to be supported by suitable lintels or the frame will deflect – cracking in the wall above may confirm the problem. A distorted door needs to be removed and trimmed with a plane. If the paint is poor the door may be damp and must be allowed to dry out before repainting.
- ■ Door hinges, knobs and fittings may be broken or worn, requiring replacement.

**Left:** Anywhere water can 'pond' will eventually be at risk of damp and decay

**Right:** A skilled joiner can repair even the most ravaged doors – common weak points are at joints and ledges at lower levels where water has become trapped

B. Mulford

# Defect: Rain blowing under door

## SYMPTOMS

Damp floors, sticking door.

**Cause** **Rain runs down the face of the door and back underneath instead of dripping off the front**

(Solution) *Fit a weather board*

Fit a weather board (with a grooved drip underneath, as for sills) to the bottom front of the door, and fit a water bar (a projecting strip of metal or plastic) into the sill underneath the door. This may require the door to first be taken off its hinges so a rebate can be cut along the bottom edge.

Ground floors should be at least 200mm (8in) above ground level

## Key points to check when maintaining doors

- Deterioration due to damp is usually very localised. Often it's just the base of a doorframe that's rotten or decay has set in along the very bottom edge of the door. Damaged wood can be cut out and patch repaired by scarfing in new lengths of seasoned timber cut to size.
- Because doors comprise quite a large surface area, they are prone to shrinking and swelling with seasonal changes in moisture and temperature. This can be up to several millimetres, enough to cause sticking in damper months. The extent of movement can be minimised by keeping the timber well decorated.
- To decorate a door thoroughly you need to temporarily remove it at the hinges so that all the edges can be fully primed and finished.

## UPGRADING SECURITY

The local police Crime Prevention Officer will give free advice on home security. Briefly, sash windows can be discreetly secured using a 'dual screw' barrel bolt that is inserted through the top rail of the lower sash into the lower rail of the sash above. This can be unscrewed to open the window. There are also window restrictors, which allow a window to be opened only a small amount; these are advisable if you have young children in the house. Modern replacement windows should have integral security locks, but it is worth checking that the beading is fitted internally so it can't be prised off and the panes removed. Doors should at least have a separate five-lever mortise lock and a deadlocking cylinder Yale lock so that if a burglar smashes a glass pane he can't just reach in and turn the latch. Additional sliding bolts are also advisable on external doors.

NB Always remember that an emergency escape may be necessary in the event of fire, so keys must be readily accessible. Consider fitting 'thumb turn' rack bolts, or cylinder locks that can be opened without a key from inside.

# External timbers

Fascias are the horizontal timber boards that, on many houses, run along the eaves at the base of the roof slopes. They may be attached directly to the brickwork or else cover the rafter ends to make 'boxed' eaves, with a soffit infill underneath. Soffits may be of timber boarding or, on cheaper houses, of lath and plaster. They are a useful place to fit grilled vents for loft ventilation. Bargeboards are similar timber facings that run along the verges

Late Victorian/Edwardian exuberant craftsmanship – needs regular painting

Half timbering is mostly decorative – but parts may be structural

Decorative timberwork to porches

fabulous carved dripping icicle bargeboards are now considered architectural gems. Elaborate Edwardian fretwork and decoratively timbered gables are period features well worth preserving. Consequently, replacement with a bit of bland 'maintenance-free' UPVC doesn't really do it justice.

'Half-timbering' and roughcast render are a common feature of Edwardian houses. Mostly this was just for show – decorative 'falsework' stuck to a brick base. So if decay sets in, they simply need replacing with new decorative strips of timber. But the posts and beams around windows can sometimes be load bearing, supporting gable brickwork plus the loadings imposed from the roof purlins.

Naturally, being so exposed to the weather makes external timberwork vulnerable, but a lot of it has proved surprisingly durable thanks to the quality of the original timber, skilled joinery, and regular decoration. There is often a contributing factor behind decay, like leaking guttering causing rot to fascias.

The solution is pretty much as described for rot to windows and doors. Localised small areas of decay can be cut out and the remaining timber treated and filled. However, any rot to load-bearing members or to exposed rafter ends (which is rare) may have structural implications and should be traced back within the roof space. Areas of more extensive rot will need replacement with treated timber. Despite the cost of having original patterns matched in new timber, the additional value that such period features confer on the property should make it worthwhile.

at gables (and were originally called 'vergeboards'). But many earlier houses simply dispensed with fascias and bargeboards altogether and instead used the traditional method of exposed rafter feet, corbelled brickwork, and pointed verges.

From around the 1880s ornate wooden eaves and heavily bargeboarded gables became popular on more expensive houses. Fascias moulded into intricate Gothic fantasy designs and

**Below:** Superb detailing to late Victorian/Edwardian gable bargeboards

## DECORATION

Owning an old house is rather like running a vintage car: all that wonderful character and charm justifies a premium in the property's value but comes at a price. To prevent deterioration, external timbers demand cyclical repainting. Recommended intervals between paint cycles vary from three years to five, depending on how aggressive and

**Below:** Poor maintenance can lead to rot to eaves timbers

exposed the location is. If your paint is crazed, flaking, or peeling, the timbers will obviously need a good seeing to.

The other option, neglect, is always dearer in the long run, as replacing original timbers is not cheap, and potential buyers will attempt to knock large sums off the sale price at the first sight of rot. Exterior painting is best carried out between June and September, when the moisture content of the wood is lower.

## LEAD PAINT

Painted joinery, both externally and internally, would normally have been finished in natural oil-based paints. These combined a base, usually of lead carbonate (to provide body and a white pigment), mixed with a binder of linseed oil to make it stick, plus pigment for colour and thickening agents like wax or soap. Varnish was similarly produced, but with resins added to the linseed oil instead of pigment.

The higher the lead content, the more durable and also the more expensive the paint. Despite their toxicity, these paints remained popular into the 1960s. Today lead paint is banned almost everywhere except for a handful of specialist uses, notably on Listed buildings

In most old houses layers of old lead paint will be lurking under later coats of modern gloss. Lead paint testing kits can be used to check if necessary, but it's safest to assume that the material is present. However, there's only a potential health risk if lead compounds are ingested or inhaled, usually from sanding. So simple precautions should be taken against ingesting dust when rubbing down. See also chapter 12.

## STRIPPING

With luck, thorough preparation may reveal a sound initial paint base only requiring spot priming and a finishing coat. Otherwise, the correct approach is to strip the timbers to bare wood, cut out and replace any decayed areas, and treat end grains with preservative prior to decorating. Although water-based strippers and flushing are preferable, solvent-based paint strippers are very effective for removing layers of modern paint. These can either be painted on or applied as a thick poultice. Older oil-based paints can be dissolved using alkaline-based removers. Caustic soda dissolved in water will often remove tar and linseed oil.

## PAINTING TIMBER

After rubbing down and preparation, the best finish for external joinery is the modern equivalent of the original lead-based variety – natural linseed oil paints. These are similar to the original but use

substitutes such as zinc in place of lead, and can be applied over existing sound lead paint, or direct to bare sanded wood.

Bare wood was traditionally treated with a primer coat or two of raw linseed oil, warmed up to about 60°C to make the oil thinner so it soaked in more easily. The same method can be used to good effect today. Any cracks can first be filled using a linseed-oil-based putty. This can be painted over before it's dried with the first coat of linseed-oil paint. Each subsequent coat should be allowed to dry and given a light sanding. Linseed-oil paints take longer to dry than today's plastic paints and should not be applied in cold damp weather.

These natural paints require greater skill to use compared to modern non-drip paints, as they are thinner (and thus drip more), becoming thicker as they dry. They need to be applied in several thin coats or they can wrinkle when dry. However, linseed-oil paints have the major benefit of lasting considerably longer than conventional modern paints – often 15 years or more. Their remarkable longevity is because the oil soaks into the wood rather than just sitting on the surface ready to flake off like modern solvent-based paints. Also the finished surface has a certain amount of elasticity so it isn't as physically hard and prone to cracking as modern paints. Above all, because they're breathable they don't trap damp behind a surface 'skin'. Although more expensive to buy, using natural oil paints should work out significantly cheaper in the long run. Alternatively, where ease of use is a key factor, modern vapour-permeable paints and stains can be a good option. Most are water-based, solvent-fee and acrylic-free with a flexible microporous finish that doesn't not seal in moisture.

Finally the last stage: sit back and admire the gleaming architectural period features and the visibly enhanced value of your home.

Natural linseed oil paints soak into the wood and last for many years

Chris Gare / Gare.co.uk

# CEILINGS

Splendid high-ceilinged rooms displaying glorious elaborate plasterwork, decorative cornices and ornate ceiling roses are one of the great delights of 19ᵗʰ and early 20ᵗʰ century houses. But suspicious bulges, bowing and cracking in old ceilings can put something of a damper on things – all the more unnerving when they're poised menacingly above your head! In this chapter we explore the ingredients of original lath and plaster, and look at the skilled techniques used to create the wonderful ornamentation found in so many homes of this era. Also, defects in more modern ceiling materials are discussed since they are commonly found in modernised older houses.

RBKC Linley Sambourne House

Rumour has it that if you slam a door in a Victorian house, great clumps of thick ceiling plaster will rain down on you with a vengeance. Indeed, the old joke about the lining paper being the only thing holding the plaster together can sometime have more than a grain of truth in it.

Victorian rooms are known for their high ceilings, typically 2.5m or higher, except for those in the more compact rear addition rooms, which were the domain of 'staff'. As a rule, the grander the house, the higher the ceilings, but even fairly small houses had relatively generous headroom thanks to standards stipulated in new by-laws. Contrast this with a typical pre-Victorian cottage, where you are forever in danger from low flying beams

One of the highlights of a typical suburban villa was an ornate ceiling displaying classical cornicing and a decorative central rose. Like much else in the Victorian home this sort of embellishment was a measure of social status. The entrance hall, no matter how small and narrow, was always given the most dignified architectural treatment with ornamental plasterwork, the division between the main front hallway and the inner staircase hall marked with an elaborate bracketed arch. In the dark days before electricity, the gas or oil lamps would cast much of their light upwards, making the ceiling one of the main focal points of the home. So visitors would also be treated to the sight of ornate plasterwork and fine detailing in front parlours.

**Above left:** Clumps of thick plaster can rain down

**Above:** Unusually light, elegant rose

**Above right and right:** Decorative cornices in entrance halls

**Below:** Ceiling rose 'smut catcher' *(Photo: RBKC Linley Sambourne House)*

**Below:** Late Victorian 'Arts & Crafts' panelled ceiling

# The meaning of ceilings – what are they made of?

## LATH AND PLASTER

Victorian houses had ceilings made of traditional lath and plaster, which has a pleasing texture that's slightly irregular. 'Laths' are thin strips of wood (about 25mm wide), which were nailed to the underside of the ceiling joists and spaced about 5mm apart. Plaster was then applied to the underside of the laths, where it was held in place by being squeezed through the gaps while still soft to create a 'key'. The plaster was made from lime mixed with sand, bulked up with horse hair for strength, and was usually applied in two or three layers of increasingly fine finish. The 'render' base coat was followed by the floating coat and finally skimmed with a setting coat of pure lime or plaster of Paris to give a smooth finish.

There are good reasons for retaining original ceilings. As well as enhancing historic interiors, traditional lath and plaster is chunkier and has far better soundproofing and insulating qualities than modern plasterboard, being substantially thicker. Unfortunately, some builders are quick to condemn old ceilings on the flimsiest evidence, such as the odd bulge, or where there's a small missing piece of plaster.

But never assume that an old ceiling needs replacing just because the surface is no longer flat and a few cracks have appeared. New plasterboard ceilings in old houses convey a flat, sterile appearance. Such advice is sometimes aimed at drumming up extra profits, but more often it is simply down to ignorance of traditional skills and materials. The fact is, traditional hair-reinforced plasters can withstand a great deal of deflection and still retain their strength. Despite appearances, they may be perfectly sound and repairable for a fraction of the cost of replacement.

To be certain what type of ceiling you have, take a look under the loft insulation, or lift a bedroom floorboard. If there are a lot of small timber laths with creamy blobs of plaster in between, the

ceiling is original. But in many older houses some areas of plaster have become 'live' and sound hollow when you tap them. The first coat may have lost its key to the lath, or the coats themselves may have separated.

## PLASTERBOARD

By now many original ceilings will have been replaced with modern plasterboard, usually with a coating of skim plaster or perhaps finished with lumpy textured paint. Plasterboard basically

consists of sheets of compressed rigid gypsum plaster sandwiched between heavy-duty lining paper. Boards are available in thicknesses of 9.5 or 12.5mm and fixed to the joists with dry wall screws driven into place using cordless screwdrivers. Clout nails are no longer used as the vibration from hammering can loosen plasterwork (sometimes badly fitted older boards have the heads of the clout nails visible – they should be driven just below the surface without breaking the paper prior to skimming). With the plasterboard in place, the joints can then be sealed with scrim tape and joint filler and the surface given a skim finish with a thin coat of plaster.

## OTHER MATERIALS

'Fibreboard' sheets appeared as a forerunner to plasterboard during the inter-war period. There's only a slim chance you might find these if your house was improved or extended around that time. It was made of compressed wood fibre in its undecorated

Lath without its plaster

state, and is tan in colour with a soft spongy feel. Defects are similar to those of plasterboard although it is flammable.

Fortunately, you are also fairly unlikely to encounter asbestos cement sheeting, which was also used as a building material during the inter-war period and beyond. However, it was widely used to repair bomb damage during the war. The distinctive light grey coloured sheets can occasionally be encountered in outbuildings and old lean-tos, plus it was sometimes used to line boiler cupboards and flues on account of its fire-resistant

Asbestos cement sheeting

properties. Asbestos cement sheeting makes a hollow sound when tapped and is thin, strong and lightweight. Unlike other ceiling materials, you can't easily stick a drawing pin or sharp point into it. In ceilings the sheets are usually supported within a thin timber framework. But the simplest check you can make is to fish out the survey report from when you bought the house. If you have asbestos, it should have been mentioned.

Asbestos cement sheeting is generally considered to be safe as long as it's left undisturbed and undamaged. In small amounts it is best left alone or concealed by a new suspended ceiling. Care must be taken not to drill or sand it, as the release of its fibres into the air could constitute a serious health threat. If you have to disturb it, specialist advice should first be obtained.

## CEILING COVERINGS
On lath-and-plaster ceilings, lining paper and wallpaper were commonly used as a finish, sometimes applied over a layer of

unbleached calico. If there are embossed decorative patterns these may well be impossible to match if a small area of plaster has been replaced. This may necessitate stripping the whole ceiling of paper, which in turn can cause old plaster to lose its key and come loose! A small ceiling job can often turn into a major project.

Unfortunately, many attempts at DIY 'makeovers' have turned out to be hideous crimes against good taste, as well as potentially hazardous. Polystyrene foam tiles and linings were once a common DIY solution when it came to disguising poor surfaces. However, these are now considered to be a fire risk, as well as being unsightly and of little practical use. They should be removed and the underlying surfaces made good.

Cladding with strips of tongue-and-groove timber was another popular DIY project some years ago. As with polystyrene tiles, its installation actually downgrades the fire resistance of a ceiling and is particularly hazardous in kitchens. The ability of flames to track across surfaces and cause ignition at some distance from the source of a fire is well known, so any potentially flammable coverings aren't a good idea.

## SUSPENDED CEILINGS
Because of the generous ceiling heights in Victorian houses, hiding a poor original ceiling by constructing a new one underneath is usually an option. This is one way that converted flats are insulated against sound and fire. Conversely, you may discover some gloriously ornate original plasterwork concealed behind a subsequently installed suspended ceiling.

Suspended ceilings are relatively lightweight because they carry no loading from the floor above and are normally constructed about 150mm below the old ceiling with a framework of 100 x 50mm joists spaced at 450mm centres. Additional support can be provided from cords or flexible hangers secured centrally from the main ceiling joists. Plasterboard sheets are screwed to the new joists, the joints scrimmed and a thin skim plaster finish applied. Packing with mineral wool quilt insulation will significantly reduce heat loss and the void is very useful for running new electric cables and concealed lighting (any pipework should have access for maintenance).

**Below left:** After the foam tiles have been stripped
**Below:** New suspended ceiling framework

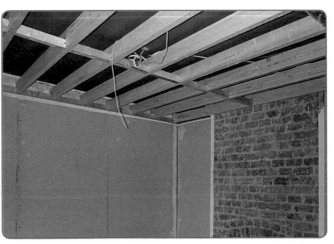

# Repairing lath and plaster

It's not unusual to come across localised ceiling damage where small clumps of plaster have fallen away exposing the naked laths behind. This type of small 'bald patch' can normally be repaired quite easily, as long as the laths are secure.

- Where the **laths have come loose**, any sound ones can be re-fixed to the joists using long stainless-steel screws with wide washers (nailing causes too much vibration). Additional support for loose laths can be provided by fixing new battens between the joists, working from above.

- Where **laths need replacing,** first cut a rectangular shaped hole to expose the joists on each side. Defective laths can be carefully cut out and removed, and replaced with new ones (available from specialist suppliers), which are screwed to the undersides of the joists. The plasterwork can then be built up in two or three coats, depending on the required thickness.

- Where the **laths are sound** but the plaster has fallen away, you can fill the damaged area from below using modern gypsum plasters (which are compatible with older lime plasters and will adhere to laths). First all the old damaged plaster must be removed and raked from between the laths. Then an undercoat of browning plaster should be applied and pressed well into the laths. After 2 or 3 hours when it has started to harden, the surface should be scratched to improve bonding. After 24 hours it should have hardened completely and a thin coat of finishing plaster can be applied using a steel trowel to achieve a level surface.

- Where an old ceiling is **very badly damaged**, the part in poor condition can be taken down and reconstructed, reusing some of the sound existing laths. The new lime plasterwork can also incorporate some of the old lime plaster, ground-up and added to the mix – another example of how Victorian buildings being incredibly 'sustainable'.

Fixing laths using headed stainless steel ringshank nails

Seanwheatley.co.uk

# Cornices and roses

By the mid-Victorian period, artistic ceiling ornamentation had filtered down the social scale, with decorative plaster cornices applied at the junctions of walls and ceilings in all but the very cheapest houses. They were traditionally formed by running a shaped metal template along the wet plaster. However, much ornamental plasterwork was actually made of cheap, lightweight, papier mâché until mass-produced ceiling roses, brackets, cornices, corbels and plaster ornaments later became widely available to speculative builders. This ready-made 'fibrous plasterwork' was produced from plaster of Paris using flexible moulds with a reinforcement of hessian scrim or a timber backing. Heavier items like ceiling roses would be nailed or screwed through the ceiling to the floor joists above, whereas lighter cornices were simply stuck in place with dabs of plaster.

Clogged with a over a century of paint, decorative cornices and ceiling roses may be hiding superb ornate detailing. Many once fine mouldings will by now have morphed into a lumpy, formless mass. Exposing the true glory of original plasterwork can often require considerable patience and plenty of TLC, using a poultice stripper, a brush and a toothpick. Minor repairs using plaster of Paris are possible, but restoration of decorative plasterwork is a skilled job. Fortunately, even where much of it has been lost, missing sections can normally be re-cast by taking a mould from the surviving original work. New cornices can be formed in the traditional *in situ* way in stages of increasing

**Below (l to r):** Original cornice showing oak leaf enrichments. Cutting the repeat pattern to use as a mould. Running new cornice ready to accept enrichments

Seanwheatley.co.uk

Adding individual enrichments

refinement. First the basic shape is 'cored out' using a rough lime and sand mix. Then, once dry, a finer mix gauged with plaster of Paris is applied. Finally, an accurate profile can be created by carefully running a fine template known as a 'running mould' along the full length of the cornice – taking care not to sneeze en route!

Modern replacement DIY plaster coving usually has a 'manufactured' look, with straight, simple patterns, and sounds hollow when tapped. But even this is better than the cheaper polystyrene foam 'fire hazard' variety.

Nearly complete – 81 individual pieces!

## Strengthening ceiling joists

If there is very pronounced dipping evident to a ceiling, lift the floorboards in the room above (or look in the attic space) to check whether the joists have deflected. Perhaps they were undersized or set too far apart when the house was originally built. Or more likely, notches or holes for pipes and cables may have subsequently been cut, further weakening old joists. A heavy load, such as a water tank, in the loft may be imposing extreme stresses. Either way, the floor structure may need strengthening. This can normally be done from above.

Where there's a loft above a weak ceiling, it's often possible to strengthen the joists by inserting a new thick timber beam (or a steel) across the top of all the joists in the roof space, leaving a small gap between them. The beam can normally be supported at either end in the main walls (or party walls). Each of the old ceiling joists is then supported by individual steel straps hung down from the beam, looping underneath them. This is a non-invasive repair that prevents further deflection in old ceilings.

Where you have rooms above a ceiling, new floor joists can be inserted running parallel to the old existing ones. Alternatively it may be possible to support an old ceiling from below, if there's sufficient spare headroom, with a new timber beam running underneath.

# Common causes of ceiling damage

### OVERLOADING
Cracks in ceilings of all kinds will result if the floors above are defective or the ceiling joists themselves are inadequate for the loading, eg when the roof space is used for storage of heavy items.

### VIBRATION
Wild parties upstairs? 40-tonne trucks driving past your front room? With original lath and plaster, vibration can eventually cause cracking and loosen plaster and cornices. Plasterboard is more resistant.

### DAMP
Where ceilings have become saturated with water, lath and plaster is prone to softening and can lose its key. The source of the damp must first be fixed, then if the plaster is still sound it must be allowed to dry out, which can take up to a couple of months.

Persistent leaks can leave a legacy of decay in concealed timbers, so allow them to fully dry and check their condition before covering them up. Damp plasterboard can warp and bow and needs to be cut out and replaced.

Brown stains are typical of old water leaks, whereas black specks indicate mould (usually due to condensation) and should be treated with a fungicide (see below). Severe staining may need to be sealed prior to decoration. Plastering is a hard skill to master, but small areas of less than a square metre should be within most DIY-ers' capabilities. See 'Patch repairs' section in chapter 9.

### ALTERATIONS AND MOVEMENT
Cracked or loose plaster is often the obvious initial symptom of ceiling defects. Where a ceiling crack is more than about 2mm wide, and where the plaster on either side is not level, it may indicate that a more serious problem has occurred elsewhere in the building. Localised cracking or bowing may be due to movement caused by structural alterations, particularly where old chimney breasts have been removed and can be the first sign of a structural problem. See chapter 11.

In fact, one of the most common causes of damage to old ceilings is from holes being cut in them to fit recessed lighting.

However, fine hairline cracks at plasterboard joints are extremely common and are often due to no more than initial shrinkage (drying out) or slight expansion and contraction of the boards. Cracks at the joints between walls and ceilings due to differential movement or expansion are also common. They can be taped and filled, but may reappear as the house adjusts to seasonal ground and temperature changes. The provision of coving around the perimeter of the ceiling can be effective at disguising small cracks of this type.

Seanwheatley.co.uk

# Decoration

Ceilings were originally lined and painted with soft distemper or limewash. These chalky, water-based distempers had a powdery finish, so modern vinyl emulsion (which began to replace them in the 1960s) doesn't adhere to them very well. They can generally be removed with a soft brush and hot soapy water, but stubborn layers may require careful use of a steamer. In most cases they will have been painted over with modern oil-based paints that can be removed with a solvent-based stripper.

When decorating or skimming over an old ceiling, if the plaster surface looks crumbly it may first need to be stabilised by painting with PVA bonding fluid using a wide brush. When dry, a thin layer of finishing plaster can be applied.

Stains from tobacco smoke can be removed with sugar soap prior to decoration. Small areas of loose plaster can often be ignored if cracks are filled and a heavy-gauge plain lining paper applied before painting.

Before and after cleaning

## Warning: removing Artex

Between the 1960s and the 1980s textured decorative paints were popular, and were often used to hide defective surfaces. They may contain asbestos fibres that would become a health hazard if breathed in, so removal needs to be done with care.

Artex can be quite difficult to remove without damaging the surface below, but you should be able to steam it off with a steam stripper: the Artex splits from the plaster and can then be scraped and peeled away.

One thing you must not do is try to remove it with any kind of abrasive tool. If in doubt leave it in place. A better alternative is to knock off the peaks (wear a mask!) and plasterboard over it. Or a skim plaster coating can be applied directly on top after applying a light coat of PVA adhesive. Or a suspended ceiling could be installed to conceal it.

# Defect: Sagging and bulging

## SYMPTOMS

Ceilings uneven; cracked finishes and bowed surfaces.

### Cause   Wet plasterboard

**Solution** Plasterboard tends to warp when very wet. Once the cause of dampness has been eradicated, the area affected can be cut out and replaced with an infill piece of board and plastered to match. In extreme conditions plasterboard ceilings will collapse, *eg* from burst pipes and overflowing tanks in lofts.

### Cause   Incorrect support to boards

**Solution** Plasterboard sheets may not be fully supported, particularly at the edges. Boards should have been fixed to the joists in a staggered pattern with their sides butted up and 3mm gaps left at cut ends, and screwed or clout nailed every 150mm (6in). Additional securing with screws will be needed to improve support.

### Cause   Loose laths. Plaster has fallen away

**Solution** Original ceilings can fail because the old plaster was mixed with insufficient hair reinforcement, or the laths were fixed too close together leaving too narrow a gap for the plaster to squeeze through and form adequate nibs. The plaster then loses its key and falls way from its backing. Other causes include overloading and vibration. The laths themselves can also fail due to damp, decay or beetle – see chapter 6.

From below, prop the area of sagged plaster with a sheet of plywood and a length of timber to push it back into position. Then, from above, clean the ceiling removing all broken nibs, and lightly spray with water (to prevent the old material sucking water out of the new plaster). Pour rapid setting plaster along the line of the gaps between the laths, to form new nibs. When dry, remove prop.

Alternatively, if laths are loose or parts of the ceilings or cornices are sagging, they can be screwed back into the joists using long stainless steel screws with large washers at frequent intervals along the joist. Screw heads are countersunk and filled - see boxout.

Where all else fails, a new suspended ceiling can be constructed below the original. In severe cases, take down the old ceiling and replace with new plasterboard.

## Defect: Cracks in lath and plaster ceilings

### SYMPTOMS
Small irregular shaped cracks and unevenness.

 **Cause** **Timber laths have come loose to a small area, or the plaster has lost its 'key' with the laths**

 **Solution** Carefully check the cracked area for lumps of loose plaster (eg by pushing very gently with a broom handle). Once

one part has lost its key or the laths have slipped, other loose areas may fall down (or even the whole ceiling). If the problem area is fairly small, cut out the loose part and then screw the laths back to the joists and fix a suitable piece of plasterboard to the joists. Replaster to a flush finish.

## Defect: Cracks in plasterboard ceilings

### SYMPTOMS
Thin straight cracks at board joints

**Cause** **Poor fixing or expansion and contraction at board joints**

**Solution** Cracking between plasterboard panels is not normally a serious problem. Joints can be raked out and filled using a covering strip of jointing tape or fabric before applying joint filler and plastering. Consider lining with a heavy-gauge lining paper to conceal the joints.

Another common defect is a line of small, round craters (about 10mm/0.33in wide) or lumps, due to plasterboard clout nails being nailed in either too far or not far enough. The solution is normally to apply a sufficiently thick skim plaster finish, having first inserted some additional screws to provide improved support for any loose boards.

## Defect: Mouldy ceilings

### SYMPTOMS
Recurrent black mould growth.

**Cause** **Condensation due to poor insulation and ventilation**

**Solution** This is a common problem resulting from poor insulation and a lack of ventilation, usually more evident to ceilings in bathrooms and kitchens, where steam is produced. The steamy water vapour hits a cold ceiling and condenses back to water. You may find this occurs particularly in poorly insulated rear extensions or flat-roofed areas.

The solution is to insulate above the ceiling with mineral wool to a depth of at least 270mm, and allow a path for ventilation above that. A new insulated suspended ceiling with foil-backed plasterboard is an excellent alternative if space permits. Improving ventilation is important, eg with extractor fans, open fireplaces and trickle vents in windows. Make sure tumble driers are ventilated to the outside, and if possible minimise indoor clothes drying and cut down on boiling food.

To get rid of the mould, clean off with water and apply diluted bleach (1:4 bleach/water) or a suitable fungicide. Finish with a coat of mould-resistant paint. See also chapter 6.

# CHAPTER 9 Victorian House Manual
# INTERNAL WALLS

There are few graver blunders that a pickaxe-wielding builder can make than mistaking a load-bearing internal wall for a simple partition. So in this chapter we explain how to tell them apart, how to locate your spine wall, and the correct way to go about making a new opening in a 'structural' wall. Common problems like cracking and noise transmission are also addressed, and we investigate some of the strange ingredients that the Victorians sometimes added to lime plaster mixes.

RBKC Linley Sambourne House

If walls could talk, doubtless they would have some intriguing stories to tell of times past. But if you know how to read the signs, internal walls can hold some useful clues to the health of your house. The first signs of structural movement are often evident here, and if any of the original walls have been removed, it is worth taking a few minutes to work out which parts of the house they were supporting before they disappeared.

Load-bearing internal walls were generally built with fairly rudimentary foundations despite providing essential structural support for floor joists, walls above and roof loadings. Construction was predominantly of solid brick or stonework, usually not much more than 100mm thick – less than half that of the main walls. And since they were destined to be plastered over and unseen, they were often built by apprentice bricklayers using cheap under-fired or misshapen bricks.

For the Victorian occupants of your home, keeping the functions of different rooms separate was of fundamental importance. So maintaining a clear division between the front drawing room and the rear parlour, and having a separate entrance hall where visitors could be greeted, were prime indicators of the property's status. The modern concept of open-plan living would have been regarded as little more than a crazed architectural fantasy in the 19th century. But to our eyes older houses can sometimes seem rather dark and narrow, perhaps not making the best use of the available light. Hence the popularity of 'opening up' and 'knocking through'. But much 'home improvement' work of this sort has been done so badly that some old structures have become unsafe as a result. So as a rule, it is best to leave walls alone. In any case, estate agents tell us that buyers like original features, so your house is probably worth more with them intact.

# Partition walls

Other internal walls merely serve the purpose of partitioning one room from another, such as where upstairs landings are separated from side bedrooms by space-saving, thin pine boarding. Partition walls were generally constructed of a lightweight timber frame studwork, comprising vertical posts (studs) and horizontal beams (plates) provided at floor and ceiling level. Timber noggins were then nailed horizontally between the studs. Like ceilings, the walls were faced with lath and plaster. Laths are small strips of pine nailed to the studs and noggins, with gaps of about 5–10mm between them for the plaster to ooze into and bond to the wall. By the Victorian period cheaper, sawn softwood laths had replaced the traditional hand 'riven' type (split along the grain) made from chestnut or oak. Lath and plaster is about 25mm deep, thicker than modern plasterboard and has better insulating properties. In some Victorian properties laths were also used to line the inner surface of main walls over vertical timber battens to form a slim air gap in a bid to protect against damp.

**Above:** Inside a timber-stud lath-and-plaster wall

**Right:** Thin bedroom / landing partition of pine boarding

**Right:** Solid wall to landing – but is it load-bearing?

**Below:** Landing /side bedroom partition part glazed to boost light to dark stairwell

## BRICK ON EDGE

The cheapest way to construct a solid partition wall using the absolute minimum number of bricks was to lay them on their sides. Building a thin 'brick on edge' wall with gaps left between all the bricks in each course, lattice style, could make even greater savings, using cheap rejected soft or misfired bricks. The gaps could then be patched with earth or clay mixed with lime and hidden behind a coat of innocent looking plaster. So where you find original walls that are notably slender – not much more than about 60mm thick, they're likely to be either 'brick on edge' or thin timber studwork construction.

# Period plasterwork

It's widely believed that the first job when renovating an old building is to start frantically hacking off all the plaster from the interior. But it's rarely necessary to remove old lime plaster. Any cracks or gaps can normally be filled, and visual flaws are easily concealed by decorating. Traditional plastered walls that aren't perfectly straight or regular are an important part of an old building's character. So don't be tempted to make them perfectly smooth and sanitised when redecorating.

The fashion for stripping old plaster to expose bare 'feature' brick or stonework is a modern phenomenon that was unknown to the Victorians. Even the humblest larder or privy was carefully plastered or at least whitewashed. Walls that were never meant to be displayed tended to be made from poorer materials with inferior-quality workmanship.

However, in many properties you may find the lower parts of some main walls have been stripped and replastered in recent years, often over cement render 'tanking' applied to conceal damp. Modern gypsum plasters normally have a pinkish-brown colour (or occasionally light grey) and a harder and flatter finish than lime. But being highly rigid and inflexible, they have a tendency to develop cracks as the old walls naturally flex in tune with the seasonal ground conditions. So ideally, for replastering an authentic lime mix should be used, especially for the main walls.

One of the joys of Victorian houses is that it's sometimes possible to recycle old lime plaster. Simply hack it off, break into small pieces and sieve into fine granules. It can then be mixed with new sand and water to form fresh lime plaster. This is about as 'sustainable' and 'green' as buildings get!

## LIME PLASTER COATS

Plaster was applied directly onto brick or stonework walls, or alternatively onto a backing of timber laths. Depending on the quality of the building, it would comprise double or triple coats with each coat getting progressively thinner. The mix differed only slightly between each coat, the weakest being the topcoat, which used finer sand for a smoother finish.

The first coat was known as a 'rendering coat' when applied to masonry, but when applied to a background of timber lath it was called the 'pricking up' coat – referring to way the plaster 'pricks' through gaps between the laths (or so we are told). Alternatively, both are also referred to as 'scratch coats' because their surface is 'scratched' in a diamond pattern to create a key for the next layer.

For better-quality work, where achieving a very flat finished surface was important, an intermediate 'floating coat' or 'straightening coat' was applied. This would be carefully worked

**Left:** Timber frame 'stud' wall with solid infill

**Below left:** Typical plastered solid brick 115mm thick internal wall

**Below centre:** Lath and plaster stud walls can be 'structural'

**Below right:** Solid 115mm thick (single brick width) interior wall with hair-lime plaster and wallpaper

| STANDARD PLASTER MIX: NON-HYDRAULIC LIME (LIME PUTTY) | | | |
|---|---|---|---|
| Used for ceilings and internal walls above ground. | | | |
| | **1 Scratch coat**<br>or 'pricking up coat' on ceilings and 'rendering coat' on masonry | **2 Floating Coat**<br>Or 'straightening coat' | **3 Setting Coat**<br>Or 'skim coat' |
| **SAND** | 3 parts sharp well-graded sand | as per Scratch coat | 1 or 2 parts silver sand (kiln-dried and sieved) |
| **LIME** | 1 part mature lime putty (thick) | as per Scratch coat | 1 part mature lime putty (thick) |
| **HAIR** | Horse or goat hair<br>2kg per tonne of plaster | None, or sometimes a finer goat hair<br>2 kg per tonne of plaster | None |

to get it dead flat. To even out bumps and hollows, particularly to ceilings, a length of straight timber known as a 'floating rule' would be passed across the surface. This intermediate coat was then 'scratched' to provide a key for the final coat, using a 'devil float' (with nails pricked through it) applied with a light circular motion. The final 'setting coat' or 'skimming coat', just a couple of millimetres thick, was a mix of lime putty and fine sand designed to create a smooth surface.

## THE MIX

Traditional lime plasters might comprise 1 part lime putty (non-hydraulic lime) to 2.5 or 3 parts sharp sand (soft sand tends to result in cracks). This is a similar mix to lime mortar, the main difference being the added ingredient of tufts of chopped animal hair – horse, goat or ox – as a binder to the undercoats. This is critical (especially for lath) as it dramatically reduces shrinkage and cracking as well as making it stronger. This can be purchased ready mixed and provides excellent flexibility, which is important in old buildings prone to seasonal movement.

But these weren't the only ingredients, if urban folklore is to be believed. To save money and bulk up the mix, additional cheap and abundant materials such as earth or clay, and dust from crushed chalk, bricks and tiles were added, along with less-wholesome ingredients such as urine, dung, fat and blood. Alarmist tales about anthrax spores present in animal hair (prior to 1895 when controls were introduced) are not supported by any evidence and are probably a bit of an urban myth.

# Load-bearing walls

## PARTY WALLS

The walls that separate you from your neighbours in a terrace or semi-detached house are generally the same thickness as the main walls, typically about 230mm. But the temptation to save on materials and labour means some party walls were built to the width of only one brick (115mm). Even worse, it was not unusual for party walls in lofts to be omitted altogether. See chapter 2.

Because the brickwork was hidden behind layers of thick plaster, party walls were often built by apprentices using cheap materials, which sometimes resulted in difficulties when it later came to joining them up to the main external walls. One solution was to embed small 'bonding timbers' in the mortar. But by now these have often rotted, as described in chapter 5. Either way,

many internal walls have little or no bonding to the main walls, with nothing more than a continuous joint behind the plaster, which inevitably will be prone to cracking.

Another common problem with party walls between houses is the lack of sound insulation. Cheap construction, with walls of only single brick thickness or with gaps and holes in them, means there is often scope for improvement. See 'Defects' below.

NB If you plan to carry out any work on a wall you share with neighbours, it is important to be aware that under the terms of the Party Wall Act 1996 you are legally obliged to first get written consent from the joint owner next door – see www.victorian-house.co.uk.

## SPINE WALLS

In a typical terraced house, about half-way through the main building there is a spine wall, dividing the front and rear reception rooms. It is a common 'improvement' to open up these two adjoining 'parlours' by taking out this wall. But the spine wall is actually a major structural component holding up the floor joists to the rooms upstairs, as well as the wall directly above (between the main bedrooms); plus it plays an important role supporting roof loadings.

Even without any subsequent alterations, spine walls were often overloaded from the word go, the design having to span across the hallway and in many houses also accommodate a large opening with double doors between the reception rooms. So it's

*Front/rear parlour spine wall with laths visible above opening*

Above: Looking back: spine wall divides front and rear reception rooms; original double doors have been removed

Below: Looking forward – from dining room to front room

Above: Poor design in cheap back-to-back housing; door frame should not support ceiling beam

Below: Floor joists above temporarily supported with acro props whilst new steel is inserted (beam ends should rest on padstones)

not uncommon for these walls to have settled. Consequently the roof structure relying on it for support (eg via ceiling joists and struts to the purlins) can follow it down, so the roof sags. Similarly floors on all storeys may slope down towards a spine wall that has settled. Even where the wall has been removed, the remaining part below ground floor level may still be providing support to the reception room floor joists.

So taking out a structural wall such as this is a major job requiring advice from a structural engineer. The removal, if done badly, can transfer loadings unevenly causing deflection elsewhere in the structure.

## IS IT STRUCTURAL?

Conventional wisdom has it that if a wall sounds hollow, it is only a partition wall and can be merrily ripped out. Regrettably, things are not always quite that simple. Despite their lightweight construction, some stud walls are actually load-bearing, with overlapping floor or ceiling joists resting on them, and may also be taking loads from walls above or the roof timbers. And, rather confusingly, some stud walls were filled in with brick rubble.

It is therefore safer to assume that all walls are load-bearing until proved otherwise. So it's not a bad idea to first consult a structural engineer, chartered surveyor or pop down to your Local Authority Building Control office for a friendly chat before starting any work.

To discover whether a wall is load-bearing, start by taking a look up in the loft. Note how the struts and ceiling joists are supported: they often rest on at least one internal wall to the bedrooms below, which is in turn supported by the reception room walls.

To ascertain whether upper floors are imposing any loads on the wall down below that you want to demolish, check the direction of the floor joists. The joists normally run the other way from the direction of the boards (so if your boards run from side to side, the joists will run from front to back), as can be seen from the lines of nailing. To be absolutely certain, lift a few upstairs floorboards over the wall below to confirm whether any joists are resting on it. But even walls whose only purpose appears to be that of partitioning adjoining rooms, with no joists resting on them, may still be structural, because there's an upstairs wall directly above. Plus it's often forgotten that a wall may also be providing 'lateral support' helping to tie-in the main walls at either end.

# Knocking through and opening up

If you were to imagine an entire terrace with the roof removed, looking from above all the rooms would appear as lots of individual boxes forming a strong cellular pattern. Removing internal walls weakens this structure and the building has to re-adjust. However, there may be situations were owners are keen to 'knock through' or 'open up' the layout. The walls most commonly demolished are the one dividing the entrance hall from the living room, and the spine wall between the front and rear receptions. If you are determined to take a wall out, before wielding the sledgehammer in anger be sure to check that your building is not Listed, otherwise an application for Listed building consent will need to be made. It's important to note that structural alterations must always be carried out with Building Control approval, which normally involves appointing a structural engineer to design the support for the remaining masonry above. Upon completion of the works a completion certificate should be obtained, as this will be required when you come to sell the house.

Gilesterrybuilding.co.uk

New opening boarded and ready for plastering

Even partition walls that do not support any of the building's structure can't just be bashed down any old way. For one thing, they may be supporting other adjoining partition walls, or may contain pipework and electric cables. Plus an unexpected avalanche of heavy chunks of lath and plaster could ruin your chances of playing in the Premiership. So caution is the watchword. Cover and protect nearby surfaces. Shut down adjacent water and electricity supplies. And start at the top, working in small sections one step at a time.

Where a new opening such as a doorway is required in a load-bearing wall, the weight of the structure above needs to be supported by a suitable lintel, usually of steel or concrete. As with demolition work, Building Control must be notified in advance and engineers' calculations provided to specify the appropriate lintel type.

Original door openings were usually spanned by timber lintels, with the masonry or studwork for the wall above built off the top. Timber door frames often have a false jamb, or a secondary subframe behind, which allowed dimensional inaccuracies in the carpentry to be corrected easily, as well as providing a surface to terminate the lath and plaster.

## PROJECT: Making an opening

**1** The wall above must be supported temporarily while a slot is cut for the new lintel. This is done by first cutting holes just above the position of the proposed new lintel about every 600mm.

**2** Sturdy timber 'needles' of about 150 x 50mm are then placed through these holes in the wall, and are supported each side by adjustable steel 'acro' props, in turn resting on a scaffold plank to spread the load. (On suspended timber floors the joists must be checked first to ensure they can provide adequate support.)

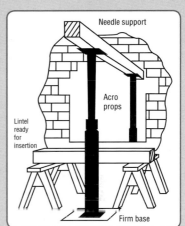

Needle support

Acro props

Lintel ready for insertion

Firm base

**3** The new lintel is inserted in the wall and must extend either side of the proposed new opening by at least 150mm (6in) 'bearing'. A hard engineering brick or stone pad is normally needed under the lintel ends to improve support.

**4** The lintel is bedded in mortar and the masonry above built up. After 24 hours or more, when the mortar is dry, the props can be removed.

**5** The new opening below the lintel is first marked out on the wall and the plaster chopped out vertically with a bolster. An angle grinder can be carefully used to cut the masonry, dismantling small stages at a time. The reveals can later be made good with plaster. Alternatively timber frame liners can be screwed to the reveal, or installed with metal frame ties bedded into small pockets cut in the reveal wall, and pointed up once the frame is square. Finally, the surrounding masonry and gaps to the floor are filled and levelled.

# Repair and renovation

## CRACKING

Small cracks (up to about 1.5mm width) will often open and close over the course of a year as the building's shallow foundations respond to the ground moistening and later drying out in tune with the seasons – see chapter 5. But cracking may sometimes conceal loose 'live' plasterwork that requires more extensive repair work (see below).

To deal with fine cracks, a standard flexible filler can be used, such as decorators' caulk, applied with an applicator gun. For small patch repairs, lime plaster can be applied direct to masonry, or over traditional timber laths or sections of reed matting. Both

are preferable to modern 'expanded metal lath' (EML), which can be prone to corrosion.

Larger cracks may indicate possible settlement or subsidence, although it's very common for old movement to have stabilised. Door openings are weak spots: cracking above a door frame, doors that stick badly, or frame linings out of square can all indicate a structural problem in a load-bearing wall. This sometimes happens where internal walls were built off a floor surface instead of having proper foundations, so if your internal doors are well out of alignment this could be a symptom of settlement to the floor.

Vertical cracking often occurs at junctions between internal timber stud walls and solid main external walls, due to inadequate fixing between them at the time of construction or differential movement between different materials. If such cracks are more than about 3mm wide, or the crack tapers with varying widths, the outside wall may have bulged out due to a lack of restraint and may need to be tied in. See Chapter 5.

Once the cause of the movement has been rectified, a bolster chisel can be used to chop away the edges of the crack and the loose debris raked out before moistening the plaster and applying filler in layers.

## PATCH REPAIRS

In many older houses some areas of plasterwork have become 'live'. One clue to this is where you tap the surface of a masonry wall and it sounds hollow. And if the plaster moves when you press against it, this shows that it's lost its key to the lath behind, or the coats themselves may have separated. However, this isn't necessarily cause for concern. Lime plaster is very resilient because the hair reinforcement binds it strongly together. This makes it act as a large sheet, even if a few smaller areas have parted company from the wall.

Small areas of loose plaster can often be hidden with a heavy lining paper before decorating. If the wall is in a bad way and several areas are loose, it is better to strip all the old plaster off and replaster the wall completely, or to dry-line the wall with plasterboard, for example where plaster at low levels has been very damp and is loose and crumbly. See step-by-step.

**Left:** Wonky walls and door frames, but movement is historic
**Below left:** Cracking often relates to defects in other parts of the building
**Below:** Exposed lath

## NEW WALLS

Modern stud walls lined with plasterboard are lighter and easier to construct than those of masonry. Although they have relatively poor soundproofing qualities, this can be improved by filling the voids in the studwork with mineral wool insulation. Acoustic 'sound-shield' board may also be worth installing to party walls etc. Specialist boards are available for humid locations like bathrooms, and pink-coloured sheets provide enhanced fire protection, useful for loft conversions.

The use of concrete blockwork for dividing walls is fine as long as there's adequate support below, *eg* when replacing a previously removed solid wall, or upgrading firebreak walls in the roof, which are located over solid party walls.

## STRIPPING AND STEAMING

One common problem is that walls may, in recent years, have been covered with woodchip wallpaper, which doesn't look great and can be troublesome to shift. To remove woodchip, use a steamer. First scrape the surface using a toothed scraper or flat cheese-grater to help the steam and water to soak in. Then apply the steamer, but do not hold it in one place for too long or there is a risk that the plaster can start to come away, turning a small job into a major project. Fortunately, modern plasters are compatible with traditional materials and can be used to patch repair old surfaces. For removing Artex finishes see Chapter 8.

## Fixing things to stud walls

Should you need to fix something to a lath and plaster wall it's important to be aware of their limitations. To support fairly light objects, such as small picture frames, screws can be inserted into pre-drilled holes through the plaster and into the laths. Pre-drill a couple of fine pilot holes – it may take more than one attempt to avoid the gaps in between them. Avoid hammering nails into the lath as the wall will bounce, with a risk of fracturing the plaster nibs.

For weightier fixings such as kitchen wall units or shelves, the screws need to pass right through the lath and plaster into the sturdy timber studwork posts or noggins behind. The traditional method was to install a stout horizontal batten fixed across the main studwork frame, with the shelves then secured to the batten. This is the same principle used for picture rails run along the upper walls.

**Right.** Wallpaper stripper gets back to bare plaster

**Left:** Modern gypsum plasters are compatible for internal walls

**Below left:** Period lime plaster gives an authentic look

**Below:** Stripped and ready for decoration

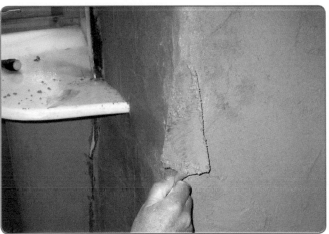

## Defect: Surface cracks in plaster

### SYMPTOMS
Small irregular shaped cracks and unevenness in lath and plaster; thin hairline cracks in plasterboard.

**Cause** **Timber laths have come loose or the plaster has lost its key with the laths due to poor adhesion of plaster or vibration**

**Solution** *Carefully check the cracked area for lumps of loose plaster*
If the problem area is fairly small, cut out the loose plaster and then screw the laths back to the joists and replaster to a flush finish. In severe cases, replace or line with plasterboard.

**Cause** **Plasterboard walls can suffer from thermal movement or shrinkage at joints or poor construction**
Hairline cracks along joints are very common and not normally a cause for concern. But joints may not have been properly scrim taped and filled when the boards were installed

**Solution** *Rake out, scrim tape and fill joints*
In severe cases, rejointing and skimming may be required, or replacement of damaged boards.

## Defect: Movement cracks

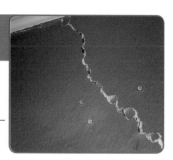

### SYMPTOMS
Cracking, typically at abutment of main walls.

**Cause** **Structural movement**

**Solution** Structural movement is more likely where cracks are tapered and more than about 3mm wide or where the plaster on each side is not level. The cracking may be indicative of more serious movement within the structure as a whole, *eg* at bays or main walls leaning or bowing out. Load-bearing internal walls may have insufficient foundation depths and could require strengthening or the construction of supporting brick piers or buttresses. See chapter 5.

**Cause** **Differential movement**

**Solution** Where dissimilar materials in the wall construction have expanded or contracted at different rates (*eg* where timber stud walls meet solid masonry walls), rake out cracks and apply flexible mastic or filler. Fit a cover strip (*eg* a batten) if differential movement recurs.

## Defect: Noise transmission

### SYMPTOMS
Sound intrusion, noise pollution and nuisance

**Cause** **Inadequate thin or defective separating walls between dwellings. Missing walls in roof spaces**

**Solution** The poor noise resistance of a separating wall between adjoining dwellings can be due to it being built of only single-width brickwork (115mm/4.5in) with gaps and holes. The solution is to block any holes and insulate the wall for sound. Build up missing party walls in the roof. See chapter 2.

Areas of poor-quality bare brick (*eg* under floors or in loft) can be rough-rendered and gaps filled with expanding foam to seal any gaps (but don't block ventilation air paths).

Ideally, construct an independent new inner wall of timber studwork and plasterboard over the old wall, leaving a minimum 50mm gap in between. Use 75 x 50mm timbers for the framework, secured at the edges to the walls, floor, and ceiling, but leaving a void so that it is independent from the old wall. Pack with a minimum of 25mm mineral wool quilt insulation. Board over with two layers of 12.5mm acoustic plasterboard with staggered joints, sealing the joints with scrim, sealant, or coving. In addition, to reduce sound via the roof void fix a new layer of plasterboard over the old ceiling or board and insulate the loft floor above and line the firebreak wall with insulated plasterboard.

Studwork partition walls between rooms can similarly be upgraded by packing them with mineral wool quilt.

# Patch repair to lath

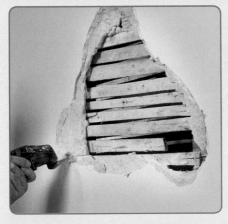

**1** Cut away loose plaster around the damaged patch until it feels solid.

**2** Any defective timber laths must be repaired or removed and replaced.

**3** Carefully cut around the surrounding edge of the adjoining plaster with a sharp knife. Leave it cut at an angle.

**4** Loose patches of plasterwork can be secured by screwing a small washer plate through to the studwork frame.

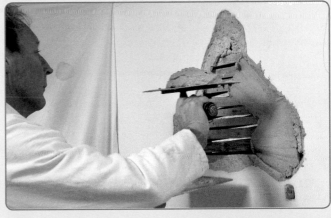

**5** After dampening the adjoining plaster edges and the laths to kill suction, firmly press in fresh plaster to the patch so it 'keys' through the lath. Special gauged plaster can be applied to the full thickness in one coat, and also sets quicker.

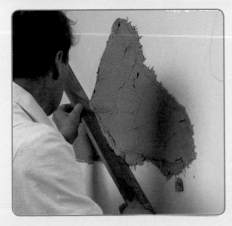

**6** The surface is 'screeded' with a board so it's flat with the surrounding wall.

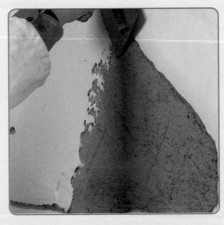

**7** Finally trowel on a fine finishing coat. Here a thin creamy mix comprises NHL2 in a 1:1 ratio with medium plastering sand, gauged with fine casting plaster. It is applied no more than 3mm thick so can appear virtually transparent.

All images courtesy
**Studio Scotland Ltd**
from
**The Master Stroke**
DVD Tutorial series.
To view DVD trailers go to
**www.themasterstroke.com**

# Dry-lining a wall

It's normally best to retain the original plasterwork as it's part of the history and character of an old building. The next best option is to replaster in matching lime. Where this isn't possible, dry-lining with sheets of plasterboard can provide a smooth new surface on old internal walls. To boost the thermal performance of cold main walls thicker insulated plasterboard can be used – see chapter 15.

By specifying specially formulated acoustic boards you can improve sound insulation. Similarly, fire resistance can be upgraded by using special fire-board (which has a distinctive pink colour). In bathrooms and kitchens there's a risk that humid, steamy air can penetrate through the plasterboard where it condenses into water causing damp problems. To prevent this select foil-backed 'vapour-check' boards – or better still, place a plastic sheet vapour barrier over the wall before plasterboarding, lapped and taped at joints. For walls adjacent to baths and showers a moisture-resistant type is best.

Dry-lining a wall can also provide a handy space to run new electric cables and pipes, but these must be suitably protected.

Before starting it's important to check that the walls are free from damp. The traditional method of dry-lining is to construct a framework of treated timber 25 x 19mm battens built onto the wall. The boards are then fixed to the framework with special non-rust screws or clout nails. Alternatively, 'quick to fit' metal frame kits are available. It's not essential to hack off the old plaster as long as it's reasonably firm and secure.

The 'direct-bond' method can be used to cover solid walls that are reasonably true. Here, the boards are secured directly to the wall using dabs of plaster adhesive, so any large areas of old loose wall plaster must first be removed.

## CUTTING PLASTERBOARD

Having measured the wall and marked the board accordingly, score along the line fairly deeply use a sharp craft knife and a

straight rule, and then repeat. Lift the board off the floor and give it a sharp knock along the cut line so that the waste part snaps away neatly. Cut off any remaining paper, trying to avoid tearing the surface. Awkward shapes can be cut with a saw.

## TOOLS REQUIRED

- ■ **Pencil, chalk, and tape measure**
- ■ **Spirit level**
- ■ **Long straight edge rule**
- ■ **Hawk and steel trowel**
- ■ **Sharp craft knife**

## MATERIALS

- ■ **Plasterboard and direct-bond adhesive**

**1** The floor and ceiling are marked with a chalked line, allowing for the thickness of board plus at least 10mm of adhesive.

**2** Next, the walls are marked with vertical lines to indicate the positions of the rows of dabs. Use a long straight-edge, spirit level and a piece of chalk, or a chalked plumb line.

**3** Mix up the special adhesive plaster and scoop some of it onto a hawk. Then use a steel float to place them on the wall. Dabs should be 250mm long and 50–70mm wide.

**4** Dabs are needed at 600mm centres, plus 50mm below the ceiling. A continuous line is needed above the floor. Apply enough for one panel at a time.

**5** While being lifted clear of the floor with a foot lifter, the plasterboard panel is pressed firmly onto the dabs of adhesive. Then it is wedged in place at the bottom.

**6** Finally, the plasterboard is tamped into place with a long straight-edge, aligning it with the marks on the floor and ceiling. Repeat the process for subsequent panels.

# Direct decoration to plasterboard

Plasterboard walls and ceilings are normally skimmed with a finishing plaster to about 2–5mm thickness, but they can be decorated or wallpapered direct instead. It is best to use tapered-edge boards (rather than square-edged joints), as the shallow recess over the joints can be filled with jointing compound (setting time approx 90 minutes) and scrim tape for an invisible joint.

## TOOLS REQUIRED

- **Steel trowel**
- **Taping knives (100 and 150mm)**
- **Feathering sponge**
- **Tin snips (cutters)**

## MATERIALS

- **Jointing compound**
- **Air-drying finishing compound**
- **Joint tape or scrim**
- **External corner metallic tape (optional)**
- **Sandpaper (120 grit)**
- **Dry wall sealer paint and emulsion**

## USING TAPERED PLASTERBOARD

**1** Apply jointing compound to the tapered joint with a 150mm taping knife.

**2** Take a strip of paper joint tape and bed it into the compound. Then cover with more jointing compound so it is flush with the board face.

**3** If necessary, the edges can be smoothed with a damp sponge to minimise sanding. Allow joint to dry.

**4** When set hard, add a second coat of jointing compound with a trowel and feather it out beyond the first coat, then smooth it off at the edges. When dry, apply a finishing coat of air-drying compound and again feather out beyond the edges of the previous coat. Allow to dry. Rub down until smooth, and it's ready for sealing and decorating with a minimum of two coats of emulsion.

**5** After step 1 above, bed the paper joint tape in the compound as before, but apply a 200mm wide band of jointing compound over the joint tape.

**6** When dry, apply a second coat of jointing compound 250mm wide extending down the edges of the joint tape.

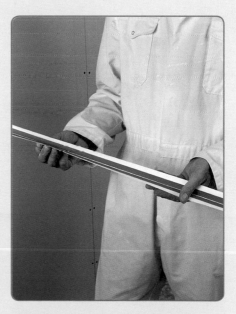

**7** When set hard, apply a finishing coat of air-drying compound over the entire joint, covering the tape and feathering out onto the plasterboard face. Repeat stage 3. Rub down until smooth, seal and decorate.

**8** For internal angles, apply jointing compound to both boards using a 100mm knife, then repeat steps 2–4.

**9** External corners need a tougher finish. Metallic 'flex tape' is cut to size with tin snips. Apply a 50mm wide band of fast-setting jointing compound each side of the corner. Bed the tape onto the corner so its metal side faces in towards the plasterboard. Then cover each side with layers of joining compound and finish as described above.

# FLOORS

Original antique pine floorboards can look great – but how do you know what dangers are lurking beneath them? If the furniture vibrates alarmingly when you walk past, is it simply because old floors were designed that way, or the result of botched DIY? Slugs appearing up through spongy floorboards may be one of the more obvious clues that all is not well below, but damp hidden floor timbers may have been quietly decaying unseen for many years, until disaster strikes. Original solid floors, often quarry tiled, can also suffer from their own special problems.

**Right:** Patterned entrance porch floor with 'encaustic' tiles

**Below left:** Entrance halls often had hard-wearing quarry tiles ...

**Below centre:** ... or geometric encaustic tiles in grander houses (Photo: RBKC Linley Sambourne House)

**Below right:** The Victorians wouldn't recognise today's stripped pine floors

Before the Victorian era, the floors in many ordinary houses and cottages comprised nothing more than a few flagstones or bricks placed directly over the soil. In some rural dwellings compacted earth strewn with rushes had to suffice. So the typical Victorian house, with its ground floors built from pine floorboards on suspended timber joists, was a major design improvement in mainstream housing.

Largely as a response to new regulations requiring better construction standards, floorboards to ground floor living areas had to have a minimum of 150mm ventilation space underneath. Hardwearing solid concrete and tile construction was used for kitchens, sculleries and some entrance halls.

As the century progressed, the introduction of machine-sawing and imported softwood resulted in a gradual reduction in sizes and boards becoming more standardised. So whereas earlier Victorian boards might be about 200mm wide, those from later in the century were generally narrower, about 150mm (6in) wide,

and typically 20 to 25mm thick. By the early 1900s slimmer boards of as little as 80–100mm (3–4in) in width were considered a sign of a good quality floor, for example the interlocking tongue-and-groove designs in beech, oak or maple found in some more expensive Edwardian houses.

# Solid or suspended?

The upstairs floors will be entirely of timber floorboard and joist construction, but if your ground floors are obscured behind layers of carpeting, the type of construction may not be immediately clear. One simple test is to heavily stamp your feet a few times (if you fall through, you know there's a problem!). A hollow sound, and a slight bounce or 'spring', will normally confirm suspended timber construction. Solid floors should, of course, feel completely rigid.

You should also see airbricks to the main walls near ground

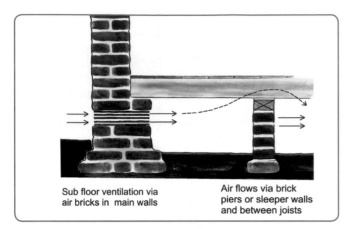

Sub floor ventilation via air bricks in main walls

Air flows via brick piers or sleeper walls and between joists

Typical hard wearing quarry tiled kitchen floor

level, indicating that the original floors were of timber. One sign of a refurbished Victorian house (particularly one 'done up' in the 1960s and '70s) is a solid concrete ground floor throughout. The floors may have become so damp and rotten that replacement with concrete was considered the best option. But if the new concrete floor was carelessly installed, it may by now have deflected or sunk – see 'solid floors' below.

For all their faults, suspended timber-boarded floors score highly for practicality and appeal. They're warmer than stone or concrete, access to pipes and cables is relatively straightforward, and they look great when stripped and sanded. And traditional floorboards can reveal clues to the structure of the house. For example the joists normally span in the opposite direction to that of the floorboards, which often tends to be the shortest width in a room, as indicated by the nail runs.

If the property has been extended in recent years you may come across modern chipboard panels, which are not generally liked as they are prone to creaking and can weaken or even disintegrate if they become damp. In addition they cannot be easily lifted or refixed should access to pipes or cables be required.

Some more expensive houses had a double floor covering where a surface of parquet or tongue-and-groove boards was laid over a base of plain floorboards.

### FLOOR COVERINGS
The Victorians wouldn't recognise your beautiful gleaming stripped and varnished pine floorboards. Clues to the original floor finish can often be glimpsed when you lift modern fitted carpets to expose an old timber floor. You may notice a pale oblong area to the centre surrounded by darker outer edges. This is because it was common practice before the days of wall-to-wall carpeting to place a large oriental rug in the middle of the room, with the area beyond stained black or given a coating of dark bitumen-based paint to protect the floor and fill small gaps – nicely setting off the Turkish rugs and assorted pelts of dead animals. In contrast, bedroom floors were sometimes scrubbed with sand and limewater for a bleached effect or fake marbling applied. Plain floorboards were occasionally stained and varnished to resemble expensive mahogany, but genuine fine-quality hardwood floors in more expensive properties would simply be waxed and polished so the quality of the timber and craftsmanship could be admired.

One of the cheapest and most popular types of floor covering in Victorian homes was oilcloth – a large rug comprising canvas sheet sealed with linseed oil sometimes decorated with a painted tiled pattern. More expensive linoleum ('lino') manufactured from linseed oil, resins and cork dust to a jute backing was also prized for its durability and the range of colours, and is today making something of a comeback thanks to its green credentials. But for many homeowners today, the natural beauty of original timber is best displayed, and the preferable finish for treating old timber floors is to simply apply a natural breathable wax or a light, natural oil that wears away gracefully and can be easily reapplied.

# Suspended timber floors

Most houses had deal floors made from imported Baltic or Scandinavian pine. Even the grandest houses, with fine polished hardwood floors in the main rooms, would have plain deal floors in the less-prominent parts of the building.

Sometimes what appears to be a traditional timber floor may actually be a more recent addition, for example where a solid concrete slab has been clad with pine floorboards nailed to timber battens resting on the surface of the concrete (raising the height of the original floor).

### GROUND FLOORS
Floorboards were normally fixed across timber joists with special flat nails called brads. Boards would terminate over the centre of a joist for support. The maximum unsupported joist span was often

Dark border around space for central rug, before the days of fitted carpets

Constructing a new timber ground floor

# TECHNICAL DATA

### Joist spans

Victorian softwood floor joists were traditionally made from 200 x 50mm (8 x 2in) timbers, the nearest modern equivalent being 195mm x 48mm. They were rarely less than 150 x 50mm (6 x 2in). Ceiling joists to lofts are typically 100 x 50mm (4 x 2in).

Ground-floor joists typically span no more than 1.8m without support, and upper floor joists 2.5m. The span that a joist can safely cover without support depends on the joist size, and how many joists there are supporting the floor.

Typical dimensions for sleeper wall timbers are 100 x 75mm (4 x 3in) and herringbone strutting /noggins 38 x 63mm (1.5 x 2.5in)

### Centres

Depending on the span, joists were spaced fairly close together, commonly at 14in (356mm), 16in (406mm), or 18in (457mm) centres, but in older buildings spacings tend to be rather haphazard. Upper floors may be spaced as close as 300mm as spans can be longer (unlike ground floors they don't have the benefit of frequent support from piers).

### Air bricks

The total surface area of all vents should be 500mm$^2$ for every 1m$^2$ of floor area, placed within 450mm of floor corners, spaced no more than 2m apart.

---

no more than 2 or 3 metres, so the joists in turn needed to be supported on thick lengths of wood called 'wall plates', in turn resting at periodic intervals on hidden 'sleeper walls' or a series of short brick 'piers'. Sleeper walls were built on rudimentary rubble 'foundations' in a honeycomb pattern with gaps in between so as not to inhibit airflow (and save on bricks). Internal partition walls similarly sometimes have vents set within them below floor level.

Lift up a floorboard, and under the coating of dust and debris you should be able to make out the (hopefully) dry ground below. The space under the floor was normally at least 300mm, but deeper 'undercrofts' to houses on sloping sites sometimes allowed this cellar-like area to be used for storage, with access via an external door. Some more expensive houses have a thin layer of 'oversite' concrete (a weak mixture of lime or cement mixed with waste 'pit ballast' or 'coke breeze') laid over the bare earth below the floor to restrict moisture rising upward. But the majority of homes just had bare earth strewn with building rubble.

Most problems with floors at ground level arise from dampness of various kinds, which can eventually cause decay in adjacent timbers. So ventilation is provided by airbricks on the lower outside walls to help moisture evaporate before it can cause decay to timbers. Fortunately, any decay is usually very localised and at worst may require replacement of a few joists or a timber floor plate. The cause of the damp can normally be traced to one of the following:

■ High external ground levels – causing lower walls to become damp
■ Very wet ground – causing damp to rise via the main walls and the sleeper walls; this is usually down to surface water from

the garden flooding through air bricks, or hidden leaks from water pipes under floors.
■ A build-up of debris in the sub-floor void forming a bridge between damp ground and floor timbers.

It is important that there are a sufficient number of airbricks on each outside wall to maintain a good cross-flow of air under the floors – typically two or three in opposite walls for a small terraced house. This should be enough to prevent stagnant corners, typically near solid

kitchen floors. Airbricks were mostly of terracotta, or cast iron grilles; where these are damaged or badly rusted they should be replaced as they provide easy access for vermin if broken.

### A VIEW FROM THE CELLAR

If you're (un)lucky enough to have a cellar, it should provide a good opportunity to observe some of the less brilliant aspects of Victorian floor construction. In particular, joist ends are usually bedded in the brickwork of the main walls, which tend to be periodically damp. In some properties any consequent decay will have been rectified by the provision of new joists with their ends

View from under a ground floor: joists rest on wall plate over (dry) supporting sub floor wall

Inside a bedroom floor – note lath and plaster ceiling.

protected by a DPC, or supported independently in joist hangers. You may also find that the original span was too long for the relatively thin joists, necessitating the subsequent provision of a brick pier or RSJ to improve support to the old floor above.

## HOW LEVEL SHOULD THE FLOOR BE?

Most floors of this age will not be perfectly true and level. Some will have a definite slope that corresponds to the general settlement of the whole building that took place many years ago, as the foundations found their 'level of repose'. Unless there are signs of recent cracking and movement to the walls, or problems with damp and poor sub-floor ventilation, this isn't necessarily a problem and should usually be within 'acceptable tolerances'. The severity of the slope can be checked with a marble or spirit level.

**Above:** Unusually severe floor slope – now stabilised

**Left:** Small electric sanders can be useful for awkward corners

## UPPER FLOORS

Upstairs floors are also of suspended timber construction but are less likely to suffer from damp problems than those downstairs. Where the joists bridge wide spans they tend to be proportionately larger and more closely spaced (typically at 300–450mm centres). Also you're more likely to find bracing – in the form of small diagonal 'herringbone' bracing struts, or noggins (scraps of wood) – between upstairs joists, at about 1.5m intervals, to help prevent bowing or twisting. In larger houses, iron tension rods were sometimes used. To level uneven joists, they would be packed underneath with thin slivers of slate or wood.

Because Victorian builders might sometimes be tempted to economise on materials, it's not unknown for houses to be built with relatively thin joists spaced a little further apart than necessary. As a result some upper floors can feel a

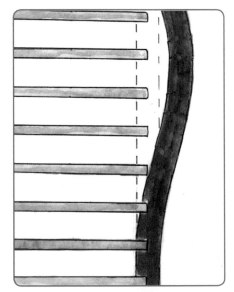

**Right and below:** Where walls have bowed out or joist ends have rotted, the joists can be re-connected to the wall by fitting steel 'shoe' joist extenders.

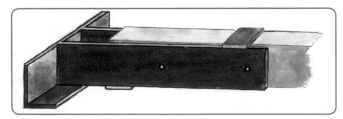

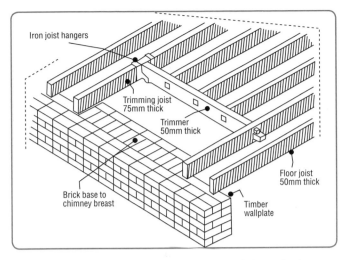

Iron joist hangers

Trimming joist
75mm thick

Trimmer
50mm thick

Floor joist
50mm thick

Brick base to
chimney breast

Timber
wallplate

little springy. Many old floors have subsequently been further weakened by having notches and holes cut in them by plumbers and electricians. This is likely to result in a certain amount of deflection, with consequential cracking to the ceilings below.

The upstairs floor joists are supported with the joist ends either built into pockets in the main walls or resting on protruding brick or stone ledges called corbels. In larger houses the joist ends may rest on stepped ledges formed as the main walls get thicker floor by floor as they descend. These methods, particularly the first, make joist ends a potential target for damp and rot, and today they have been superseded by metal joist hangers. However, it's important to remember that the quality of the timber used in old buildings can generally withstand such conditions far better than the modern equivalent, as long as it's well ventilated. Also, by not fitting the joist ends too tightly into the masonry air could flow around them allowing any moisture to evaporate harmlessly away. Where joist ends have rotted or walls bowed out, steel joist extenders can be fitted – see page 178.

Upstairs floors often rely on internal walls below for additional support, which is one reason why their removal can pose potential dangers – see chapter 9. Structurally, the upstairs floor joists do more than just hold the floor up: they also help tie in the main walls. If the joists run from front to rear, then the bowing of side walls can be a problem where the walls aren't tied in – see chapter 5.

In some cases the upper floor may also provide support for a partition wall in an upstairs room where there's no wall below it. Here, the supporting joists would be strengthened by being 'doubled up' and bolted together directly below the wall.

The floors in most rooms also had to accommodate chimney breasts and fireplaces with a projecting hearthstone in front (raised hearths are a modern practice). So that the timber joists didn't terminate under the hot hearth, they were cut short and connected instead to a trimmer beam run parallel to front of the hearth (ie at right angles to joists). To form the hearth itself, 'deafening boards' were fitted and packed with cinders or clinker and the hearthstone laid on top. The floorboards were then laid flush with the hearthstone. A similar method would be used in landings, with a frame constructed around the large stairwell opening by butting the ends of the floor joists with thick 75mm-wide 'trimmer' timbers.

## FLOORBOARD REPAIRS

Most problems can be remedied with a few straightforward repairs. Where the originals can't be reused, to help retain the building's period character, reclaimed boards are widely available, or it should be possible to have matching new ones made from seasoned timber.

Weakened boards can be reinforced and strengthened by discreetly doubling them up with a length of new board fixed underneath, spanning between the joists. Alternatively, additional support can be provided by fitting new battens underneath, screwed to the adjacent boards on either side, as long as they're in sound condition. Split boards can be glued and cramped back together. Where you've got localised damage to corners, this can normally be repaired by splicing in small strips of new timber.

Lifting old boards without damaging them can sometimes be difficult, particularly the tongue-and-groove type where the interlocking tongues generally need to be cut. But even plain boards were sometimes 'secretly' nailed in their end edges over a joist with the nail driven into the joist at a 45° angle so the next board concealed the nail head.

One common cause of damage is where it's necessary to cut across a board above the side of the joist, rather than above its centre. This facilitates lifting but means that when it comes to reinstating it, a new batten will need to be screwed to the side of the joist, to support the cut end of the board. For re-fixing boards, it's best to use brass screws rather than

**Right:** When replacing floorboards fit a new batten to support cut ends

nails. This allows for possible future lifting and also protects delicate plaster ceilings below from hammer vibration.

Most old houses will show signs of past woodbeetle activity, something that's very common but normally superficial. So it's important not to overreact to old boreholes in the floorboards. Only in extreme cases would the odd length of board be sufficiently worm-ridden to need replacing (see page 123).

## DIY DISASTERS

Sadly, many houses have suffered from inappropriate alterations over the years. Probably the most serious cause of harm to old floors is from generations of plumbers and electricians carelessly prising up old boards and cutting chunks out of joists. Then there are various DIY 'improvements' such as fitting laminate flooring of a non moisture-resistant type in bathrooms and kitchens. Or ceramic floor tiles laid on suspended timber floors where the flexibility in the floor structure has caused the tiles to crack – worse, damp may have seeped down through the joints, causing hidden decay to the floor timbers.

To be suitable for tiling, timber floors must be very stable. The tiles should be placed over an oil-based hardboard layer laid with the rough side up (to provide a key for the tile adhesive).

It's not unknow for an upper layer of floorboards to have subsequently been laid over the original ones. To spot this check if the surface boards butt against the skirting rather than under it, or if floor levels are unequal from one room to another. This can sometimes be a recipe for trouble if the reason for covering them over may have been to hide rotten or beetle-riddled old boards. The problem is, you can't tell their condition without lifting the new ones (unless visible from the cellar below). If there's no significant amount of spring to the floor and the walls aren't damp and the airbricks are clear, you may be OK. Otherwise the painful process of lifting the newer boards and replacing defective timbers may need to be carried out.

## INSULATION AND SOUNDPROOFING

A good flow of air under timber ground floors is essential to disperse damp and pre-empt the risk of decay. But this can make

### TIP

When lifting floorboards, start by lifting one of the boards close to the skirting along one side of the room. The first board running parallel to the skirting usually extends underneath it, and can be left in place. When nailing or screwing boards down, be aware of plumbing or electrical cables concealed just below floor level. A metal and cable detector is a very useful gadget.

things a bit draughty where there are gaps between the boards and at skirtings. Victorian floors were not insulated, so heat loss can be a problem if, like most people, you want to retain the original boards. One solution, if you are willing to go to the trouble of lifting and refixing all the boards, is to place rigid polyurethane or quilt insulation between the joists (supported on battens). Alternatively floors can be insulated from below by lifting a few boards to gain access to the sub-floor void.

If you plan to fit a carpet, a thin layer of plywood or hardboard can first be fixed on top of sound boards to smooth out irregularities and provide extremely effective draught-proofing, although the slightly raised floor level may then necessitate minor adjustments to doors and fittings. Solid floors can be covered with suitable laminate or wood flooring over a thin layer of underlay, although floor levels will again be raised slightly.

Attempts were sometimes made by Victorian builders to deaden sound transmission to floors, especially between the main bedrooms and servants' quarters in attics. This was done by filling voids to upper floors with 'pugging' – a coarse material such as ash or sand, laid on boards (pugging boards) placed between joists, or simply filling the space with layers of woodshavings, felt or straw.

Lath-and-plaster ceilings have a mass greater than modern plasterboard and can absorb a greater degree of sound. Noise transmission can be reduced either by fitting modern acoustic floor panels over existing surfaces, or by constructing an insulated suspended ceiling below – see chapter 15.

Butchery! Joists severely weakened by criminally excessive cutting –floor at risk of collapse

Mineral wool quilt between joists can accommodate pipe runs

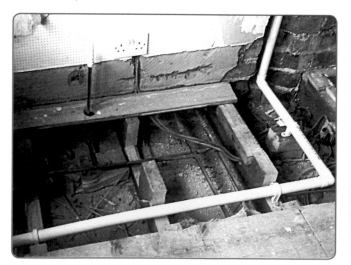

Patrick Stevenson

# Defect: Sloping floors

## SYMPTOMS

Uneven floors; floor surface out of true; gaps to skirting boards; a distinct ridge to floor; the floor in an upstairs bay significantly out of kilter with the room.

### Cause  Settlement

Internal walls that support floor joists often had little in the way of foundations and consequently often settled in the years following construction, causing floors on all storeys to slope downwards in the direction of the wall. More recent movement may have occurred where additional loadings have been added to an old building, such as after a loft conversion.

**Solution**  *If the settlement is old, no attention may be necessary. A typical example is the spine wall that divides the main reception rooms, sometimes with double doors between. See chapters 5 and 10.*

### Cause  Floor joists twisted or bowed as shrinkage occurred in the timber

Joists may have warped or deflected under the weight of enormously heavy furniture over many years. Shrinkage is more common where fresh new timber has been used for repairs or new floors; original joists were normally made from well-seasoned timber.

**Solution**  *Fit noggins (bracing struts) between joists at mid span. Screw down any loose boards*

### Cause  Structural movement to main walls

– eg parts of the building with shallow foundations on clay subsoils are more likely to move in drought conditions, or movement to adjoining bay windows etc.

**Solution**  *Once any structural defects to the walls have been remedied, localised making good to affected joists and boards may be required. See chapters 5 and 7.*

### Cause  'Humping' of a floor surface, often as a result of poorly planned alterations

– eg due to a steel beam in the floor structure, where the joists have settled either side of the beam. May also be a symptom of old wartime bomb damage where air pressure from blasts caused whole floor structures to 'jump up' before resettling unevenly.

**Solution**  *This is not normally a serious problem but can be improved by lifting the boards and packing the joists to make them level*

# Defect: Excessive floor 'spring'

## SYMPTOMS

Pronounced deflection to the whole floor in a room when walked on, so that nearby furniture and ornaments vibrate. Floors may sag.

You expect a small amount of spring in most suspended timber floors, particularly upstairs. Apart from rot-related causes (see below) there are two main causes of excessive springiness:

### Cause  Design fault

Inadequately sized timbers are more likely to be found in smaller terraced houses of 'cheaper character'. This may have been aggravated by the tremendous weight of Victorian furniture. Joists may have warped or deflected under the weight but failed to recover when the load was removed.

### Cause  Joists cut / poor DIY

Original undersized or unbraced joists may have been further weakened by notches cut in them for pipes and cables. The depth of a joist gives it greater stiffness than its width, so a basic rule for cutting is that it should not exceed one-eighth of the depth of the timber (normally cut from above) and not within the first 300mm or so from the joist ends. Waste pipes should always be run above floor level, not through joists. For running electric cables small holes should be drilled centrally (or no less than 50mm from the top of the joist).

Another problem is incompetent removal of structural walls, which may have left floors unsupported. See chapter 10.

**Solution**  *Fit timber 'noggins' as bracing between joists (typically 50 x 38mm).*

Overly springy and weak floors can normally be stiffened by wedging blocks of wood known as 'noggins' between the joists at mid span and every 1.5m or so. Or it may be necessary to install additional joists run between or bolted alongside the existing ones. Weak points in floor joists (eg holes and notches cut for pipes and cables) can be stiffened by attaching straps or metal plates alongside. If any supporting walls have been taken out, structural repairs may be needed.In severe cases settled sleeper walls supporting ground floor joists may need to be rebuilt.

# Defect: Rotten or infested floor

## SYMPTOMS

Soft, damp, or spongy floorboards in localised areas; damp smells; slugs coming up through ground floor.

Dampness aggravated by a lack of sub-floor ventilation to ground floors is the most common cause of decay or woodbeetle in timber floors. See chapter 6.

**Cause** **The joist ends may have been built into the brickwork of damp external solid walls.**

Joists were traditionally built into pockets in the main walls (something that's no longer permitted – today joist hangers must be used in both main and party walls). Where the walls were particularly exposed to the elements, or poor maintenance allowed damp to penetrate, these embedded joist ends could eventually start to rot.

Rotten joist ends in main walls are usually found in ground floors, but damp can penetrate walls even at higher levels, particularly in locations that are very exposed to wind and rain, or where the pointing is badly eroded, or cement mortar has trapped damp. Leaking downpipes are a common cause of damp to upper walls. These conditions can cause an outbreak of rot that can spread, as upper floors were not designed with through-ventilation.

**Solution** *Lift boards over joist ends and check condition* If walls are damp, joist ends must be protected with a DPC, or rehung from steel joist hangers. But first check for rot in the timbers by prodding with a screwdriver. Attend to the causes of dampness and treat any decayed timbers and joist ends with preservative. Improve ventilation – see chapter 6.

In extreme cases, lifting the floorboards adjacent to such walls may reveal that joist ends have been completely eaten away. Consequently there may now be a gap between the wall and the remaining body of the joist. A similar problem is sometimes found at higher levels where the roof has pushed the upper walls out, so that they've parted company with the joist ends. A useful solution in both cases is to fit steel 'shoe' joist extenders that re-establish connection between the joists and the wall (see page 174).

To carry out these works, the floorboards directly above the joist ends need to be lifted, as well as a small strip of the ceiling below. Once the joists have been extended and reconnected to the wall, they are concealed within the floor void, so you don't have to worry about such repairs looking pretty.

Where a weakened joist or beam end is visually important, there are two main methods of achieving an 'invisible repair'. The traditional approach is to make a 'scarf repair', replacing the defective section in new matching treated timber.

Alternatively, joists can be tied back in and strengthened by inserting a steel bar through the wall from outside. This involves first drilling a suitable hole through the wall and into the joist. A threaded stainless-steel 'helical bar' with a fabric sleeve is then inserted, and the sleeve filled with resin to form a tight connection between the bar and the timber and masonry.

**Cause** **Damp sleeper walls**

Sleeper walls or brick piers help support ground-floor joists.

There should be a DPC, usually of slate, to keep the timber joists dry and free from decay, but there may be none. Such defects are not common, as the underfloor area should be dry. The sleeper walls themselves are usually trouble free, but on rare occasions damp or subsidence can cause them to disintegrate.

**Solution** *Check the condition of the ground-floor sleeper walls by lifting floorboards near the centre of the room* If there is no DPC, the joists may have to be raised so that a plastic DPC sheet can be placed under the timber wall plate. If the brick sleeper wall or pier has disintegrated it will need rebuilding. Improve ventilation and fix the source of damp – see below.

**Cause** **Dampness from hidden leaks**

Beware kitchens and bathrooms with suspended timber floors. Hidden leakage from pipes and sinks, or from extreme condensation, can cause rot under units that is only revealed when the units are replaced or the floor collapses while you're doing the washing up!

Water supply pipes are often run in from the street under the house to a kitchen or bathroom at the back. Hidden central heating and waste pipes can also pose a risk. A leak in your supply, or next door's, may go unnoticed for years, causing the earth to become saturated, affecting the stability of sleeper walls and structural walls alike.

**(Solution)** *Lift a few boards and check for any dampness under the floor*

Repair leaks in water supply, CH and waste pipes. Leaks are often close to the sanitary fittings. Even quite minor leaks can cause severe rot if they remain unnoticed over time. Cut out and replace defective timbers and treat the immediate area as described above. Check also that outside surface water is draining away from the house, particularly near doorways. Downpipes discharging by the walls should be redirected. External surface water from the garden may require an additional gulley to improve water dispersal.

**(Cause)** **Blocked airflow**

There may be insufficient numbers of airbricks for air to flow

freely, or they may have been blocked or rendered over in a misguided attempt to prevent draughts. It is important that airbricks are kept clear because a lack of ventilation can permit a build up of dampness, eventually leading to fungal decay and woodboring beetle attack. Where the house has an extension with a modern concrete floor, there should be ventilation ducts extending through the new floor to the new outside wall.

**(Solution)** *Clear blocked vents, replace damaged vents, or fit additional terracotta or plastic airbricks (typical size 215 x 140mm)*

Vents are needed in all walls to prevent 'dead spaces' with no airflow near adjoining solid floors. Soil from flowerbeds must be kept clear of vents. If the concrete floor in your extension has blocked off the old airbricks, it may be possible to improve airflow by fitting extra vents to a side wall. Alternatively a 'periscope' vent can be fitted and channelled to the exterior using extractor ducting.

# Defect: Loose or damaged floorboards

## SYMPTOMS

Loose, twisted and shrunken boards; draughts from gaps in floorboards; creaking boards; uneven floors.

**(Cause)** **Poor workmanship or timber shrinkage**

Floorboards that have been cut for DIY fitting of cables and pipes may not have been refitted too cleverly. If a board has been cut at the side of the joist rather than in the centre, there may now be nothing supporting it, so it dips alarmingly when trodden on. Properly fitted boards should have few gaps.

**(Solution)** *Fit 50 x 38mm noggins between joists to support boards*

- Where the board ends aren't supported, lift the defective board to reveal the joist. Firmly nail or screw a batten alongside the old joist to extend it under the unsupported board. Replace and screw down the floorboard.
- Lift the boards and check their condition and screw them securely – woodscrews are more effective than nails at pulling a warped board back against its joist.
- The surface of twisted boards can be levelled by planing or by using a floor sander once the nails have been driven well below the surface.
- You don't want gales blowing up your floorboards, but don't try to solve this one by covering up the airbricks – they are there for a purpose.
    Gaps between boards and at skirtings are a source of draughts and can be sealed with a bitumen-based mastic (gun applicator) or by cutting small timber wedges to fit the gaps. Or fit a strip of timber beading along the base of skirtings.

**(Cause)** **Incorrect nailing after lifting boards**

Joists settling can also cause annoying creaks

**(Solution)** *Creaking boards usually just need screwing down or nailing in with brads*

Make sure the boards are secured at every joist – screws are preferable if there is movement between the fixing nails and the timber. If the surface is very poor, the boards can be replaced with matching new ones. Uneven floors can be levelled by covering with hardboard sheets prior to carpeting.

**(Cause)** **Rot or beetle infestation can cause floorboards to become soft**

Woodworm is very common and is recognisable by its distinctive

boreholes. In extreme cases boards can collapse underfoot (very unusual). See chapter 6.

**(Solution)** *The source of damp must first be eradicated and badly affected timbers cut out and replaced*

In most cases the damage is very localised and can be treated as described in chapter 6. Good airflow to the boards and warmth from central heating will reduce moisture content in the timber, helping to defeat the woodworm beetle, which thrives in damp, unventilated conditions.

# Overhauling and sanding a timber floor

Once the cause of any decay has been remedied (*eg* damp and lack of ventilation) the floor timbers will need to be overhauled. The boards can then be sanded to produce an attractive, hard-wearing, natural finish.

## TOOLS REQUIRED

■ **Claw hammer and jemmy to lift boards**

■ **Screwdriver**

■ **Spray for timber-treatment fluid**

■ **Hire a belt sander and edge sander for floor stripping plus dust mask, gloves, goggles and ear protectors**

■ **Jig saw or hacksaw**

■ **Tape measure and spirit level**

## MATERIALS

■ **Rot and woodworm fluid (wear protective clothing)**

■ **Floor lacquer or varnish**

**1** Damage to floors often occurs near external doorways and walls. Start by removing all rotten or damaged floorboards to at least 300mm beyond the area where decay is visible. The lines of the joists below can be seen from the nail runs. A good place to start lifting is at an existing joint between the ends of boards. Insert a jemmy or claw hammer and prise the boards up. With tongue-and-groove boards, the tongues can be cut using a circular saw along the length of the board. Remove any nearby skirting that obstructs access.

**2** If there are no convenient joints at board ends to start from, a board can be cut. Ideally it should be cut directly over a joist, so that relaying it will be easier. But it is normally simpler to cut it at the side of the joist and when relaying to fix a batten to the side of the joist to support the board end. Beware of pipes and electric cables, which often run under the centre of floorboards.

**3** Now the extent of any hidden damage to the floor timbers can be checked. If joists have to be removed, cut them just wide of the wall plate supporting them.

**4** If there is any rot to the wall plates under the joists, these must also be removed. This can be done by supporting the joists and levering the wall plate out in sections, cutting nails with a metal saw as necessary.

**5** Remove old rotten timber. Clean the area and carefully apply rot fluid to treat nearby timbers.

**6** New timbers should be pre-treated and pressure impregnated, or wood preservative applied and allowed to fully soak into the grain.

**7** Prepare a strip of plastic DPC and place it under the new wall plate (*ie* on top of sleeper wall or pier).

**8** Having positioned the DPC, slide in the new wall plates, making sure they are level. To adjust the level, use a wedge to prop up one end and pack any gaps (*eg* with slate or mortar). Remove any temporary joist supports.

**9** Fix joists to wall plates by skew nailing through the joist sides.

**10** Overlap new joist ends next to the existing ones for extra strength. Skew nail joints or bolt overlapping joists together.

**11** The new floor structure should extend well past the initial area of rot. Check there is a good through-flow of air under the floor and clear any blocked air vents; if necessary fit additional new airbricks.

**12** Refix sound old boards or replace defective ones. Use brads or 20mm No. 8 screws, and reuse old nail holes to help prevent hitting a cable or pipe.

**13** If small areas of replacement board are too thin, pack with some sheet wood.

**14** Gaps to old boards of more than about 3mm caused by shrinkage can be sealed with tapered timber strips (wedges) glued in place. Use a flexible floor sealant or papier mâché for smaller gaps.

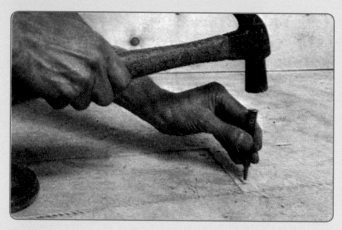

**15** Punch in all protruding nail heads before sanding and fill small holes with wood filler, or they will tear the sanding belt.

**16** Large belt sanders can level rough floorboards, but can be difficult to control. Ensure the abrasive sheet is tightly fitted. Tilt the machine back slightly, switch on and gently lower it to the floor. Work in a straight line in slow diagonal sweeps across the grain. Start with medium or coarse sheets.

**17** Make the second run with fine or medium sheets at 90° to the first, then finish off with a third slow run along the grain using fine sheets.

**18** Use an edge sander to get into corners and for warped boards. An orbital sander can be used to finish.

**19** After vacuuming all loose dust, apply two or three coats of varnish, the first coat thinned and sanded when dry.

# Solid floors

Floors to some entrance halls and most kitchens/sculleries and ground floor WCs were normally of solid construction, comprising a weak mix of concrete laid over a bed of hardcore or rubble (often including building rubbish such as smashed bottles, broken bricks and tiles).

## FLOORING MATERIALS

Plain quarry tiles held in place on a bed of lime mortar were probably the most common solid floor finish. In eastern counties, traditional Norfolk 'pamment' (or 'pammet') terracotta floor tiles were produced in a variety of different colours. But it was the Victorians who developed floor tiling into an art form, with exotic geometric and decorative encaustic tiles echoing medieval monastic designs. In the best middle-class housing there was a vogue for parquet and exposed wood floors from the 1870s, laid like a veneer over conventional floorboards. By the Edwardian era woodblock flooring had become enormously popular in better-quality houses.

Flooring in the more visible parts of the houses also performed the socially important function of impressing visitors combined with subtle demarcation of status; this can be seen where lavish geometric tiling in entrance halls stops short at the boundary between the 'polite' part of the house and the service quarters to the rear.

Concrete floors in kitchens and sculleries were sometimes surfaced with 'granolithic', a mix of cement with stone chippings. Cheaper materials included traditional flagstones or thin slabs of slate, laid over a roughly level surface of compressed earth. Hardwearing York stone was popular, but artificial stone was also used (made from cement and coke breeze), or even a sprinkling of surplus under-fired bricks. In smaller poorer houses simple lime-ash ground floors were common, comprising a stiff mix of

Above and below: Floor art; exotic blend of encaustic and geometric tiling

Edwardian woodblock

**Below:** Norfolk pammets
**Bottom:** servants quarters with hard wearing limestone slabs

Gleaming 'terrazzo' hallway.
*(Photo: southwest flooring)*

Marble floor in rectory entrance hall　　Edwardian tiled hall　　Minton encaustic tiles

lime, wood ash and sand as a covering to bare earth; but by now may have degenerated to the consistency of crumbly sand.

In the most expensive homes polished marble flooring would be used to provide the required top-quality finish. Inevitably such high-status materials would be emulated further down the social scale where what appears to be expensive marble may actually be *scagiola*. This was a cheaper material made from glued gypsum with a polished surface of coloured stone or marble dust, skilfully stained to imitate the real thing; or it might comprise a more expensive material like *terrazzo* – marble chips set in mortar.

## DAMP PROOFING

It was not generally considered necessary to incorporate effective damp proofing into solid floors until well into the 20th century, even though most houses had DPCs built into walls from the mid 1870s onwards. However, in better-quality homes a layer of asphalt might be laid over a base of concrete before levelling with a thin screed prior to tiling.

To keep any dampness at bay, most Victorian solid floors relied on the fact that quarry tiles, flagstones or slate slabs were impervious and water resistant, and any moisture in the ground could freely evaporate through the joints between them. As long as the ground wasn't marshy, and the floor was raised up

a few inches higher than external ground level, then floors should remain comfortably dry.

Fireplaces sucking in currents of air helped any resulting damp to evaporate harmlessly away. But many older properties have since been sealed up. And with today's lifestyles, cold solid floors in steamy kitchens and bathrooms can attract condensation. However, the concept of floor insulation wasn't unknown to Victorians builders – sometimes glass bottles were embedded within lime concrete floors to form a surprisingly effective air gap.

## FLOORS OVER CELLARS

Some grander houses had solid ground floors constructed over brick cellars built like small railway arches, in the form of vaults or tunnels. As floor loadings are applied, the bricks or stones in the arches are compressed and, acting in unison, they transfer the loadings by pushing out sideways. This is why the cellar walls at either side sometimes need to be propped with buttresses. So before cutting or altering such floors and walls, it's important to check the implications with a structural engineer, as it all works interdependently.

# Modern solid floors

A modern solid concrete floor will be substantially thicker than its Victorian equivalent, as well as incorporating insulation and a damp-proofing membrane. After excavating the topsoil, a typical solid floor might comprise a layer of at least 100mm of well-compacted hardcore. This is then levelled with a layer of sand ('blinding') before being covered with another 150mm or so of concrete to form a thick 'slab' base. Sandwiched above or below the concrete is a damp-proof membrane or 'DPM' (a thick 1200-gauge plastic sheet) to block any dampness from the ground. The membrane should extend up the sides of the slab and join up with the DPC in the walls. Thick insulation boards (eg 100mm polyurthane) are incorporated above the slab, or can sometimes be used in lieu of a finishing screed. Finally the slab is finished and levelled with a weak sand/cement screed typically 65mm thick.

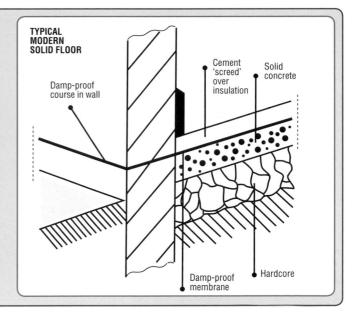

TYPICAL MODERN SOLID FLOOR

Damp-proof course in wall

Cement 'screed' over insulation

Solid concrete

Damp-proof membrane

Hardcore

## EXPOSING ORIGINAL FLOORS

Where a thin screed has been applied over an old floor surface using a weak cement mix, it's possible that the original tiles or stone flags have survived, particularly where it was laid over a plastic sheet DPM. One indication that a concrete screed was simply poured on top of the old floor rather than excavating it is where the current floor level is significantly higher than the original.

## TILED FLOORS

Floor tiles were traditionally made from clay in plain single colours (typically reds, buffs or blacks) and produced with square edges and unglazed. The popularity of these 'quarry tiles' ('carre' is French for square) peaked in the Victorian era. They are commonly found in kitchens and sculleries on account of their remarkably durability. Most are 6 x 6 inch.

However, Victorian floors are especially well known for 'geometric' tiles and exotically patterned 'encaustic' tiles. Geometric tiling comprises lots of small straight-edged plain tiles laid in patterns of two or more colours with a very fine grout line between. Commonly found on front garden paths and in entrance halls, geometrics were made in shapes such as triangles and rectangles and in a range of natural clay colours, from off-white through to red and brown to blue-black.

Encaustic tiles (literally meaning 'burnt-in') revived medieval techniques of inlaying clays of different colours to create patterns within each single tile. Decoration was achieved by stamping a design into the body of a plain clay tile before firing, and filling it with liquid clays of contrasting colours. Being relatively expensive, encaustics were often combined with quarry and geometric tiles to cover large areas at less cost.

The most common problem today is where individual tiles have come loose or have broken or are uneven. Usually they can be prised out carefully with a knife and relaid. More extensive damage is usually the result of movement in the base in which they're bedded. However, unless tiles have already come loose or are damaged, it's best to avoid trying to lift them as they tend to be fairly brittle and prone to breakage. Where the tiles have been laid in a modern cement base (rather than soft lime), taking up and relaying them isn't practical as many will suffer damage

either in the lifting-up process, or when cleaning off the mortar residue. If repairs can't be made without wholesale removal of tiles, there may be no choice but to replace the floor with a modern replica.

## RELAYING

As with tiles, it's quite common for individual stones or bricks to have come loose, rocking disconcertingly under the pressure of foot traffic. In most cases these can normally be re-bedded fairly easily with a dab of hydraulic lime mortar. But in severe cases, there may be no alternative but for the entire floor to be relaid. So that the floor can be correctly reinstated, as each stone or brick is removed it should be numbered in chalk and its position marked on a plan. If possible try to avoid disturbing the base (substrate). If the base has to be replaced, the procedure is to build up a new layer of hardcore and then relay the floor on a new bed of weak sand lime mortar. With many solid floors the stones or bricks were butted close and weren't pointed. Where pointing is

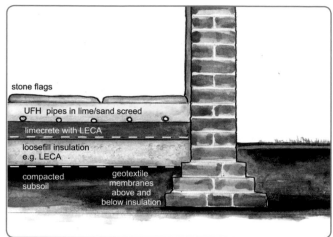

required, use lime mortar to allow evaporation of moisture.

Where a new solid floor is required, limecrete is ideal. This is much favoured by conservationists as a natural alternative to concrete. Limecrete typically comprises one part hydraulic lime (NHL 3.5), one part sharp sand and two parts lightweight aggregate or insulating material (eg lightweight expanded clay). It is laid 100mm thick and forms a breathable floor slab, which has a certain amount of flex. It's fully compatible with underfloor heating and normally laid within a 75mm sand/lime screed.

---

## Defect: Sinking, cracked, or bulging concrete

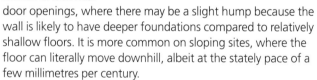

### SYMPTOMS

Uneven floors; cracked tiles; floors sound hollow when stamped on; doors stick; walls out of true; gaps below skirtings.

**Cause** **Poor mixing of the original concrete, or screed that is too thin**

Clues to defective floors include internal walls that are badly cracked due to being built off a floor instead of having their own foundations, and internal doors that are out of alignment. Gaps at skirtings are another indication of problems. Look for irregular dips in the floor of more than about 15mm, which may indicate localised settlement.

**Solution** Some unevenness and slight settlement is not unusual in older solid floors. If the settlement is not excessive, the simplest solution is to lay a levelling screed over the existing concrete. However, a severely dropped floor slab may need complete relaying. If the problem is acute but localised, concrete can be pumped into the gaps that have formed in the floor by 'pressure grouting' through holes at 1m spacing in the slab – an expensive, specialist job.

---

**Cause** **Poor compaction**

As well as being of dubious origin, the hardcore was not always compacted too well before being covered in concrete. One problem that can occur is that parts of the hardcore base under the concrete can start to compact many years after construction,

so the floor sinks and cracks. Typically this is evident around the edges or corners of rooms. Check near door openings, where there may be a slight hump because the wall is likely to have deeper foundations compared to relatively shallow floors. It is more common on sloping sites, where the floor can literally move downhill, albeit at the stately pace of a few millimetres per century.

As a rule, once you get cracks over 5mm, or large localised gaps below skirting (15mm or more) where the floors are obviously sloping, these are likely to require significant remedial work, as described above.

---

**Cause** **Sulphate attack**

This is a chemical reaction that can occur between cement-based concrete and some types of hardcore (eg old clinker from fireplaces or shale), causing the concrete floor to expand and push upwards with cracks and a bulge in the middle of the floor. This is more likely to occur where floors have subsequently been relaid because traditional Victorian lime-based materials are not normally prone to sulphate attack.

**Solution** This is not a common problem. It shows by the floor humping upwards in localised areas. In severe cases the concrete will need to be broken up and renewed.

In the diagram labels:
stone flags
UFH pipes in lime/sand screed
limecrete with LECA
loosefill insulation e.g. LECA
compacted subsoil
geotextile membranes above and below insulation

Because it's breathable, water vapour can escape – a useful feature in period properties. The floor finishes must also be breathable, so it suits natural flagstones, tiles and wood coverings. It can even meet current Building Regulations by incorporating natural insulation material.

## MAINTAINING FLOORS

Most tiled or flagstone floors require very little maintenance, just occasional light scrubbing with a stiff brush and hot water with a dash of washing-up liquid. To prevent flooding the floor with excess water and avoid soaking, mop and rinse as you go. Victorian floor tiles were generally unglazed, and over time tend to develop their own natural sheen through the polishing action of wear from shoes. Even so, tiles may in some cases have become stained. Specialist tile cleaning products can be used sparingly to remove stubborn surface dirt, but must be rinsed off. Avoid detergents, caustic soda and acids. Never use abrasive cleaning methods such as rubbing with scouring powder.

To bring out the colour of old stone flags or clay tiles, applying a smear of beeswax can be highly effective, but keep the joints clear of wax, so as not to block evaporation.

Do not use modern sealants as they trap moisture, which should be free to evaporate out. Even exterior tiling (common to Victorian entrance porches and front garden paths) should not be sealed as it reduces its resistance to frost.

When it comes to maintaining marble floors, gentle cleaning and polishing is required because, despite its hard image, marble is easily chipped and damaged.

## WOODBLOCK FLOORING

Woodblock floors were bedded in a layer of hot asphalt bitumen adhesive as a basic moisture barrier. But after a long period of time, individual blocks sometimes come loose, the bitumen base becoming brittle with age. However, having them reset should be relatively straightforward, although any associated damp must first be resolved. Woodblock/parquet floors are sometimes restored by sanding and varnishing although conservationists generally prefer less harsh methods. For example, built-up layers of old polish can be removed using a suitable solvent, and the floor given a light manual sanding before applying a traditional finish of natural oil or wax polish. This method allows for occasional rewaxing or oiling.

# Defect: Damp floor

## SYMPTOMS

Wet floor surface or damp under vinyl; floor tiles lifting or cracked; tiles sound hollow as damp collects below the surface of flooring.

### Cause  Dampness in the ground around and under the house

**Solution** To minimise dampness in the ground, remove adjoining concrete paths and replace them with gravel next to the walls to allow moisture to evaporate. The ground around the house should be at least 200mm lower than the indoor floor level, and should slope away from the walls, to disperse surface water away from the building.

### Cause  Impermeable floor coverings over traditional solid floors, which need to breathe and perspire

**Solution** To allow the floor to dry naturally, the first step is to expose an old solid floor by removing carpets or other coverings. Once exposed there may initially be a smell of trapped damp and it can take several weeks for floors to fully dry out so you need to thoroughly ventilate the room. Once the house is warm and ventilated, any slight moisture from the floor will be free to naturally evaporate without trace. Joints between tiles should be of breathable lime mortar rather than cement.

To accommodate the need for old floors to 'breathe', loose rugs of natural fibres like wool or cotton (without a rubber backing) are well suited. The ideal floor covering is natural matting of seagrass, jute, sisal or coir, with a border of exposed floor left around the perimeter of the room.

Where excessive moisture has been present, white hygroscopic salts can be present, attracting moisture from the air, making floors appear damp. Although normally harmless, it may look unsightly. Loose salts can be brushed off or vacuumed, but shouldn't be washed because the water re-dissolves the salts sending them back into the floor. In severe cases special pastes can be applied to draw out the salts.

### Cause  Leaks from defective plumbing

Copper pipes embedded in concrete floors are very prone to corrosion and leakage unless protected with tape or run in ducts. Leakage can occur for a long time before failure is detected. Screeds that are too thin over buried pipes will break down (more likely to apply where modern cement screeds have been applied)

**Solution** *Replace pipes with new surface-run pipework* To avoid damaging any historic tiles, the disconnected defective old pipework can be left *in situ* and replaced with new surface-run pipework along wall surfaces.

### Cause  Condensation

Cold floors, particularly in kitchens and bathrooms, can attract condensation as warm moist air and steam condenses. This may look like leakage or even rising damp.

**Solution** *Install extractor fans to remove humid air and improve ventilation.*
A less cold surface covering such as wood flooring can be placed over a layer of rigid insulation material to make the surface warmer (but raised floor heights may necessitate adjustments to units and fittings etc.). Floor mats of a breathable material such as jute or seagrass may help.

# FIREPLACES AND FLUES

Here we come to what are surely the crown jewels of many a period home – stunning antique fireplaces in marble or cast iron. But the Victorians were masters of deception, so things may not be quite what they seem. A Victorian house just isn't complete without its fireplace – the most important fitting in the architecture of the interior. But behind those beautiful period grates and surrounds there may be deadly hidden dangers.

# Fireplaces

You can judge the status of the rooms in your house by the quality of their fireplaces. Starting with marble for the front drawing room, they diminish in quality and size as you proceed through the house, until in the servant's room you find a mean little fireplace designed to burn the minimum amount of coal. But in Victorian houses appearances can be deceptive. That awesome marble fire surround may actually turn out to be well disguised cast iron or cheap slate cunningly painted to look like marble.

**Above:** Marble mantle surround with cast iron arched insert c.1870
**Above right:** Small cast iron job in back bedroom

## THE REAL THING

What if the originals were ripped out long ago? The best guide to tracking down the beauties that may have formerly graced your residence is to find an original fireplace in a neighbouring house of similar age. The old openings from which they were removed may also provide some clues – shadowy lines in the plaster on the chimney breast can indicate the proportions of a missing surround. If you find any surviving pieces of the old grate within the opening, a

1  Early rustic fireplace with simple wood surround
2  Modern repro of mid Victorian arched topped cast iron combined insert & surround
3  Later splayed cheek insert with modern plain wooden mantle
4  Cast iron tiled cheek repro fireplace
5  Arts & Crafts style polished cast iron combo with inset tiles
6  Hardwood Edwardian with mirror and tiled insert and hearth

Victfires.com

fireplace specialist or an architectural salvage yard may have matching parts in stock.

When choosing an appropriate replacement fireplace, it should be suitable both for the type of house and for the room in which it's to be installed. An elaborate, oversized Louis XIV grate will not suit a small cottage and would be equally out of place in a maid's bedroom.

## CONSTRUCTION

Fireplaces are a complex part of the building's construction, with each one needing a separate flue to safely discharge combustion gases all the way to the top. The most economical and efficient method of construction was to bunch flues together within a large chimney breast built into the party walls, which helped insulate one house from the next as well as saving on materials.

Maybe it was down to the smouldering memory of the Great Fire of London, but the Building Acts specified construction standards surprisingly rigorously – including the method of construction of chimney breasts, the widths of flues, angle dimensions for turns within flues, and minimum chimney heights (at least 0.9m/3ft above the roof). The Building Acts also stipulated that all fireplaces have a fireproof hearth measuring a minimum of 450mm (18in) to the front and 150mm (6in) to either side. This was usually

achieved with a weak concrete infill supported over brick arches spanning between the wall and the floor trimmer joist.

Given that this was the age of steam, it's not surprising that the Victorians are credited with a number of important developments in fireplace efficiency. Combustion was improved by a reduction in the width of the fireplace back to one-third that of the front opening. The problem of uncontrolled draughts was addressed with the 'register grate', with its integral iron fire basket and adjustable iron plate (a small hatch) that could regulate the airflow. This restricted the throat enough to retain the maximum amount of heat in the room whilst allowing smoke to escape. Advances continued with smoke hoods and vented ash-pan covers ('frets') to improve combustion.

It's often possible to date a property by the style of its fireplaces, although in mainstream housing speculative builders adopted a

cautious approach to changes in fashion. From the 1870s arch-topped cast-iron fireplaces started to be superseded by square openings with tiled 'splayed cheeks' designed to reflect more heat into the room. In contrast, the Edwardians had a fondness for elaborate, tall, polished hardwood surrounds with glazed tiled inserts and hearths.

# Removed chimney breasts

Taking out a chimney breast to make more living space is fairly common, though generally inadvisable, structural alteration. The trouble is, this kind or work is sometimes carried out without proper support provided to the remaining masonry and the stack above – with potentially lethal consequences. Structural alterations like this must be approved by Building Control.

Consider this: your party wall with next door may be only 230mm (9in) thick – sometimes even less. Balancing over it is the full chimney stack structure, a huge amount of masonry, supported below by the chimney breasts on either side. Suppose one owner decides to take out the upstairs chimney breast. And then the neighbour does the same. Unless properly supported, the stack and the masonry below it could be unstable, just waiting for the wind to blow.

## BREAST CHECK

To see if any chimney breasts have been taken out, start by taking a look outside at the chimney stacks. Most houses have at least one shared stack over the party wall and another on the rear addition serving the kitchen/scullery and the back bedroom. In the roof space the main stack usually divides into two chimney breasts that continue into the bedrooms below and down into the front and rear reception rooms.

If the chimney breast is missing in any of these rooms, the question is, what's supporting the remaining masonry above? Look for telltale signs such as a leaning stack, or a cracked, badly patched ceiling next to the wall where the chimney breast used to be. Or brown ceiling stains where rainwater has dribbled down the old stack. These are clues that there might be

**Above:** Danger! Unsupported masonry in loft after chimney breast removed in bedroom

**Below:** Proper job – pair of steel gallows brackets support masonry (although Building Control sometimes stipulate a steel beam)

tonnes of unsupported masonry balancing menacingly just above your ceiling.

The minimum requirement is support in the form of metal 'gallows brackets', or a steel joist that (if it's there at all) should be visible in the loft or within the floor above. Unless consent was obtained at the time of the works, Building Control will need to inspect and retrospectively confirm the adequacy of the method of support. Apart from the obvious health-and-safety issue, unless satisfactory approvals have been obtained, when you come to sell the house, the surveyor and solicitors will raise awkward questions, likely to scare away buyers.

# Opening it up

One of the great home-owning experiences is the joy of excavating an old boarded up fireplace to hopefully reveal a splendid, original cast iron grate. But more likely all you'll find is an old opening half filled with rubble, soot, dead birds, and mounds of old sandy lime mortar.

To open up fireplaces that have been boarded over with studwork and plasterboard for many a long year may require the application of a little brute force. More challenging are openings blocked up with solid masonry (normally concrete blockwork) for

which you may need to hire a Kango type power breaker. But don't get too carried away: take care not to demolish any of the brickwork that forms the 'builder's opening' (the large square opening in the brickwork of the chimney breast into which

**Left:** Stripped back 'builders opening' – but Victorian chimney breasts were always plastered

**Below:** Curved cast iron bracing bar under arch supports masonry and should be retained

the fireplace is fitted). The top of the opening normally comprises a brickwork arch with a discreet supporting curved iron bar embedded underneath; remember these are supporting substantial loadings of masonry to the chimney breasts above.

If you propose to reuse an old flue, then before starting you'll need to check that: (1) the chimney breast in the room above is still there; and (2) that the chimney stack on the roof hasn't been taken down. It may have been capped or cowled to protect it from rain, but it should not have been completely sealed.

If you take a look up the flue from the fireplace (wear eye protectors!) you may be able to see, amongst the sooty old brickwork, how the flue narrows past the throat (aka the 'gather'). It should look like an upside down funnel, that's reasonably smooth as far as the eye can see. If you see crudely stepped brickwork this may inhibit the flow of exhaust gases and cause combustion problems for open fires. To check, light a twist of paper in the opening. If it burns well and the flame is drawn inwards and upwards, the flue is clear. But you still need to get a chimney sweep to clean and inspect the flue before lighting fires. If all you can see is a flat concrete slab with a hole, the fireplace most has most likely been adapted for use with a gas appliance, and will need to be removed for reversion to use as an open fire.

A word of warning: if you harbour a plan to use the 'builder's

opening' to the full extent of the brickwork – perhaps as large open Tudor style 'inglenook' with a small traditional fire basket - it may prove difficult to get the new fire to draw properly. These openings were built to take a fireback and grate, or a fitted range or stove, with flues designed to suit the relevant appliance.

Pointing fireplace masonry prior to lining flue (right) and re-plastering chimney breast (below)

Woodlouseconservation.co.uk

**Above:** Reclaimed period fireplaces can be sourced from salvage yards
**Right:** Modern period repro with living flame effect gas fire. Will need suitable flue liner. Part J of Building Regs also stipulates minimum hearth sizes (eg 13mm thick x 305mm deep x 686mm wide) depending on opening

## REFITTING A FIREPLACE

Having sourced a delightful replacement 'chimney piece', before attempting to install it some key checks need to be made. First, check whether the chimney breast is plumb and the floor in front is level, as fitting a fireplace perfectly straight and true on an old chimney breast that isn't level can look a bit odd. If there's a difference of more than about quarter of a bubble on the spirit level, it's probably better to fit the fireplace at a slight angle to match. Bedroom fireplaces were often fitted off-centre to allow for flues from lower floors rising up inside the same chimney breast.

Installing a new fire or heating appliance means first checking how this will affect the flue, because a flue liner designed for an open fire, for example, may not be suitable for a gas-fired appliance. By law, gas-burning appliances must be fitted by a competent Gas Safe registered engineer.

Also be aware that urban areas are normally smokeless zones, where burning wood or coal on an open fire is legally prohibited – this can be checked with your Local Authority.

Not unreasonably, current Building Regulations stipulate a number of safety measures, such as the need for fireplaces to have a sufficient supply of air, and to be constructed from non-combustible materials, with safe escape of exhaust gases to protect the building from catching fire and the occupants from becoming asphyxiated.

The hearth to the floor at the front of the fireplace should be made of non-combustible material (brick, stone, or concrete) and be at least 125mm thick. These are usually intact in old houses, but some may be cracked due to settlement of the floor and in severe cases may require hacking out and reforming with new concrete. To comply with Building Regs hearths need to be raised at least 13mm above floor level – see caption.

# Flues

A flue is simply a funnel, or vertical exhaust pipe, designed to safely transport smoke from the fire out to the external environment. It is based on the simple principle that warm air generated by fires naturally rises, and in the process disperses the smoke and combustion gases.

At its simplest a flue comprises the space enclosed by the chimney breast masonry and the stack. But in order for smoke to rise efficiently, a smooth internal surface had to be created. This was achieved by lining the internal masonry walls with 'parging' - a coat of lime plaster sometimes reinforced with a little cow dung added to the mix. This also helped seal the masonry to prevent leaks as well as smoothing over internal ledges where soot might have built up.

Construction became more complex with the provision of additional fireplaces in bedrooms and adjoining reception rooms, with multiple flues needing to be accommodated with a single chimney. This was achieved by building thin internal partitions of brick, slate or stone known as 'withes' or 'mid-feathers'.

To prevent rain and sleet falling vertically onto the fire, and to check downdraughts, flues were built with a gentle curve of corbelled brickwork, so you normally can't see daylight when peering up a flue. The standard flue size for an ordinary fireplace was about 230mm square (9 x 9in), whilst for those serving kitchen ranges 337 x 230mm (13.5 x 9in) was common.

**Above and below:** View down an unlined flues – eroded joints can allow fumes to escape

**Above:** Some fireplaces to outside walls have an external ash door

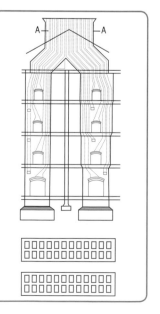

Over the years rain from unprotected chimney pots has often combined with soot and acidic gases from coal fires to degrade the mortar joints inside the flues. Loose parging and lumps of mortar can fall off and collect on ledges, and if the flue isn't swept to remove this debris it can act like a sponge and soak up polluted water that eventually seeps through the chimney breasts. Symptoms of 'parging failure' include debris in the fireplace, blocked flues, and smoke escaping.

Feeling sleepy? You may be inhaling the poisonous fumes from next door's gas fire. Leaks can sometimes go undetected for long periods, particularly within lofts or to adjoining houses. The thin wall of brickwork between flues can deteriorate, enabling smoke and fumes to pass from one flue into another. And if it becomes blocked, a build-up of poisonous combustion gases will blow back, re-entering the room. The dangers are graver still where old unlined flues are used for gas or oil-fired appliances that produce deadly, odourless Carbon Monoxide, which can prove fatal. Therefore it's essential to check the condition of flues before lighting fires.

Special brush system and vacuum used to sweep stove flues

It's a fairly simple task to test airtightness using smoke pellets. First create an upward draught (or 'draw') by burning some crumpled newspaper in the fireplace. Then light some smoke pellets placed on the hearth before taking a look in upstairs rooms and in the loft for signs of smoke leakage. CCTV cameras can be used to help detect defects, such as masonry flaws, faulty linings, fire damage and birds' nests. The solution to most problems of this sort is for the flue to be professionally lined. In fact, lining is necessary for all active flues of this age, particularly those serving gas fires and boilers, and will also help to prevent other problems, like staining on chimney breasts.

## STAINING AND DAMP

Ugly stains on chimney breasts are generally the result of old acidic, tarry dampness in flues seeping through eaten-away mortar joints and into the plaster. Dampness on upstairs chimney breasts may also be due to problems at roof level such as defective flashings,

Hot gases in unlined flues can condense on cold walls staining brickwork

rain coming down pots, or eroded brickwork and pointing to the stack. See chapter 3.

But there's another, more insidious kind of dampness. Over the years, the unholy alliance of water vapour from internal condensation and rainwater in flues, combined with the toxic combustion gases from coal fires (tar acids, ammonia, sulphates etc.), can cause damp to find its way into the surface plaster of the chimney breast. And the plaster may be contaminated with our old friends hygroscopic salts (which can appear anywhere that damp has previously come through, such as old leaks from defective flashings).

Damp soot and combustion gases leach through brickwork with staining evident along the path of the flue

Even when you've solved the cause of the damp, it can be infuriating to see patchy dampness reappearing on the plasterwork. Each time the air in the house becomes a little humid the salts promptly absorb it and liquefy, doing an excellent impersonation of penetrating damp. But when the atmosphere is dry (or when you get the builder round to check it) the damp has mysteriously disappeared!

When plaster is heavily stained or contaminated by salts, the best remedy is to hack off all the affected plaster, wash down and seal the exposed brickwork (which may contain further salts), and replaster with a 1:3 cement/sand base coat and a 'Multifinish' plaster topcoat to at least 300mm beyond the old staining. Alternatively, stained plaster may be sealed and dry-lined with foil-backed plasterboard.

## AIR SUPPLY

The provision of a sufficient amount of oxygen to ensure the efficient combustion of fires and appliances is a key part of compliance with current Building Regulations. The main reason for this concern is to ensure that occupants have enough air to breath and aren't asphyxiated in competition with large open fires that can consume more than 260 cubic metres of air per hour (in addition to burning the fuel, a large amount of air is needed to create a draught that flows over the fire and up the flue).

## WOODBURNING STOVES

Not only are these stoves very effective at heating rooms, they also boast impressive green credentials. However, if you live in a 'smoke control area', you're not allowed to burn fuels such as wood or coal that emit smoke, either in stoves or fireplaces (although

## Disused flues

In most old properties today, some of the original fireplaces will be disused and boarded up, particularly those serving bedrooms or bathrooms. But behind the scenes all may not be well. Unless chimney pots have been capped off, rain can come trickling down redundant flues causing damp in disused fireplaces, or birds may have taken up residence. Damp patches on chimney breasts can also be caused by moist air condensing on the inside of cold disused flues. If unused chimneys are boarded up but not ventilated they can suffer from condensation and damp staining.

In the 1970s the craze for boarding up fireplaces and fitting airtight double glazing resulted in some homes becoming so hermetically sealed that serious problems subsequently developed with condensation and toxic mould. See chapter 6.

This is why redundant pots should not only be capped off but also ventilated (*eg* with an air brick or purpose-made vent). And to encourage a healthy through-flow of air, a vent should also be fitted to the blocked-up former fireplace in the room below. Disused fireplaces should be ventilated top and bottom to get a through flow of air to disperse unhealthy stagnant moist indoor air. There are two things to check:

**1** If fireplaces are boarded up there should be an open vent or airbrick built in, to allow a flow of air up the stack (unless the stack has been removed or taken down into the loft).

**2** The redundant chimney pot should be protected with a hood, so that rain doesn't get in, but with sufficient draught to ventilate the flue. There are several suitable pots and caps available (see chapter 3). An alternative method is to reduce a redundant chimney in height, fit airbricks in the sides of the brickwork, and cap off the top with a suitable concrete paving slab bedded on mortar. Flues should first be swept to prevent the risk of soot coming into the room through the vent.

---

smokeless fuels such as coke briquettes are permitted). Fortunately, some 'cleanburning' stoves approved by DEFRA are exempt. These produce low emissions when woodburning, and some are also approved for use with coal. Your local Council should be able to advise whether you're located in a 'smokeless zone'). N.B. Building Regs require CO detectors to be fitted in rooms with woodburning stoves.

### VENTILATION

Victorian fireplaces can be a little draughty in winter but are an important source of ventilation to help remove moist humid air from the home. Some even had ingenious 'passive ventilation' systems in the form of a small vertical air duct built alongside the flue. When warmed by the fire, the air in these ducts would naturally rise, drawing stale air out of the room, but could be closed when necessary with a fabric flap over the grille to stop downdraughts – the only snag being that, where these terminate in the roof space, pumping in lots of warm moist air can cause condensation in lofts, so it needs to be vented externally instead.

Fresh air was sometimes fed direct to the fire through a 50mm iron pipe from outside, to alleviate the problem of fires sucking gales of icy air into the room through gaps in doors and windows, resulting in frozen backs and scorched faces.

### TECHNICAL DATA

See Building Regulations Part J. See also the Solid Fuel Association website www.solidfuel.co.uk

The ratio of the flue diameter to the fireplace opening cross-section should not be more than 1:8. Most grates are no more than 450mm wide by 500mm high, and are suitable for a flue as small as 185mm.

---

## Defect: Smoky fires

### SYMPTOMS

Smoke doesn't go up the chimney but billows into the room.

Fireplaces rely on the principle that hot air rises. So you're off to a head start with a regularly used flue (or a flue adjoining one that's regularly used) because this helps keep the escaping smoke warm. The next best type of flue is one that's well insulated. Some chimneys are inherently too cold to draw well, particularly those on outside walls.

There are several possible causes of smoky fires, but sometimes a smoking fire is simply down to the weather – a change in wind direction can distort the air pressure differences between the fireplace and chimney pot that normally give the smoke an extra push up.

**Cause** **Air starvation – insufficient air to carry the smoke into the flue**

A lack of air is the most common cause of smoky fires. An open fire needs at least six changes of air in the room per hour to burn well. Air is drawn into the fire from the room, which in turn needs to be replaced.

**Solution** *Boost air supply*
Check by opening a door or window – if the smoke clears, the problem is likely to be air starvation. Install wall vents or underfloor ducts. Rooms must not be completely airtight.

**Cause** **Blocked chimney**

If smoke blows back into the room It could be because the chimney has become blocked – *eg* by debris and soot. One clue to nests blocking flues is the appearance of straw and feathers in the grate, or nesting birds depositing large quantities of twigs in Spring. Decomposing dead birds tend to attract swarms of flies.

**Solution** *Ensure chimney has been swept*
Some old flues have voids and bends that can harbour birds' nests, which drop off and cause blockages (hence the need for regular sweeping). Debris may be from damaged parging or from slipped bricks or stones that separate one flue from another (known as 'bats' or 'mid-feathers'). The flue will need to be lined. Excessive soot may be due to using unsuitable fuel. To check the flue is clear, a metal 'coring ball' can be lowered down the flue.

**Cause** **Unsuitable size of flue**

Too large a flue (more than about 230mm square) can cause smokiness because it takes a long time to get warm enough to help smoke rise (larger kitchen flues with cooking ranges were in constant use). Conversely, too small a flue (less than 185mm square) can choke the fire. Flues should be the same size all the way up.

**Solution** *Line large flues with suitable flue liner*

**Cause** **Downdraught causes smoke to blow back**

Downdraught blows smoke back into room. There are several

causes of downdraughts. Typically the chimney is not high enough above the ridge, or wind currents near chimney tops create a 'high-pressure zone', caused by high buildings etc. nearby. Stacks that are too short or overshadowed by surrounding buildings or the proximity of a large tree or hill etc. on the windward side will divert air currents causing a downward draught.

**Solution** *Extend chimney with 'long tom' pipes, or build up the stack*
Fit a special draught-inducing cowl or circular deflector. Raising the height of the stack, *eg* with a pot extender can help.

**Cause** **Adverse flow conditions due to poor design or combustion**

Any obstacles the smoke and gases are likely to encounter as they pass up the flue will affect the aerodynamics of the flue gases leaving the chimney. Clearly the less resistance,

the faster the flow. Hence the importance of smooth flue walls – or a flue liner. Defective construction, such as any sharp bends in the flue, can cause poor chimney draught, as can air leaks at mortar joints. Also, if the throat over the fire is too large, the escape of smoke and gases is slowed. Or if the fireplace opening itself is too large in relation to the flue (*ie* more than six times larger), or too high (more than 610mm), smoke may enter the room.

**Solution** *A chimney sweep can check the run of a flue with rods. If severe, open up the front of the chimney breast and rebuild defective areas. Check the size and position of fireplace throat and lintel.*

**Solution** *Fit a throat restrictor*
A throat restrictor will reduce the entry zone to about 100 x 250mm (4 x 10in) – or wedge a thin sheet of metal across the front of the throat. To lower the fireplace opening, fit a metal canopy or place a thin piece of metal 75–100mm (3–4in) high across top of the opening to reduce the height to 510–560mm (20–22in).

**Cause** **Unsuitable chimney pot**

A badly fitted or partially blocked pot can cause an obstruction to the flow. Cowls or guards on pots may look clear from ground level but could actually be sufficiently blocked to affect the burning of the fire.

**Solution** *Replace with suitable chimney pot for flue. Check cowls are clear*

## Defect: Fire burns poorly

**SYMPTOMS**

Fire becomes choked and struggles to burn.

**Cause** **As described above for a smoky fire, but with less obvious symptoms**

A fire may burn well for a while, but the fire then chokes on exhaust gases that are not fully dispersed. This may be due to a poor airflow or because of a downdraught. Problems with indoor air supply are common in houses where draughts have been totally sealed up. So there's just not enough air being sucked into the fireplace. Downdraughts are complex because the house itself can act like a big chimney, being full of warm air that naturally rises, sometimes with immense suction.

**Solution** *As for a smoky fire*

If there's a lack of room ventilation, try opening the door to the room to see if it improves the draw. If it does, fit an underfloor grille in the hearth or insert air vents through the external walls. Poor airflow may also simply be due to draughty upstairs windows or a poorly fitted loft hatch competing with the chimney to suck air (and smoke) upwards. Similarly, there is competition from extractor fans, cooker hoods and other fires, which all need air to work; if they have a stronger pull than the chimney, they win! In very windy conditions, an air vent on the downwind side of the house can create 'negative pressure' that literally sucks the smoke back down the chimney. The solution may be to simply relocate the offending air vents or extractors.

## Defect: Smoke and fumes leak into other rooms

**SYMPTOMS**

Obvious smoke leaks. Invisible fumes and carbon monoxide cause drowsiness and can be fatal.

**Cause** **Air leaks**

Air leaks through defective brick joints or cracked parging can allow toxic smoke and fumes to seep into rooms. They also cool the exhaust gases, reducing the draught.

**Solution** *Line the flue*

Use a smoke bomb (available from DIY stores) to trace leaks. Make good defective joints and line the flue (see opposite).

## Defect: Birds entering the house via the chimney

**SYMPTOMS**

Feathers, sticks, straw and soot in grate; bird ingress.

**Cause** **Unprotected chimney pots allow bird access and nesting**

**Solution** *Fit a special protective 'bird guard' to the pot*

## Defect: Chimney fires

Fires in flues may burn unnoticed until structural damage is caused or fire spreads to the main house.

**SYMPTOMS**

Hot chimney breasts; smell of burning

**Cause** **Excessive soot due to inefficient combustion**

Soot and tar can build up on ledges inside a flue, and may eventually ignite causing hidden chimney fires that are hard to extinguish. This is more common where the fuel being burnt is green unseasoned timber as it generates excessive tar. But soot deposits can also be a result of inefficient combustion caused by poor air supply etc (as discussed earlier). A chimney fire can burn at over 1,000°C, causing metal liners to collapse.

**Solution** *Sweep flue and improve air supply*

Flues need to be swept at least annually to remove combustible soot deposits and blockages, especially if burning wood or peat, which are particularly aggressive fuels and produce a rapid build up of tar and deposits. Sweeping with a brush and rods is still the best method. Sweeps should be NACS or HETAS approved and ideally also members of the Guild of Master Sweeps.

*For further information visit the National Fireplace Association at* **www.nfa.org.uk**

# Installing flue liners

If you want to use an old fireplace the flue has to be impervious to poisonous combustion gases. And a flue that's more than a century old won't be. So old flues must be lined, to prevent aggressive combustion gases and damp from condensation from attacking the lining and eating away at the mortar and brickwork.

Inserting flexible steel lining with cowl fitted over pot

This is normally a specialist job because there can be problems associated with installing flue liners in older houses, such as unstable masonry, plus working at height requires safe access.

Where a flue is already lined, you still may not be home and dry. Old flexible steel liners eventually suffer from corrosion, and need renewal every 10 to 15 years, depending on how regularly the fire is used. It is advisable whenever an appliance is changed to renew any existing metal liners. In addition, gas and oil appliances need special chimney-top terminals to prevent blockage from birds or debris, and to help disperse gases. New flue liners also need to be insulated, particularly if the chimney is against a cold outer wall. There are three basic types of flue liners: stainless-steel liners, poured concrete linings, and rigid clay or concrete pipes.

## STAINLESS-STEEL LINERS

Flexible single-skinned liners (class 1) are used mainly for gas fires and for oil- or gas-fired boilers, whereas hardier double-skinned liners (with a smooth inner skin and a corrugated outer surface) are required for burning solid fuels and for multi-fuel stoves. In terms of size, a 150mm or 200mm diameter liner is generally required for a wood-burning stove, but larger inglenooks with open hearths need wide-diameter flue liners, 300mm minimum, which are relatively expensive.

They are installed by being pulled either up or down the old flue by an attached rope. The ends are then fixed in position – the top is secured by a sealing plate bedded on the stack top and a metal terminal added. To reduce the risk of condensation, the space around the liner can be back-filled with loose fill mineral insulation such as vermiculite or Leca.

Wood-burning stoves need special rigid metal liners (minimum of 1mm thick), as they generate extreme heat in the flue. They are installed in sections with joints secured by clips. Again, the space around the liner is insulated by being back-filled. Oil-fired boilers also need a special insulated lining due to their high operating temperatures.

To replace an old gas fire with a solid-fuel stove or an open fireplace normally means having to also change the flue liner.

## POURED CONCRETE IN SITU LININGS

Linings can be made on site by pouring concrete down the flue around a special inflatable thin rubber tube, which is later deflated and removed to leave a smooth new flue. Although cheaper to install than other methods, the drawbacks are that it can require some opening up of chimney breasts at bends to position the tube correctly, and that the large amount of water used can cause dampness. There's a risk that installing a heavy mass of inflexible concrete inside a crumbly tube that was once able to move gently could result in localised cracking and damp problems over time.

## SOLID LINERS

Rigid pipes of clay, concrete, or terracotta come in sections that slot together, but installing them is only really feasible where the chimneys are large and fairly straight or where they are being rebuilt. Old kitchen boilers often have flues lined with asbestos cement pipes.

## Fitting a flue liner

Fitting a flexible steel flue liner should be a fairly simple operation. Liners are inserted from the top down which means you need two people. Proceed as follows:

- First have the chimney professionally swept to remove accumulated soot and debris.
- Select a liner of the correct type and diameter for your appliance.
- Arrange safe access to the stack with scaffolding
- Attach a long cord to the end of the liner and from the top of the chimney feed the cord down the flue until it reaches the fireplace below.
- At the fireplace one person gently pulls the cord down, while the person up top gently feeds the liner 'snake' down in small stages.
- To close the space between the liner and the top of the chimney a 'top plate' can be fitted underneath the chimney pot, with a 'top clamp' to take the weight of the liner.
- At the bottom, a support bracket holds the lower end of the liner in place. A flex adapter can be used to join the liner to the flue pipe from a stove or appliance.
- Once in place, the gap between the outer skin of the liner and the masonry flue should be insulated, which can be done using non-combustible loose fill (eg leca) poured down from the top. Alternatively, you can fit special insulation wraps or sleeves that just clip around the liner as it goes down the chimney.

Finally a 'rain cap' should be fitted to the top, because unlike traditional masonry flues, steel liners won't absorb rain and can act like a 'rain slide', leading to puddles on the hearth.

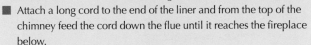

# INTERNAL JOINERY, KITCHENS AND DECORATION

If you enjoy a spot of *Time Team* archaeology, there's no better place to start than in a Victorian home. It's surprising how much of the history and status of your house can be revealed from its internal joinery and finishes. If you know where to look, old joinery can also harbour hidden clues indicating the presence of potential killer defects which, if spotted in time, can be safely eradicated.

RBKC Linley Sambourne House

As you walk through a Victorian or Edwardian house, you will notice changes in the style of the joinery. The skirting, stair fittings and architraves all become progressively cheaper and smaller as you move to the rear and upstairs, as do the cornices and fireplaces.

Everything was geared towards impressing visitors, so the entrance hall and front parlour would shine with the most lavish workmanship and materials affordable, to create an aura of comfort and wealth. It is these features that give the house its historic character and add to its value. Unfortunately, in some properties misguided home-improvers have lopped tens of thousands off the value of their houses by converting them into sanitised replica 1970s boxes – or even into mock-Jacobean inns, complete with plastic beams and injection-moulded woodworm! The advice is simple: you own a design classic – keep it that way by conserving the original features. It's also worth retaining as much old joinery as possible for the simple reason that it was of far higher quality than today's kiln-dried off-the-shelf variety.

If the interior mouldings have been stripped out of your house there may still be clues as to what was originally there, such as old bumps in the plasterwork, redundant fixing holes and sheets of hardboard concealing doors and banisters. Look at neighbouring houses of the same age to see whether they retain their original features.

It's sometimes possible to track down suitable replacements, with a little detective work. However, to obtain an exact match for original skirting boards and architraves for example can often be impossible because of the infinite number of local variations in times past. If replacement or repair isn't possible, it may be worth having new matching items crafted using materials of a similar quality.

In a Victorian property the internal joinery and finishes directly reflected the status of the house. Softwood was regarded as a cheap and inferior material, despite being of far higher quality than most equivalent timber available today. Pine (red or yellow deal) would be disguised by being painted or was sometimes camouflaged with skilled 'fake-graining' to resemble more expensive woods. The modern fad for stripped pine interiors

would horrify any self-respecting Victorian. Only superior-quality hardwood joinery would be left unfinished in all its naked glory, or a simple wax polish or oil finish applied to enhance its natural beauty for all to admire. Thus you often find mahogany banister rails, and painted pine newel posts and balusters (the thin vertical spindles that run from the handrail down to the tread). Or the newel posts might also be made from mahogany or another fine quality hardwood. Most woodwork, including doors, architraves, skirtings and picture rails, would be grained in a dark or medium tone or else painted a dark colour such as brown or green – cream and off-white becoming popular later in the century.

# Stairs

Even in the plainest houses an original staircase can be an absolute delight, with beautifully crafted newel posts, banister handrails and balusters (spindles). But that doesn't mean they're always easy to live with. Many old stairs will be considerably steeper than we're used to today, with an alarming lack of

**Above:** Exuberant late Victorian balustrades
**Left:** Plain 'stick' design to attic room
**Right:** Same house as below left, has hard wearing stone steps for servants to semi-basement

headroom and stair treads that bear little resemblance to the dimensions of modern feet. But Victorian staircases were actually a highly functional design, optimising the available space, with a handrail to provide support and the balusters arranged to let the light through so you could see the way. Making the best use of natural light was important, since the illumination provided by candles, gas jets, oil lamps, and fireplaces was so dim.

In poorer 'cottage' houses, particularly those with no entrance

*Delightful mid period cast iron balustrades with polished mahogany banister c 1870s*

hall it was common to place the staircase between the two principle rooms on each floor, with a small lobby at the foot of the stairs. This was economical to build, with less timber required (for balustrades etc.) and support provided by the brickwork walls on both sides (or one wall of brick and one studwork). But the classic Victorian terraced house design has the staircase located to the inner hall against the party wall to provide an elegant and imposing feature.

Larger houses would often have two or more staircases – a main one for the family and a discreet back one for servants made from hardwearing stone (or sometimes cast-iron spiral stairs) plus additional plainly designed flights serving basements and attic rooms.

In the earlier Victorian period when all the components had to be made by hand it was common for decorative joinery to be limited to the main staircase, with

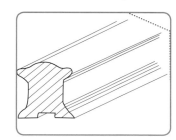

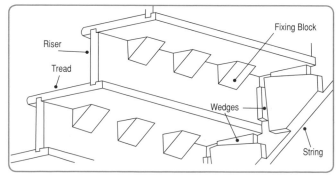

Riser

Tread

Fixing Block

Wedges

String

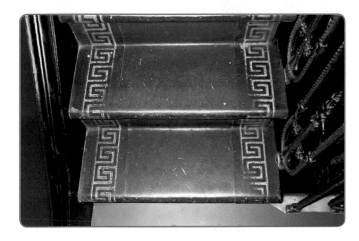

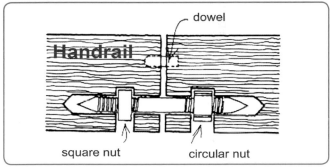

Handrail

dowel

square nut          circular nut

In a typical Victorian house there will also be a couple of steps down to the back rooms on each floor to accommodate the split-level layout with the rear addition built at a lower level.

Even if the existing stairs are in fairly poor condition, you need to think twice before ripping them out because installing a new staircase requires Building Regulations approval, and it's rarely possible to fit new stairs into an old building without making major alterations. The main difference between a modern one and its 19th-century ancestor is the pitch. Some originals employed narrow treads and high risers to achieve what would now be an illegally steep angle. So it is generally far better to repair existing Victorian staircases than replace them. Even if you succeeded in sourcing one of the right dimensions, obtaining authentic reproductions of elaborately carved original components can be difficult.

Today, the design of new stairs is controlled by the Building Regulations which stipulate that they must have a clear minimum width of 800mm, and must not be steeper than 42°. The 'goings' (which are the stair treads without their small projecting front 'nosings') must be not less than 220mm deep and risers no more than 220mm high. There must be clear headroom of at least 2m and an unobstructed landing space at the top and bottom, so you cannot have a doorway directly adjoining the stairs. A balustrade at least 840mm high is required at the side of the stairs and the gap between spindles must be less than 100mm. The full Building Regulations can be seen at www.victorian-house.co.uk.

Although the staircase is a complicated exercise in joinery, most problems today simply relate to normal wear and tear, such as the odd creaking tread. Fortunately it should be a fairly straightforward job to carry out basic maintenance such as replacing a damaged or badly worn tread or refixing loose blocks or wedges under treads. However most repair work will be a lot easier where access can be gained from below as well as from above, eg via a handy understairs cupboard. Where access can't easily be gained from underneath because the underside is plastered, try to minimise the area of damage and if possible reinstate any removed lath and plaster (lime plaster can be ground up and remixed). The occasional broken spindle is another common maintenance issue, and it's normally possible to effect a neat repair by gluing and temporarily clamping the two parts together. For additional strength a split spindle can be reinforced before gluing with a dowel drilled into the broken end of each piece.

Another problem affecting old timber stairs is that the treads, particularly the nosings, become worn away underfoot. Slight dips can be packed with felt or rubber underlay beneath the carpet. The packing must be fixed in place, eg with a thin layer of plywood nailed over it. Worn nosings can be prised off or cut away from the front of the tread and replicated in new bevelled timber and then glued and screwed in place. If the damage extends further back, the whole tread may need to be replaced.

Old staircases can sometimes be susceptible to attack from beetle or fungal decay where understairs spaces are poorly ventilated and there's a damp adjoining wall or floor. Evidence of old beetle attack is fairly common and is not normally serious. If it is still active, with signs of wood dust around the bore holes, it will require treating – see chapter 6.

plain square section 'sticks' used as balusters on secondary flights to basements etc. But from mid century, steam-powered woodcutting and moulding machinery made it possible for builders to buy large bundles of turned balusters at much lower prices.

Staircases were generally well built from good-quality timber and formed, as they still are today, from four basic elements: treads, risers, strings and carriages. The steps (treads and risers) slotted at each side into rebates cut into the strings (the long lengths of timber running top to bottom); a close fit was ensured by driving in thin hardwood wedges at the joints with the strings, and small triangular blocks were glued under the treads at the joints with the tops of the risers. Finally, additional support was provided by two or three thick lengths of timber running along the underside known as carriages.

The cheapest form of handrail was the 'mopstick' of plain deal. Most, however, were of polished hardwood (usually mahogany) and oval or 'toad's back' in section. Some had a metal core concealing bolted joints.

The newel posts and balusters were normally of solid wood, but were sometimes of cast iron. These could be of quite complex designs, scrolled and twisted into floral and Gothic shapes or carved into intricate fretwork. Occasionally, matching painted iron balusters were inserted amongst wooden spindles for added strength. Towards the end of the century simple plain 'stick' balusters became more fashionable.

The main staircase normally had just one straight flight, often with a single dog-leg turn usually at the top with radiating stair treads ('winders') or a small landing. In earlier houses of more than two storeys, simple narrow staircases would lead to the upper floors and attics where the servants' bedrooms were usually separated from the landings by thin pine-boarded partition walls.

## Defect: Wobbly banisters/balustrades

**SYMPTOMS**
Loose handrails and spindles.

**Cause** **Loose joints**

*eg* between newel posts and handrails, or loose balusters.

**Solution** Handrails are joined to newel posts with mortise and tenon joints, which can be dismantled and re-glued. If the joints between different lengths of handrail are loose, look on the underside of the rail for a small timber 'plug', which conceals a bolt or screw that can be tightened to pull the rails together.

If balusters are loose, the simplest repair is to re-glue the loose joints and diagonally screw through the end into the

tread or handrail. It can be impossible to obtain a perfect matching for Victorian balusters off the peg, but exact replicas can be custom made.

## Defect: Loose treads and risers

**SYMPTOMS**
Squeaks, creaks and soft treads.

**Cause** **Loose fixing blocks under treads or loose wedges**

Because the timber has shrunk or the old glue has weakened, some of the joints between stair components may have opened.

**Solution** *Securely refix loose blocks and wedges with glue*
Check how treads are supported and refix with timber packing if necessary. Where stairs are plastered underneath, the plaster will need to be removed to gain access. To avoid this, it may be sufficient to simply refix the little triangular blocks under the nosings (the small projecting front leading edge of each tread overhanging the riser). Worn nosings can be cut or prised off the front of the tread and replaced in matching new timber, glued and screwed into place.

## Defect: Gaps to the walls, movement, sagging stairs

**SYMPTOMS**
Unstable stairs, loose newel posts.

**Cause** **Inadequate support to staircase. Staircase has come loose from wall**

**Solution** *Check how staircase is supported*
A failing joint between a string and a newel post at the top or bottom of the flight can cause the staircase to sag.
The main string may be inadequately fixed to the wall. Refix with plugs and screws or use masonry bolts.

**Cause** **Structural movement to the wall (bulging, leaning etc.)**

**Solution** *See chapter 5. Repair wall and once stabilised refix staircase.*

**Cause** **Dampness leading to rot or woodworm**

**Solution** *Identify and remedy the cause of damp and treat affected timbers supporting staircase*
Check for rot, especially at the base of the newel post at ground floor level, and for woodworm under the stairs. Cut out and renew affected timber and treat remaining timbers. See chapter 6.

Classic stripped 4 panel door, but originally they would normally be painted (bottom right)

High quality hardwood doors might be varnished

Kitchen door with coloured glass and brass kickplate

# Internal doors

The Victorian panelled pine door is a genuine design classic and was almost universally used although traditional cheaper 'ledge and brace' cottage doors were sometimes fitted to kitchens and sculleries, and were common in rural properties.

The six-panel doors popular in the early Victorian period were later superseded by the classic four-panel – always with taller upper panels and shorter lower ones. Panelled doors were made from a solid frame of vertical stiles and horizontal rails (typically about 32mm thick) with thin solid timber infill panels (about 8mm thick) enclosed by decorative mouldings (omitted on doors to the less important rooms). Latches were sometimes mortised into the outer edge, but to avoid weakening the door, surface-mounted rim sash locks were commonly fitted.

From the 1850s and '60s doors in the hall and kitchen might incorporate decorative coloured glass panels. Bathroom doors (if there was a bathroom) typically featured twin glazed panes of opaque or etched glass. Double or folding doors were sometimes used to separate the main front and rear reception rooms.

Softwood doors were invariably painted, either in imitation wood grain or in dark colours, some with highlighted mouldings. If you plan to strip the doors, cold dipping is preferable to the hot-dip process, which can loosen joints and split panels (flame guns can leave ugly burn marks).

Common issues are similar to those described for external doors in chapter 7, *ie* warping, split panels and doors sticking in frames. Serious defects are rare, as the doors are not usually exposed to damp. Panels can normally be removed quite easily by prising off the beading or moulding that holds them in place and replacing the damaged panel.

If the bottom of an old door has been cut too short, perhaps to accommodate old carpets since removed, a matching piece of wood can be cut to size and spliced in using wood filler to mask the joint.

Doors that stick can be remedied by adjusting the door stops or removing the door and planing, unless the problem is due to deflection in the frame caused by structural movement in the wall. See chapter 9.

If the hinges or the inner edge of the

Traditional 'ledge' door in rural Victorian cottage

Flame guns can leave scorch marks and are not ideal for stripping paint

Original doors can be upgraded to fire resistant with special intumescent coatings

door bind on the frame, this can often be solved by sanding the corners to a slightly bevelled finish. Hinges should be positioned about 150mm from the top and about 200mm from the bottom. The screws shouldn't all be driven home until the door closes correctly.

In properties with recent loft conversions, the original internal doors may have been replaced to comply with fire regulations. If so, they may have been stashed away in an outbuilding or basement. Sometimes such sacrifices are made unnecessarily, because it's normally possible to upgrade the fire resistance of original doors using special paint or boarding and fitting intumescent seals to door linings.

Original reclaimed doors are still widely available, as are modern reproductions. However, selecting replacement doors

Over-clad door panels improve fire resistance whilst retaining original timberwork

can be a little tricky, as old openings may be of non-standard sizes and doors were sometimes purpose-made on site.

A rough guide to typical original Victorian door widths is:

Landing cupboard          610mm (24in)
Bedroom doors   686mm (27in)
Ground floor doors          762mm (30in)

Heights varied considerably but 1,981mm (78in) was fairly common. The most common metric sizes today are 724 x 1,968mm (28.5 x 77.5in) and 813 x 2,032mm (32 x 80in).

A joiner can make new custom doors, at a price.

Replacing internal doors in older houses is not always as easy as it looks, because over the years the frames may have adjusted to settlement in the building, for example there may be a slight bow at the head of the doorframe. This can be planed, or the top of the door trimmed. Holding the new door up to the frame allows you to pencil around the perimeter so any undulations in the frame can be allowed for by planing the door (working inwards from the corners, to prevent splitting). The gap at the bottom should be about 3mm above the floor covering.

Door linings were traditionally secured to the surrounding masonry or timber studwork by nails or screws in pairs to prevent twisting. To achieve a straight edge, thin packing strips would be placed behind the lining timbers. Once fitted and the plastering completed, the door would be hung and finally the gap between the wall and the door frame concealed with moulded ogee style architraves mitred

Out of true door frame caused by old settlement to building - accommodated by trimming door and building up door linings

and nailed over the joints. Fitting a new door lining is rarely required unless you're knocking through a new opening (taking care not to damage the surrounding old plasterwork).

## DOOR FURNITURE

Handles always took the form of door knobs (never modern lever handles) made from plain brass, turned wood or white porcelain, and later cast iron. The finely grooved brass design (sometimes called a 'beehive') is particularly suitable for houses of the1880s and '90s.

Beehive door knob on rim lock

Fingerplates were sometimes fitted above doorknobs to keep dirty finger marks at bay. These later became unfashionable and are often missing. However, a closer inspection may reveal old screw holes or impressions

in paintwork where such items of original door furniture have since been removed, providing sufficient clues to enable matching replicas to be sourced.

Similarly, it should be possible to repair and retain pleasingly stout old door locks, for which new keys can be made by locksmiths. Reproduction brass surface-mounted rim sash locks and ironmongery are widely available, or salvaged originals can be acquired for a more authentic feel. It's not unusual for old hinges to have become worn or damaged, but rather than simply replacing them it can be a better option to leave them in place and supplement them with additional new matching ones.

# Other joinery

## SKIRTING

Early Victorian skirting boards tend to be deeper, becoming more stunted in stature by the 1870s with the fashion for wall surfaces divided into three zones – the lower area under the dado, the central 'filling' and the uppermost frieze above the picture rail and ceiling.

Skirting boards were either made from timber mouldings or constructed from beading applied to plain boards so it could be bent to form curved shapes, *eg* at the top and bottom of the

*Painted marble effect. (Photo: RBKC Linley Sambourne House)*

stairs. As with architraves, matching original skirting exactly is often impossible today due to the infinite number of local design variations of the basic 'torus' pattern, some as high as 350mm. The skirting was fixed into wooden plugs set into masonry walls, or nailed to studwork.

Draughty gaps are commonly found at the base of skirting boards. These can be sealed with rubber strips (see page 246) or timber beading applied over the floor/skirting junction for a neat cosmetic finish. But be aware that large gaps may sometimes be an indication of problems with the floor itself – irregular dips of more than about 14mm in the floor may indicate settlement.

Defects are usually limited to localised outbreaks of rot or beetle resulting from dampness near outer walls, or poor DIY alterations – look for signs of filler at joints.

### SHUTTERS

The Georgian fashion for internal window shutters persisted to varying degrees through to the Edwardian era. Shutters were generally made from high-quality imported softwoods although in more expensive houses oak or mahogany would be the material of choice. The standard format comprised panelled leaves hinged

to the window jambs, designed to fold neatly into timber boxes at the sides of windows. An alternative Victorian system comprised vertically sliding wooden sash boards. These were concealed within panelling beneath the windows and could be partially raised to provide security whilst admitting a certain amount of light.

Being remarkably unobtrusive these may appear to be an integral part of the wall panelling, particularly when disguised by numerous layers of paint, hence many householders today are unaware that their house has shutters. If you're lucky enough to still retain the originals, they're well worth reinstating to their former glory as they're very effective at keeping in the heat as well as deterring burglars.

## PANELLING AND OTHER JOINERY

There was a clear hierarchy of wall finishes for the Victorians. One up from cheap painted brickwork was the distemper coated plaster found in utilitarian rooms like sculleries. Wallpapered plasterwork was better regarded and was the most common wall covering. Superior to this was timber wainscoting which was widely adopted in larger Victorian houses to provide a partially panelled wall finish to heavily used areas such as hallways and staircases. The lower wall was lined with continuous wooden

Plate rack and frieze

panelling inserted between an upper 'chair rail' and the skirting, the walls above being plastered and wallpapered.

Wainscoting was commonly made from softwood and painted or given a faux woodgrain appearance in imitation of more expensive hardwood.

Because traditional wood panelling was too expensive for most mainstream housing, cheaper 'matchboarding' was applied to the lower walls of many Victorian hallways and sculleries. This comprised vertical-planed tongue-and-groove softwood boards, similar to that used for flooring, topped with a horizontal run of moulded dado beading

Where the walls were not panelled, horizontal dado rails became very popular from the 1870s to around 1900. Dados were most commonly fitted in hallways (to protect wallpaper from damage caused by people rushing through) and in dining rooms (to protect expensive wall hangings), the height of the rail being determined by

the point at which the chair-backs would brush against the wall. Dado rails on staircase walls would match the height of the banister handrail opposite. Combined with picture rails running above door level, as noted earlier, the walls were divided into three decorative zones, the uppermost 'frieze' adjoining the ceiling. In some houses a wooden shelf about 80mm (3in) wide – known as a 'plate rail' – was installed, supported on shaped brackets and used to display ornaments.

From a practical perspective today, any wood adjacent to the main walls, such as matchboarding, timber panelling, or shutter boxes, is potentially at risk where walls are damp and there's a lack of ventilation. Timber fittings in basements must always be suspect for the same reasons.

When exploring Victorian houses you occasionally encounter other intriguing pieces of defunct joinery and relics of times past. For example in homes with servants, which included many quite small houses, battery powered bells in wooden boxes were introduced from the 1870s, replacing the complicated wire-pull systems, which you may still encounter when lifting floorboards.

Built-in cupboards were a feature of bedrooms and reception rooms, as well as kitchens and sculleries, making use of the alcove space to either side of the chimney breasts.

# Kitchens and sculleries

In the poorest homes separate kitchens were unknown, since only the living room would have been heated with an open fire, which was also used for all the cooking. But most houses had a kitchen and a scullery to the rear addition. This was the domain of the 'staff' where much of the hard work, cooking and cleaning was carried out. In case of leaks or flooding, floors were of solid construction and built at a lower level with the characteristic step down from the adjoining hallway.

Kitchen furnishings were functional and quite sparse, mainly comprising a scrubbed wooden table on which food was prepared in the centre of the room. Storage was provided by built-in cupboards and sometimes a separate larder. Large freestanding Welsh dressers were used for storing the plates, jugs and cooking utensils. And of course even the most humble terraced houses would have a cast-iron coal-fired range for cooking – see chapter 13.

intensive, also requiring a 'dolly tub' for soaking dirty clothes overnight, a wooden tub for scrubbing, and a heavy cast-iron mangle to squeeze out water.

Modern kitchens bear little resemblance to the rooms familiar to the original occupants, most having long since been 'knocked through' to embrace the scullery into one larger room. By now many will have second- or third-generation fitted units installed. But modern flat-packed kits are not always of the highest quality. Common defects include poor joints and upstands to worktops, loose hinges and fittings, hidden leakage or damp from condensation behind units, and poor DIY workmanship generally. It is often simpler to replace complete units with standard-size new ones than to contemplate lots of minor repairs.

**Top:** Back to basics – table, range and water pump.
**Above:** Every home had a 'copper' for washing clothes

In more expensive properties, the butler had his pantry where the crockery, silver, and wines and spirits were stored, and the housekeeper had a substantial store cupboard, or larder, for food, in which a constant war against vermin was waged. Flour and grain were kept in containers hanging from shelves; sugar loaf was wrapped in paper and hung from a hook; and tea was kept in a lead-lined chest. The pantry and the larder would be located to the rear, off the scullery, but in smaller houses these were not much more than large cupboards.

The scullery was used for preparing vegetables and meat, washing up after the meal, and for the family laundry. The typical 'Belfast' sink was white or brown glazed ceramic stoneware or fireclay with a single cold tap and a draining board. A large metal tub called a 'copper' was used for washing clothes and heating large quantities of water; in smaller houses this took the form of a 'set pot' – a brick-enclosed fireplace where the copper could be lifted out. Doing the laundry was enormously labour-

# Decoration

The Victorians liked their interiors to look more expensive than they really were – embossed wallpaper dressed up as plaster or wood, cheaper grained wood masquerading as expensive timber, and slate or cast iron pretending to be marble.

As noted earlier, the timber in Victorian interiors was always painted, unless made from expensive hardwoods such as oak or mahogany. Dado and picture rails, doors and architraves would be grained in a dark or medium tone or else painted dull green, brown or pale cream. Today's fashion for all-white joinery started from around the late 1880s with the introduction of semi-glossy cream-coloured enamel paint. Skirting was similarly painted dark colours like green or brown, or else varnished or marbled. Highly skilled 'graining' could make cheap timber look like expensive wood by imitating its grain using steel combs drawn through paint to achieve the desired effect, then sponging down with a mixture of beer and whitening (distemper) before varnishing. A recently revived variant of graining is 'marbling', used to imitate marble on such areas as wainscoting and skirting.

Wallpaper is the most important element in the decoration of Victorian and Edwardian rooms. Cheap enough for universal use and available in an enormous variety of styles and patterns, many reproductions of these papers are available today.

Ceilings and the frieze area above the picture rail were lined with embossed paper such as anaglypta that mimicked the look of expensive plasterwork, painted either off-white or a tint of the wall colour. Stencilling became very fashionable. But in dining rooms – the male domain – colours remained sombre and rich, such as dark maroon and sage green.

The later Arts and Crafts movement pointed the way to lighter, brighter interiors. The trend for less clutter and more simplicity continued into the Edwardian era with plainer

**Above:** Layers of historic paint stripped from staircase matchboarding – but beware lead!
**Below:** Authentic colour scheme adds period charm

patterned wallpapers or painted plaster walls. Dados were dispensed with, except for halls, stairs and landings although picture rails soldiered on, albeit at a reduced height.

In contrast, the walls to kitchens and sculleries were of functional whitewashed or distempered plaster. The lower walls may have had vertical matchboarding painted with washable gloss paint, or later, hardwearing hygienic ceramic tiles or glazed brick.

As well as adding considerable aesthetic charm, glazed tiles served a practical purpose solving the problem of clothing getting marked from brushing against chalky, whitewashed surfaces. Tiles were even manufactured in the form of glazed dado rails. Some entrance halls had tiled or glazed brick finishes for easy 'wipe down' maintenance, green being a popular colour. Lavishly tiled porches and fireplaces became increasingly fashionable even in artisans' dwellings. Bathroom tiles were usually white or blue, around a tiled dado, with white painted ceilings. But because tiles are made of a rigid material, they can sometimes develop hairline cracks as old buildings tend to move slightly. Fortunately, there are many good reproductions of old pattern tiles now available, and using a mildly flexible adhesive and grout should help accommodate any minor movement in future.

### DISTEMPER AND LIMEWASH

Distempers were a popular type of whitewash used for internal decoration until eventually superseded in the 1960s by modern vinyl and acrylic emulsions. Soft distemper was cheap and easy to make from powdered chalk or lime, bound with an animal glue size or casein (a resin with good adhesive qualities derived from solidified milk– hence the traditional name 'milk paints'). This produced a thin, inexpensive paint that dried quickly to a gentle matt white finish known for retaining its brightness on dry surfaces. It could be tinted in a wide range of pastel colours, but was most often white.

Soft distemper was widely used for ceilings and decorative plasterwork. Ceilings required regular redecoration because of the extensive amount of smoke from sooty candles and open fires, plus of course nicotine staining from pipes, cigars and cigarettes. But instead of just slapping a new coat of paint on top of the existing layers, the dirty coat was washed off with warm water

## What's in lead paint?

Lead was the key ingredient in traditional oil-based paints until it was finally banned in Britain in 1992 (it was banned in the US in 1978). Lead paint normally comprised 'white lead' (lead carbonate) to provide body and a white pigment, mixed with a binder of boiled linseed oil to make it stick, plus thickening agents like wax or soap. The higher the lead content, the more durable and also the more expensive the paint.. It was also coloured with various pigments and dyes, sometimes including small amounts of toxins such as arsenic (greens and whites), cadmium (yellow), cyanide (blues) and cinnabar (the parent ore of mercury, used for reds).

Linseed oil (squeezed from the seeds of the flax plant) was an essential ingredient because it hardened into a tough film. Its one dramatic downside was that it's extremely combustible – a single pot of linseed oil could, in the presence of an open flame such as a candle, ignite spontaneously – and was the source of many devastating house fires.

Oil paints were originally pretty glutinous and difficult to use, a bit like spreading tar with a broom. Only the discovery of turpentine, a natural thinner distilled from the sap of pine trees, made the paint easier

to apply, and painting became smoother. Adding turps (turpentine) also gave the paint a matt finish.

Painting was especially skilful because, until the first ready-mixed paints appeared in the mid to late Victorian period, painters had to create their own colours by grinding their own pigments. Paint had to be mixed in small portions and used at once, so the ability to make matching batches from day to day was a real skill. They also had to apply several coats, since even the best paints had little opacity.

Blues and yellows were two or three times more expensive than duller colours like off-white and stone, so they tended to be used only by the wealthier classes. Most expensive of all was bluish-green verdigris, made by hanging copper strips over a vat of horse dung and vinegar and scraping off the oxidised copper. Rich colours generally denoted expense, since you needed a lot of pigment to make them.

However, there were two very basic colours that didn't exist until well into the 19th century. The brightest available white was a dull off-white, and it wasn't until titanium oxide began to be added to paints in the 1940s that really intense, long-lasting whites became available. Also missing was a strong black. Permanent black, distilled from tar and pitch, wasn't popularly available until the late Victorian period, so glossy black front doors, railings, gates, lampposts, gutters, downpipes and other fittings that are such an element in many restored properties today are actually quite a recent phenomenon. In Dickens' time almost all ironwork was green, light blue or dull grey.

before each redecoration. This had the benefit of preventing a build up of thick paint over many years obscuring the detail on intricate mouldings. Unlike modern emulsions it did not trap moisture within walls, but allowed the structure to breathe.

Limewash was a solution of powdered lime and water with similar properties but was more widely used externally, and is available from specialist suppliers today. 'Brilliant White' paint was not available until the 1950s, so an off-white or cream colour will give a more traditional appearance. But when it comes to redecoration, whatever colour is used, a matt or silk finish will appear more authentic than modern high-gloss finishes.

## PAINT REMOVAL

Because the quality of Victorian joinery was generally very high, stripping it can be a good way to show off the original workmanship. Removing coatings of modern paint and exposing original hardwood can bring back to life beautiful features that have been hidden for generations. But when stripping old oil paint some simple precautions must be taken as there's a potential health risk if lead is ingested, absorbed through the skin or inhaled.

Abrasive methods such as sanding with power tools and wire brushes should be avoided because they release toxic dust into the air, as well as potentially damaging the surfaces being stripped. Rubbing down by hand is preferable but you still need to wear a mask or respirator and keep children well away. Heat applied using hot air guns softens paint so it can be scraped off. But these can generate fumes that may carry lead particles. Great care is also needed to avoid scorching timber, leaving unsightly black marks, and also to avoid cracking adjacent glass in doors and windows. Naked flame blow torches should never be used because of the fire risk.

The best advice is to use a poultice type paint stripper.

Solvent-based paint strippers are very effective for removing layers of modern paint. These can either be painted-on or applied as a thick poultice. Older oil paints can be dissolved using alkaline-based removers. Water-based, solvent-free 'eco strippers' are non-toxic and are

Victorians wouldn't recognise 21st century makeovers

## Poisonous wallpaper

Wallpaper became popular in the Georgian period in more expensive houses, only becoming affordable for use in ordinary homes from around the 1840s. Prior to the Victorian era it had been very expensive, thanks in part to the Wallpaper Tax (introduced in 1712 and not lifted until 1830). It was also very labour intensive to make, especially flock wallpaper, which was made from dyed stubbles of wool stuck by hand to the surface.

But the thing that's truly shocking today is the toxicity of some pigments used to create desirable colours. Large doses of arsenic, lead and antimony were used in the manufacturing process. After 1755 emerald green became a hugely desirable colour, created by soaking the material in an especially insidious compound called copper arsenate (mixed with copper acetate). Rich arsenic greens were enormously popular, and by the late 19th century it's estimated that 80% of British wallpapers contained arsenic, often in very significant quantities. When damp was present it's said the wallpaper exuded a peculiar musty smell reminiscent of garlic, and slow arsenic poisoning from this source was not unknown. People noticed that bedrooms with green wallpapers were remarkably free of bedbugs.

So if you plan to strip or salvage historic wallpaper today the best advice is to take suitable precautions. Above all avoid licking your fingers!

Embossed 'relief' wallpaper c 1870s

RBKC Linley Sambourne House

also simply brushed on. As a precaution, where lead paint is being removed, wash down the stripped surfaces with detergent after completing the work, vacuum the area and bag up all clothes and wash them separately from other clothes.

When it comes to stripping doors, it's common to send them (and other detachable items of internal joinery) to be stripped in caustic chemical baths. But there are potential risks particularly with hot dipping, such as the timber becoming either bleached or darkened. If the chemicals dissolve animal glue used in joints, it can allow parts to work loose, or may lead to shrinkage and distortion.

Scraping off innumerable layers of wallpaper using a steamer is part of the renovation experience familiar to many – the steam penetrates the paper and softens the adhesive, as described in chapter 9. But be wary of disturbing old lining paper covering lath-and-plaster walls and ceilings. Steam wallpaper strippers can weaken lining paper that's helping to bond old plaster. If stripping is essential, a simple hand scraper tool is usually sufficient to get the emulsion to part company from underlying original distemper, but take care not to gouge the plaster.

## TECHNICAL DATA
To find out about original wallpapers and decor see www.bricksandbrass.co.uk and www.periodproperty.co.uk.

# THE SERVICES, BATHROOMS, PLUMBING AND HEATING

Most of the services and creature comforts that we now take for granted in the home simply did not exist when your property was built. Nonetheless, there are some important issues with the services that current owners need to be aware of, such as lead piping, hidden cable runs and inaccessible drains. Homes of this period can also contain intriguing remnants of long-defunct fittings, such as old gas pipes and networks of wires that were once connected to servants' bells, offering a fascinating glimpse into lifestyles from bygone eras.

Catchpole & Rye

The legendary Thomas Crapper and his fellow Victorian toilet pioneers bequeathed a fine inheritance to the world. After all, where would we be today without the invention of the 'pull and let go' flushing water closet? The development of services to residential properties during the 19<sup>th</sup> century progressed very rapidly, and by the later Victorian period significant numbers of homes were supplied with piped water and connected to a public sewer. Gas became increasingly popular from the 1890s for lighting, although electricity remained an elusive concept for all but the wealthiest households.

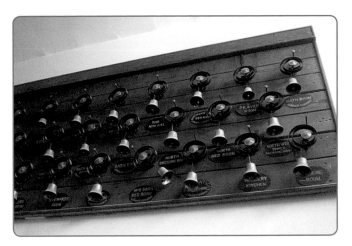

# Electricity

Electricity for domestic use was slowly introduced from around 1900, but the vast majority of houses remained without electricity until the 1920s. Early cables were insulated in rubber, sheathed in cotton braid and run in wooden trunking or lead tubing. Socket outlets had two round holes. If you come across old, redundant fittings it's worth bearing in mind that they're part of your property's history. Although not strictly Victorian, and certainly not compliant with current safety standards, once disconnected, antique switches and fittings should if possible be left on display as interesting features.

Modern plastic-insulated cable and 13-amp square-pin plug sockets appeared in the 1950s, but earthing was only introduced from the mid-1960s. Modern white or grey PVC cabling has the live wire coloured brown and the neutral wire coloured blue (respectively red and black in older cables). The bare earth wire needs to be covered with yellow and green striped plastic sleeving when connected.

The main supply entering the house terminates in a fuse near the meter, which belongs to the electricity company, and should not be tampered with – everything else is yours. Another pair of cables then runs from the meter to your consumer unit (fuse box), often located at high level in the hall or under the stairs, or else in an outside box. Most supplies come in underground, but some are from overhead cables and are connected to the house via insulated terminals at high level.

If the fuse box is made of metal, wood or bakelite (hard dark plastic), or has old rewireable fuses, it means the system is very dated and will need replacing. Modern consumer units have MCBs (miniature circuit breakers) for each individual circuit. These are special fuses that automatically switch off when they sense a fault or overload, thereby potentially saving lives (a major improvement on old rewireable systems where you had to try and replace fusewire, often in the dark). Modern 'split load' consumer units provide additional RCD (residual current device) protection for the more vulnerable circuits (showers, outdoor sockets, external lighting, etc).

When refurbishing a property, constructing a new false ceiling below the existing one can be a handy way to conceal lots of new cabling as well as improving fire and sound insulation.

**Above:** Rewirable fuses – now obsolete.
**Below:** early 20th century fusebox – still in use!
**Left:** Edwardian light switch switches.

# Electric circuits

Power circuits supply the socket outlets in the walls and are arranged in a loop or 'ring main' normally with just one or two circuits per floor. These are usually run in thick 2.5mm$^2$ 'twin & earth' plastic covered cable (usually white or grey), which connects from the 32-amp MCB (or 30-amp fuse) in the consumer unit, looping round all the 13-amp socket outlets on one floor in series and then back to the MCB.

The kitchen, being a high-demand area, has its own circuit for all the wall sockets. In addition, electric cookers and immersion heaters have separate 45-amp and 20-amp MCBs respectively but use 'radial circuits' – where a single cable is run from the MCB direct to the unit. You need one circuit for each 100m$^2$ floor area (an average terraced house has about 50m$^2$ area per floor). Cables run to outdoors should have separate RCD protected circuits.

Lighting circuits use thinner 1.5mm$^2$ 'twin & earth' cable which connect from the 6-amp MCB (or 5-amp fuse) in the consumer unit, running to each light in turn, usually at least one per floor. Each room switch is normally wired directly to the room's ceiling rose via a single cable. To get access to the cables you normally need to work from the floor or loft space above.

At least a thousand house fires each year are due to electrical faults

Similarly, dry-lining the main walls with insulated plasterboard can create useful ducts for cables and wall lights while simultaneously boosting thermal insulation. Otherwise to avoid damage to the fabric of the building, cables can be surface-run and contained in neat plastic conduit. If there's no other option, holes through floor joists should be drilled approximately half way down, and located no closer together than 3 x diameter of the hole.

## HOW MANY SOCKETS?

Depending on room size, a modern household needs about three or four double-switched socket outlets (DSSOs) for each bedroom, five or six each for kitchens and living rooms, and a couple for halls and landings. It is normally possible to add an extra socket by extending from one of the existing sockets on the ring using suitable cable. This is called a 'spur' – but no more than one spur is normally permitted from any socket.

## DANGER SIGNS

Electricity is the biggest killer in the house. At least a thousand fires and more than 50 deaths each year are due to electrical faults. Wiring usually lasts only about 35 years before starting to deteriorate and should be tested every ten years. Most houses have electrical systems that are unsatisfactory in some way, either due to badly executed alterations or just

Farewell cruel world! Criminally incompetent wiring

Mark Harding

through sheer age. Be concerned if you see any of these: old fuseboxes with rewireable fuses, roundpin plugs, radial circuits to sockets, rubber-sheathed cable, frayed flex, brittle PVC cable, or burn marks. Also check for DIY surface-mounted sockets on skirtings, unprotected surface-run cables, power points in damp walls, insufficient numbers of sockets (evident from lots of multi-outlet plug adaptors), and any switches and sockets positioned close to sinks, basins and showers, or with loose covers.

Cables should be run in protective plastic trunking, or, when laid within plastered walls, in oval plastic conduit tubing or behind a flat steel shield. Any supplies run outdoors should be of exterior quality cable and run in special protective conduits. Contact with polystyrene insulation (eg in lofts) should be avoided, as it can react with and soften the PVC sleeving.

Electricity will always head for earth, taking the easiest route (along its circuit). But in the event of a fault, leaking current could pass to earth through your body and down through the building, particularly via anything wet or metallic. So to protect against electric shock, all metal components should be connected with an earth wire (green and yellow sleeved) so that they can't retain any current with potential for dangerous shocks. The requirement is to bond metal items such as incoming service pipes (eg mains water, gas, oil), metallic waste pipes, central heating pipes, hot and cold water pipes and metal baths etc. Pipework run completely in plastic shouldn't normally need to be bonded. Bathrooms and kitchens are high risk areas, so nothing electric should be touchable where a person could simultaneously be in contact with water.

You can buy 'plug testers' which plug into sockets and have indicator lights that may help identify some deficiencies. But if the consumer unit doesn't have a sticker showing when your system was last tested, it means it should now be checked by a qualified electrician, as should any non-professional electrical alterations.

A 'Duplex burner' double wick paraffin lamp, common in the late century. The pulley allowed the lamp to be pulled down low over a table to provide a bright pool of light, or raised to illuminate the whole room.

## COMMON PROBLEMS

Before tackling any electrical work remember to switch the power off at the consumer unit first! Where there is no power to some lights or sockets, check the MCB at the consumer unit. These can be very sensitive to a lightbulb blowing and you may simply need to flick the MCB switch to restore power. Also, it's not unknown for individual MCB s to become faulty and need replacing. Check that all fuses are of the correct capacity and that all connections are secure.

More serious symptoms such as fuses that keep blowing, sparks at switches, scorch marks around plug sockets, or plug flexes getting hot may simply be due to the switch, plug, or socket itself being faulty, perhaps with a loose cable/connector inside the socket or the appliance has been incorrectly installed. Overheating may be caused by cables being overloaded with too many 'spurs' (socket extensions), so the cable is too thin for the load, or damaged cables (eg from rodent attack) in which case the defective length of cabling must be replaced, or in severe cases a new ring main fitted. Power cables buried in thick layers of insulation can also be prone to overheating. Check there are separate circuits for lights and power on each floor, and for the kitchen, the cooker and any electric heaters etc.

## LIGHTING

Early Victorian houses were lit by nothing more than candles and the light of the fire. Then came oil lamps, which were marginally brighter than candles. Later paraffin lamps and plain gas jets produced a better light than oil flames, but not dramatically so. Gas lighting was not common in urban homes until quite late in the century. The reason for this lack of popularity wasn't simply down to the fact that early 'fish-tail' jets didn't produce a very good light; they also had the drawback of generating a great deal of heat and emitting fumes, which killed off plants (though not the popular *aspidistra*). Various attempts

Gas lights in kitchen later converted to electric

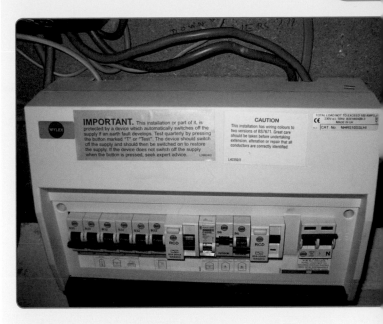

# Electrics and the Building Regulations

All new electrical work must comply with Part P of the Building Regulations, which restricts DIY electrical work on grounds of safety. However, you are still allowed to carry out some work yourself without notifying Building Control. Minor repairs and maintenance are permitted, as well as 'like for like' replacements, such as changing existing sockets, switches and ceiling pendants or even replacing damaged cables. As long as the job isn't within a 'special location' such as a kitchen, bathroom or outdoors, you're also allowed to install additional new light fittings, switches and even add a single fused spur to an existing circuit (a 'spur' is a new cable and socket run as a branch from an existing socket on the ring main). New power sockets must be positioned no lower than 450mm above the floor, and light switches no higher than 1200mm from the floor.

In bathrooms only special low-voltage safety fittings are allowed and no power sockets are permitted. All electrical circuits within bathrooms must be protected by Residual Current Devices (RCD) not exceeding 30mA and there are rules that limit fittings to defined safety zones within the room. Light switches should be of the pull-cord type or else located on the wall outside the bathroom.

All major electrical work, such as installing complete new circuits or changing a fuseboard for a consumer unit, is classed as 'notifiable work'. This requires a Building Regulations application to be made in advance so the work can be inspected and checked. However, in most cases Building Control won't need to be involved because electricians can normally self-certify their work if they're registered with a body that gives them the necessary 'registered installer' status (also known as '*competent persons*') such as the **ECA** (Electrical Contractors Association) or **NICEIC** (National Inspection Council for Electrical Installation Contracting).

Upon completion of the job, it is a legal requirement for the electrician to test the new system and hand over a signed BS 7671 electrical safety certificate. In addition, you should be sent a Building Regulations compliance certificate for all notifiable work by the operator of the registration scheme.

were made to overcome these problems, and as early as the 1860s some fashionable townhouses had large ceiling pendant light fittings known as 'gasoliers' which incorporated several burners to generate sufficient light. These gas-powered chandeliers would hang centrally from the ceiling in the main reception rooms. Ceiling pendants were often of a sliding type that could be lowered when lit to make the most of the limited light. To help disperse the smoke and fumes rising up from the burners a 'smut catcher' ventilation grille would be concealed in the decorative ceiling rose. Fresh air inlets were sometimes also provided in the corners of ceilings connected to vents on the outer walls. Where vents were fitted to walls the indoor faces were sometimes covered with silk stretched over them to prevent dust getting blown into the room.

Gas 'wall brackets' were a more popular form of lighting. These were commonly built into chimney breasts in place of traditional candle-holding sconces. Because wall brackets were sited at a lower level than ceiling pendants the flames could throw a better light. You may still come across the original gas pipes as thin as 6mm embedded within the plaster, often found at eye level on either side of the fireplace. The brass or iron supply tubes had a tap for switching the gas on and off and the burner was shielded by a glass globe or shade to diffuse the light. More exotically, some staircases were lit by lights attached to newel posts. Where such curiosities survive it may be worth exploring the possibility of making them safe and reusing them.

After the introduction of penny slot meters, gas lighting became enormously popular, with about 2 million gas consumers in Britain by the time the incandescent gas mantle was developed in 1887. The gas mantle transformed lighting, providing the first source of really bright light, while consuming a quarter of the amount of gas previously required. Mantles were woven from cotton and impregnated with chemical salts that glowed white hot but gave out a greenish light. The actual flame needed was much smaller than for old fish-tail jets, producing less heat and fumes. They could even be suspended upside down for more efficient ceiling pendants.

Today, when fitting new lighting to a Victorian house bear in mind that transformers for low-voltage lighting give off heat and can potentially present a fire hazard in confined spaces where there may be accumulated dust or old dry timber. LED lights are preferable as they consume less energy and run at lower temperatures. Also try to avoid fitting recessed downlighters in old lath-and-plaster ceilings, because even cutting small holes will weaken them, as well as breaching the building's fire resistance. Downlighters are ideally suited to modern plasterboard ceilings (such as in kitchen extensions). If all else fails it might be possible to construct a suitable suspended ceiling below the original.

# Gas

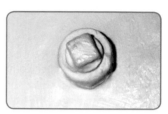

Although gas lighting had become popular towards the end of the Victorian era, gas didn't become established in homes for cooking or heating until well into the 20th century. As well as a widespread fear of explosions, customers were at the mercy of unpredictable supplies. Before the introduction of gas meters homes were subject to pressure periodically dropping to unusable levels causing flickering lights – or rooms could fill with dangerous levels of gas if flames cut out and the supply later recovered. Gas companies also had a habit of turning supplies on at dusk and off again before midnight.

Today, many Victorian homes will have original mains supply pipes of cast iron. But iron can rust or clog up, causing poor pressure, so in recent years suppliers have undertaken a major programme of replacement with modern yellow polyethylene plastic underground pipes, which should normally be trouble free. The gas supplier should replace the pipework free of charge up to the point where the incoming mains service pipe terminates at the stop valve by your gas meter. Meters are often tucked away in understairs cupboards (which should be well ventilated and free from damp). A better arrangement is to have a brown coloured external box discreetly installed at ground level – this is preferable to having your home disfigured with a large white plastic meter box stuck on the front wall. From the meter, above-ground supplies are normally run along the walls near ground level. Gas pipework should be run externally (normally in 22mm copper pipe). Any internal piping must be kept more than 25mm away from electrical cables.

Natural gas is not poisonous; the main

## What to do if you smell gas

The first thing to do with a suspected gas leak is to turn off all gas appliances and the supply at the meter. Open the windows and doors to disperse the gas, but do not switch on the lights. Then phone your gas supplier. Sometimes a leak may be traced simply to an unlit pilot light or faulty gas-cooker burner. It is advisable to have all gas appliances serviced annually to prevent any such risks.

New incoming yellow gas supply pipe routed under stairs connecting to 22mm copper pipe to meter.

risk is the potential for explosions. Hence all work on gas appliances and pipework must, by law, only be carried out by qualified Gas Safe registered engineers. However, there is the associated risk that combustion fumes can be poisonous. Which is why any rooms with open-flued gas appliances that take their combustion air from within the room, such as living-flame gas fires, will require additional room ventilation to comply with Building Regulations. This may necessitate cutting 230 x 230mm air vents into the main walls, which must be carefully planned in advance to prevent damage. Gas fires are only suitable for properly vented reception rooms, and not bedrooms etc.

# Water

In early Victorian houses (pre 1850s) drinking water still had to be carried in buckets from a nearby pump or well. Water butts were widely used to collect rain for use in sculleries for laundry and for jugs of water to flush toilets. In larger houses rain was collected via pipes to reservoir tanks in cellars, and then pumped to a loft cisterns for flushing WCs. But pressurised cold-water supplies in cast-iron pipes were widely available for urban housing from the 1860s. Water was normally piped to a single cold tap next to a stoneware sink (white or brown) in the scullery where most of the washing was done.

Pumping water – indoors with stone sink (left) and elaborate design located outside back door (right)

# Cold water systems

As a general rule there are two types of domestic water systems: stored or mains-fed.

### Stored systems

In a stored, or low-pressure system, the cold water coming into the home fills a large storage tank situated as high up as possible, usually in the loft. The water that comes out of the taps is fed from these storage tanks, so turning off the mains cold water at the stopcock doesn't instantly stop water coming out of your taps (which can take up to 30 minutes) – so look for a valve that can turn the water off coming down from the tank.

However, in most houses the cold water to the kitchen sink is supplied direct from the mains while all the hot water is from a stored supply.

### Mains-fed systems

Here all cold water comes direct from the mains at a fairly high pressure. This mains cold supply is also used to push your hot water out of the taps (eg via an unvented hot water cylinder). So turning off the cold-water stopcock should almost immediately stop both the cold and hot water running.

However, supplies were not terribly constant; water companies were criticised for being more concerned with paying shareholders than providing a reliable service! Because of serious concerns that the mains could be contaminated by waste drawn back from appliances, the 1871 Water Act stipulated that WCs must be served only by water stored in cisterns, with 'no pipes direct to WCs, boilers or urinals'. But it was the mass laying of piped water supplies that ultimately made provision of modern bathrooms and WCs feasible.

## PLUMBING AND COLD WATER

Water is piped into the house from below ground level via a stopcock in the pavement or front garden, but terraces often had a shared service with one stopcock for all and a single pipe to each dwelling. As with most services, pipework within your

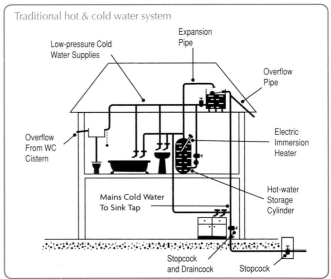

Traditional hot & cold water system

Low-pressure Cold Water Supplies

Expansion Pipe

Overflow Pipe

Overflow From WC Cistern

Electric Immersion Heater

Mains Cold Water To Sink Tap

Hot-water Storage Cylinder

Stopcock and Draincock

Stopcock

Incoming main supply pipe of lead, with typical bulbous joint – pipes should be insulated

boundary is the responsibility of the homeowner. Underground water pipes should be at least 750mm below ground level to avoid damage from frost (although it's not unknown for them to be buried at considerably shallower depths).

Incoming mains water supplies were originally made from cast iron or lead. Cast-iron water pipes can suffer from corrosion, with the rust restricting the flow and turning the water an unpleasant brown colour. Lead pipes are a potential health issue in soft water areas although in hard water areas they will have become lined internally with protective limescale. If you still have original pipes, contact your supplier, as replacement with modern heavy-duty blue polyethylene plastic pipework may be advisable. There should be a stopcock inside the house, typically in the downstairs bathroom or kitchen, for shutting off in an emergency. Supply pipes should be lagged to prevent freezing and bursting, particularly if run under floors, through cellars, or along cold main walls. The mains supply normally runs through the house direct to the kitchen/scullery sink (suitable for drinking water), with the 'rising main' continuing up to supply a cold-water storage tank in the roof space.

## COLD WATER TANKS

Most homes still use traditional systems with a cold water tank in the loft. The purpose of the cold water tank (storage cistern) is to store water for washing and for WCs in case the mains supply is cut off or suffers low pressure. When needed, this stored water flows back down by gravity, so the water pressure at the bathroom taps is governed by the height of the tank (known as 'head'). However, some older houses, particularly those with no

Corroded old galvanised steel tank (left) and an old asbestos cement water tank (right)

upstairs bathroom, have no tank, just a direct mains supply (an arrangement that today has come full circle with new systems now supplied direct from the mains).

Tanks typically store at least 230 litres (50gal) and need to be located high up, usually in the roof space. Like toilet cisterns, overflowing is prevented by a ball valve that automatically shuts off the incoming water supply when the tank is full. As the water level rises, so does the floating plastic ball (or of copper in older tanks), pushing the water inlet valve closed. Just in case this fails, water tanks and older WC cisterns have overflow pipes sticking out through the walls.

Redundant old tanks of riveted galvanised iron or steel or asbestos cement can be too bulky to easily remove, and once empty are often left sitting in lofts, which shouldn't be a problem. Tanks full of water can typically weigh more than 250kg

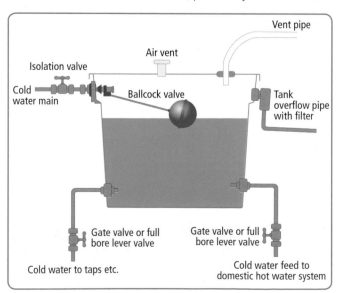

Vent pipe

Air vent

Isolation valve

Cold water main

Ballcock valve

Tank overflow pipe with filter

Gate valve or full bore lever valve

Gate valve or full bore lever valve

Cold water to taps etc.

Cold water feed to domestic hot water system

## Water tanks

In a typical traditional layout there are two water tanks in the loft – a large one feeding most of your cold-water taps plus the hot water cylinder, and a smaller 'feed & expansion' (F&E) tank feeding the CH system. The F&E tank plays a twin role, acting as a reservoir providing water for the boiler/CH system; and giving the water in your CH system somewhere to expand into when it gets hot. A typical CH system containing about 100 litres will expand by about 4 litres when heated, moving up the pipe and into the F&E tank (hence the tank should be half empty when cold so it doesn't overflow)

Points to check with tanks are:

- Is it made of modern plastic (usually black)? If not, it's likely to be of a considerable age and may be overdue for replacement.
- Is it insulated? If not fit an insulation jacket to prevent condensation causing damp.
- Check the ball valve – if it looks a bit worn, consider renewing it
- Is there an isolation valve on the supply pipe leading to the ball valve? If not it might be an idea to fit one in case of emergencies.
- Does the overflow pipe look in good condition, or is it sagging or blocked?
- Is the inside of the main cold water tank clean, and is there a properly fitted lid?
- Is it properly supported on a strong plywood deck over four or more large bearers over an internal load-bearing wall?

(equivalent to three adults) and must be well supported, ideally on a thick exterior-quality plywood deck over four or more timber bearers designed to spread the weight over several joists to a supporting wall below – remember, these ceilings weren't designed to support big heavy tanks when originally built.

Tanks need insulation jackets to stop condensation forming and soaking down to the deck and the ceiling. Loft insulation below tanks is usually omitted to help take the chill off the immediate area and prevent freezing. But a better idea is to insulate beneath the tank or deck itself so loft insulation doesn't have gaps in it. Also, make sure your tank has a fitted lid as it's not unknown for decomposing pigeons moulding away in water tanks to spread disease.

## INTERNAL SUPPLY PIPES

Most modern pipework is run in either 15mm or 22mm copper or plastic pipe. But in older houses the norm was iron or lead. Lead was also used as a constituent of solder for some types of copper pipework until it was finally banned in 1987.

Steel tubing was occasionally used in the 1970s as a cheaper replacement for copper, but if steel and copper pipes are in contact, they can corrode and may need to be replaced. Any unlagged pipes in the roof space will be at risk of bursting during winter freezes (as the frozen water expands). Pipe lagging should be provided as it is a cheap and easy way to prevent the damage and mayhem caused by such leaks.

Today, most new work is run in plastic with push-fit or press-fit joints, even for hot-water pipework, and can be safely used with other materials. When installing new hot or cold pipework in old buildings, modern materials offer some significant advantages. Flexible plastic pipes with push-fit joints are easier to thread through awkward spaces reducing the need for cutting. Even so careful planning of new cable and pipe runs is key to minimising damage, eg by avoiding cutting holes in old lath-and-plaster ceilings and walls. It's very common to find that floor joists have been weakened by having notches cut in them for pipes. But it should be possible to route pipework so most of it is surface-run. Where pipes have to be routed through floors they can often be run parallel to joists. Where cutting is unavoidable, notches should not exceed one-eigth of the depth of the timber because structurally the depth of a joist provides greater stiffness than its width. Cutting should always be made from the top and not within the first 300mm of either end. Fitting a metal plate over the open notch, that covers the top of the pipes, helps reduce the weakening effect and also protects pipes from any subsequent clumsy nailing to floorboards.

**Left:** Incoming lead water pipe, unusual at such a high level

**Right:** Original (leaking) incoming supply pipe to hall floor now replaced in plastic, with new stopcock.

**Above:** Old lead pipework (plus cat playing under floorboards!)
**Below:** Mixed metals can react causing leaks

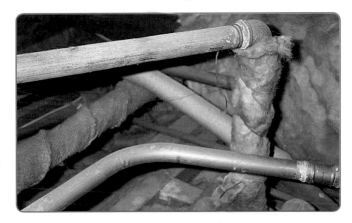

# Replacing lead pipework

If your incoming cold water supply is in lead pipework, contact your water supplier to check whether they offer a replacement service free of charge. If not, and you can't get rid of lead pipes, consider installing a suitable water filter fitted to a single drinking water tap (usually next to the kitchen sink). Alternatively, get into the habit of running the taps for a while before drinking out of them, or stick to drinking bottled water.

But in most properties it should be feasible to replace nearly all the incoming lead pipework with plastic or copper. Start by tracing the lead pipe back as far as possible.

**1** Take a hacksaw, and having made sure the mains supply into the home is turned off and the pipe is empty, cut through it with as clean and square a cut as possible.

**2** Clean the cut end of the pipe to receive a new fitting using a flat file to get rid of the worst of the blemishes and dirt.

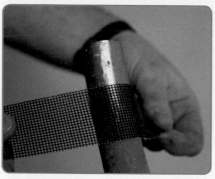

**3** Use an abrasive strip to get the pipe round and really clean.

**4** Use a round file to gently clear the inside of the pipe of any burrs and deformities. But try to retain the coating of limescale inside the pipe that is sealing it from contact with the water.

**5** You now need a fitting called a Lead-Lok that matches the size of pipe (the internal diameter). Lead pipes are in imperial measurements, normally ½, ¾ or 1 inch. When converting, ½ in normally corresponds to 15mm copper, ¾ in to 22mm and 1in to either 22mm or 28mm).

**6** Put the nut onto the pipe followed by the split olive with its flat face pointing towards the fitting.

**7** Push the rubber O-ring onto the pipe followed by the fitting itself, ensuring that the pipe is fully inserted into the fitting.

**8** Tighten the nut to the fitting by hand before using an adjustable spanner to give it an extra turn – at most one full turn should be required as lead is relatively soft.

**9** The other end of the fitting is a standard copper compression fitting that can be continued as normal to either copper or plastic pipe.

# Sanitary fittings

Although by the end of the Victorian era most urban houses had a water supply to the kitchen and perhaps to a downstairs WC, most households still washed themselves from small washstand basins or jugs and bowls in the corner of the bedroom. Ironically, some of last properties to have plumbing installed were those who could most easily afford it – the rich

**Left and above:** Modern repro 'Victoriana'

had little need for piped water supplies as servants continued to perform this function well beyond the Edwardian period. So in many wealthier homes small portable tubs, like hip baths, would periodically be set up in the bedroom, filled from cans or buckets of hot water laboriously carried up and down the stairs from the kitchen range. But for many working families housing conditions were chronically overcrowded, with a 'Saturday night bath in the laundry tub' or a cheap bathtub of zinc, tin or galvanised iron placed by the fireside. Large numbers of the working-class population would still have to share privies, wash in cold water at the scullery sink and take baths in portable tubs until well into the 20th century.

(Photo: www.bricksandbrass.co.uk)

**Left:** Lower middle class bathroom c 1910 with a cast iron enamelled bath, earthenware basin and glazed fireclay WC.

**Below:** Saturday night is bath tub time!

## BATHROOMS

Bathrooms were virtually unknown before the middle of the 19th century, only starting to make an appearance in some wealthier households from the 1860s.

Early bathrooms were just that – a room with a bath – fitted in wherever space permitted, usually taking the place of a small bedroom or jemmied into odd corners. The fashion for hot baths came relatively late, but one pioneering appliance in the 1850s was the self-heating bath or 'bath heater' with a gas burner underneath that rather excitingly shot jets of flame around the metal bathtub! From 1868 primitive new wall-mounted gas water geysers appeared, providing hot water when and where it was needed, but these were notorious for emitting noxious carbon monoxide fumes.

By the 1870s bathrooms were starting to become a standard fixture in more expensive new houses. A decade later virtually all middle-class villas were being built with a small upstairs bathroom, the

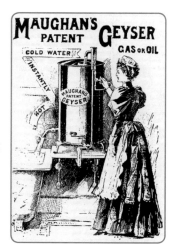

Hot water gas geyser c 1895. Geysers fitted in rooms without sufficient ventilation led to some bathers being asphyxiated by CO fumes.

A gas bath of 1873 that took only 35 minutes to heat the water. Some had gas rings directly underneath.

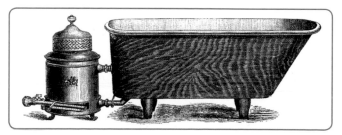

plumbing simplified by locating the bath and any upstairs WCs to the rear above the kitchen/scullery. In many urban areas improvements in mains water supplies meant there was now enough pressure in the pipes to get water upstairs. But this still left much of the population without a bathroom even by the end of the century (many had to wait until legislation in 1918). Few speculative builders of workers' terraces even by the early 1900s thought it worth including bathrooms as such luxuries wouldn't command much higher rents. But ultimately middle-class sentiment prevailed – owning a bath showed that the family cared about cleanliness and became a statement of respectability, plus the newly discovered scientific link between dirt and germs encouraged bathing as a means of preventing disease.

To cope with steam and to make the bathroom hygienic, bathware became impervious and sanitary. From the 1870s hardwearing classically styled tiles in clay, marble or slate started to appear, some moulded to imitate dados and skirtings. Bathroom tiles were usually white or blue and ceilings were painted white. Small bathrooms in cheaper houses had bare walls painted with oil paint.

By the later 19th century, water was heated by back boilers in many larger houses. Located behind the coal-fired kitchen range, these supplied hot(ish) water through a system of cast-iron pipes via a hot water tank (in an upstairs cupboard or in the roof space), or a cylinder above the kitchen range. But the cast-iron pipes could be subject to corrosion or fracture, occasionally resulting in dramatic explosions. Room heating was still dependant upon coal fires, but bathrooms without fireplaces might be warmed by paraffin heaters (popular from 1870s), or oil stoves (from the 1880s).

## BATHS

The archetypal Victorian enamelled cast-iron roll-top bath and the popular ceramic baths were introduced from the 1880s. But because the enamel was thin and easily chipped, they were usually enclosed by a wooden casing. More upmarket versions were panelled in mahogany, a close grained timber that can withstand humid conditions, with protective hardwood rails around the rim.

Freestanding baths became fashionable from the 1890s with their distinctive 'ball and claw' feet, panelling now considered unhygienic because it could harbour dirt. This period also saw the advent of modern mixer taps in brass or more expensive nickel plate. The development of tough porcelain enamel fireclay in the Edwardian era helped popularise sturdy free-standing baths,

C 1870s bath with mahogany surround

Top of the range 1893 Shanks' 'Independent' spray shower and plunge bath with ornamental feet

many remaining in use to this today having been successfully re-enamelled. The most luxurious of all were canopy baths with integral showers that appeared from around the turn of the century.

## BASINS

Even in homes with a fixed bath, most people continued to wash their hands and faces from a traditional large dish and cold-water jug in their bedrooms. By the early 20th century washbasins were being added, though they were somewhat confusingly known as 'lavatory basins' (from the French *lavage*). The choice was either a plain wall-mounted 'lavatory on brackets' made of white enamelled earthenware, or a fixed basin on metal legs, often with gunmetal taps. Occasionally basins were plumbed into existing slate or marble topped washstands to make 'vanity units'. In the Edwardian period substantial four-square pedestal basins became popular, featuring a lip to prevent water slopping out at the front.

## SINKS

The kitchen was used only for cooking, so it was rare to find sinks located in kitchens before the 20th century. Instead, washing-up, laundry, scrubbing vegetables and all the messy activities that involved water were done in the separate scullery (by servants). Sinks were of enamelled cast iron, glazed stoneware, or fireclay, and were supported on brackets, brick corbels, or pedestals rather than on cupboard base units as they are today. Adjacent draining surfaces would be of hardwood (typically teak) or impervious materials such as slate slabs. Along with Victorian baths and basins, these hardwearing ceramic 'Belfast' sinks are today very much in vogue.

## THE SMALLEST ROOM

A typical Victorian house today will have a WC somewhere 'out the back', where the original outside loo has been incorporated into a more recent extension, often enlarged to form a downstairs bathroom. Alternatively, you may have an original upstairs bathroom located near the top of the stairs if yours was one of the more expensive houses built with one. In smaller terraced houses the rear bedroom has often been converted to a bathroom.

Either way, in most homes today the original Victorian fittings will be long gone, and have probably been replaced several times over. If you do come across originals, they're well worth conserving. But unlike robust modern vitreous china sanitary fittings, Victorian sanitaryware was made of fragile glazed earthenware or fireclay. This makes it very difficult to remove old lavatory pans without damage, particularly due to the hard-setting red lead putty used to fit them.

Modern sanitary fittings of course aren't entirely free of problems. For example it's surprisingly common for small leaks to have persisted unnoticed over a long period of time. As well as leaks from hidden pipework, the chief offenders are acrylic shower trays and plastic baths which commonly develop leaks at the seals around the edges allowing water to quietly seep into the fabric of the building. Power showers are notorious for injecting water under pressure through even the tiniest of cracks, including poorly grouted wall tiles.

The resulting timber decay to floors and stud walls may only become evident as a result of a pervasive damp smell.

So that maintenance work can be carried out to individual taps and fittings without having to turn off the supply to the entire house it's a good idea to fit isolation valves to hot and cold pipes serving baths, basins and WCs. And when installing new shower trays etc., bear in mind also that it's important to provide access for maintenance to concealed pipes, cisterns and traps. With heavier new fittings such as baths or water tanks it's essential to carefully consider how the weight will be supported. A Victorian cast-iron bath, for example, can weigh up to 200kg, and hold at least 150 litres of water, with a combined load (when in use) equivalent to 5 or more adults. With all the loadings transmitted via tiny claw feet, the pressure per square centimetre can be more than a Challenger tank. Additional support and strengthening is likely to be required to floors.

# The amazing history of toilets

In the early Victorian period, most houses had some form of outside privy, or a primitive ash or earth closet. Yet by the end of the century, the fruits of a sanitary revolution meant that most of the nation's housing stock was

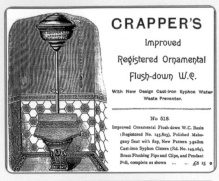

equipped with hygienic modern flushing loos and Britain led the world in advanced plumbing technology. A typical Victorian house today will have witnessed considerable changes in the way that different generations of residents went about answering the call of nature. So let's take a quick look at some of the remarkable events that made this progress possible.

## Privies

If you house dates back to the first half of the 19th century, the original occupants would probably have made do with the traditional privy-midden, which took the form of a cesspit dug nearby in porous ground, usually in the back yard or garden. A brick outhouse or wooden hut would be erected with a wooden seat placed over the pit. With no means to dispose of waste, leaking and overflowing cesspools were a major threat to health, with liquid seeping through porous ground contaminating nearby wells. Some types of privy were built projecting over a river or small brook, that might also be used as a source of water.

In wealthier households privies were sometimes installed indoors with a cesspool in the cellar or dug below the ground floor that had to be emptied periodically by 'nightsoil men' (who sold the contents to farms as manure). However, many such vessels were badly built or remained

unemptied and leaked into nearby domestic wells or into rivers providing drinking water. Many fatal illnesses of the period such as cholera and typhoid were the result of such practices. Today, where long-forgotten sealed-up cesspits are unearthed in old houses, there can be a potential risk from any trapped highly explosive methane gas.

## Dry privies and earth closets

In the mid-Victorian period many major urban local authorities, such as Newcastle and Manchester, opted for dry closets ('netties') as a cheap alternative to water systems, instead using freely available ash or earth. These were an improvement on primitive privy-middens, and made a lot of sense because every house had large amounts of ash to dispose of, as a by-product from coal fires. Ash was the perfect deodorising material, 'solidifying and fossilising the faeces' to convert it into dry solids devoid of smells – although to be on the safe side, a dash of quicklime was sometimes added; the modern equivalent today is the eco-friendly composting toilet.

Simplest of all was the 'pail system', which was widely used in rural areas (some Councils collected them weekly until the 1960s). These comprised a simple iron bucket placed under a wooden seat, with a pull handle releasing dry earth or ashes from a 'cistern box', cascading down into the bucket, the end product being a useful form of manure (even Queen Victoria suffered the indignities of earth closets at Windsor Castle). Ultimately, however, as we all know, the water closet triumphed, later becoming universally adopted thanks to the provision of piped water to homes and the technological development of flushing WCs.

## Legislation, sewers and the 'Great Stink'

The 1848 Public Health Act was a turning point, requiring every house to have access to some form of fixed sanitary arrangement, in a bid to eradicate contamination from cesspools. The subsequent Nuisances Removal Act of 1855 facilitated legal proceedings against owners of filthy privies. But the reality for much working-class housing was an insanitary toilet (if there were any at all) in an outbuilding attached to the house or at some distance in a separate block. In the older 'back to back' districts poorer households had to make do with a shared communal toilet in the yard, sometimes with only one toilet for every 20 or more houses, and a single tap supplying water for several families.

The popularity of public lavatories installed at the Great Exhibition of 1851 (with more than 827,000 people paying to use one of the 'necessary conveniences') stimulated demand for flushable water closets (WCs), which soon afterwards starting being installed in more expensive houses. The better off expected toilets to be located inside the house – one upstairs for the family, and one downstairs or outside for the servants. But the introduction of water closets had the unintended consequence of turning a private nuisance into a major public one – because, once flushed away, waste might only travel as far as an overflowing cesspool in the back yard, or would run down street. At best it might be dispersed via a short brick drain that poured waste into the nearest stream or river. This ultimately lead to the infamous event that finally prompted Victorian legislators to take action – the 'Great Stink' of 1858. That summer, the combination of unusually warm weather and an unbelievably polluted Thames (from which drinking water continued to be extracted) resulted in an appalling stench from the river, halting ministerial proceedings at the House of Commons and making it necessary to hang sacking soaked in deodorising chemicals at the windows. Legislation swiftly followed instigating the construction of London's sewer system. Subsequent legislation required all new houses to have running water and internal drainage. London's sewer system, completed by 1865, carried foul waste to settlement tanks located on the banks of the upper Thames where it was pumped into the river at high tide. Urban sewers were constructed in most other big cities shortly after London, and methods of treating raw sewage began a decade or so later.

The fact that contagion and disease didn't respect class boundaries may also have helped spur progress; only ten years after Queen Victoria's husband Prince Albert's typhoid-related death in 1861, her son, the Prince of Wales, suffered a similar illness, but survived to later ascend the throne (no pun intended) as King Edward VII.

## The rise of the water closet

Despite enormous progress in the development of sanitary fittings and plumbing, many people soldiered on with the familiar time-honoured comforts afforded by chamber pots, or commodes (a chair with a removable inset chamber pot), which were emptied the next day into an outdoor toilet. Such reluctance to abandon tradition may have had something to do with the fact that early water closets often didn't work well. Until the development of an effective integral water-sealed trap, every toilet bowl acted as a conduit for smells of cesspit and sewer. The backwaft of odours, especially in hot weather could be unbearable, hence the custom of locating privies outdoors. Worse, some primitive early WCs could backfire, filling the room with what the horrified user had hoped to dispose of.

The most common earlier type was the 'valve closet', which comprised a simple funnel dropping straight into the foul drain below via a mechanical valve, a jug of water assisting dispersal. However, the valves could be prone to jamming, allowing water to seep away and noxious gases back up. Nonetheless, valve closets were popular with the better off, with cheaper basin-and-trap WCs (hopper closets) reserved for servants.

Pan closets were another popular design, able to withstand rough usage. These comprised a small funnel with a pan at the bottom. When the handle was pulled the pan tipped the contents down into a larger vessel below, and via a 'D' trap into the foul drain below. However, pan closets suffered from inaccessible parts that were impossible to clean and had a reputation for becoming filthy.

Meanwhile major improvements were being made in the dispersal of waste thanks to the development of smooth saltglazed stoneware pipes that replaced Roman-style brick drainage channels. The arrival of flushable water cisterns from the late 1870s also greatly improved matters. Cisterns were made either of cast iron or lead-lined softwood (deal) and fitted at high level to gain maximum water pressure and operated via ballcocks and 'pull and let go' chains (although quieter modern 'combination closets' with combined low-level cisterns were developed in 1893 they didn't start to replace the noisier high-level type

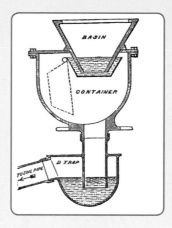

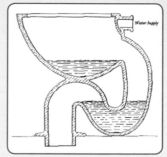

Left: Pan Closet
Above: Early wash-down WC, forerunner of the modern loo

**Above:** Invictas Washdown with ample seating
**Below:** Crapper's valveless waste preventer – father of the modern flushing cistern

Deluxe mahogany WC circa 1870 with rare low level cistern

until well after the First World War).

Modern ceramic flushing 'wash down' WCs with pedestal bases and enclosed U-bend traps that kept sewer gases at bay became popular from the 1880s. The familiar design comprised an all-ceramic pan with a flushing rim and a pedestal base enclosing the trap. Popular models included the Humpherson 'Beaufort' in white or cane colour fireclay and Twyfords' best selling 'Deluge'. Thomas Twyford also pioneered the first trapless toilet, the 'Unitas' of 1885, a one-piece, free-standing unit on a pedestal base. This eliminated the problem of leaky joints and foul odours.

These glazed earthenware late Victorian beauties with their pedestal bases, flushing rims and pull-chain cisterns were rapidly adopted in preference to the 'washout' type lavatory popular in Germany with its 'waste collecting shelf' that was (hopefully) cleared by a torrent of flushed water. British Crappers, Humphersons and Twyfords, complete with user guides, were commonplace by the turn of the 20th century. The finishing touch would be a choice of seats in mahogany, cedar, walnut, or cheaper plain deal. The luxury of cork-tipped seats was reserved for the rich. By the Edwardian era, as with baths, WCs were no longer boxed in behind mahogany or deal panelling but freestanding, made from white earthenware or enamelled fireclay with hygienic lifting seats as standard.

## COMMON PLUMBING PROBLEMS

It's a good idea to make sure you know how to turn off and drain down the water in your home. To minimise damage in the event of a serious leak first turn off the mains stopcock. Then open all your taps to reduce the flow rate. Turn off stored water coming from tanks and the hot water cylinders. If the leak still hasn't stopped, it's probably water from the CH (black and smells of chemicals) so you need to drain your CH system. If all else fails, emergency repair can be attempted using a clamp repair kit or special leak repair tapes. A proper repair can then be made.

■ Low water pressure at the taps has a number of possible causes. Low water pressure to the kitchen cold tap may be due to a leaking or corroded supply pipe or because it's an old supply shared with other properties. One symptom is where the pressure drops at times of peak demand. Low pressure to upstairs taps may be due to insufficient height of the water tank above the bathroom.

■ Leaking overflow pipes are normally due to defective ball valves in tanks or older WCs, usually requiring the ball valve to be replaced.

■ Dripping taps are generally due to faulty or worn washers, which need to be replaced, or in some cases the complete mixer/tap unit may need to be renewed.

■ Knocking noises in cold or hot pipes can be 'water hammer' due to high pressure or insufficiently supported loose pipework. Pressure in pipes from the cold water tank can sometimes be remedied by fitting a larger float of at least 150mm diameter. Loose pipes can be secured with additional clips.

■ Leaks: The main causes of leaks are:

### Frozen pipes
Uninsulated pipes run through cold spaces (*eg* lofts, porches, garages, conservatories, larders) are vulnerable to freezing and bursting. While everything is frozen the burst pipe is filled with ice and no harm is being done – until it thaws! The best advice is to lag your pipework. If you leave your home empty in winter, turn off your hot and cold water (the CH system will still run). Note that modern condensing boilers have small plastic 'condensate' waste pipes that if run outdoors can freeze, causing boilers to shut down (pipes should be lagged after defrosting with a kettle).

### Ball valves failing
Ball valves control the supply of water to WC cisterns and to the water tanks in the loft. In case of a malfunction these all have overflow pipes. But if these become damaged or blocked, the tank will overflow into the house.

### DIY
This is the most common cause of household floods. Most pipes are routed just under the floorboards, so lift the board first to check. Use fairly short screws or nails (floorboards are typically 19mm thick). Metal detectors can be worth using, but won't detect plastic pipes. And remember that pipework is sometimes embedded down walls. If you do drive a nail or screw into a pipe, don't immediately remove it – it's better blocking up the hole!

**See the Haynes *Home Plumbing Manual* for detailed solutions to all plumbing problems**

**Right:** Heart of the home - the kitchen range was kept going 24/7

**Below:** Bedroom heating - small cast iron fireplace

**Below right:** Edwardian gas fire

# Heating and hot water

Traditionally, the only heat in the home was provided by open fires or kitchen ranges. With just the living room or kitchen heated, keeping warm in other rooms was simply a matter of 'putting a jumper on'. Although some of the warm air would rise upwards, most of the heat would disappear up the chimneys or straight through uninsulated roofs and walls. To keep the fires going, a typical house would need to store about two tons of coal, in the cellar or under the stairs. Portable paraffin heaters became popular from the 1860s, and cast-iron gas grates later made a brief appearance.

## HOT WATER

Traditionally, hot water was produced in a large pot called a 'wash copper' (often made of cast iron) set into brickwork over a fire and capped with a wooden lid. Here the weekly laundry wash was done as well as food preparation and boiling Christmas puddings. The copper was usually located in the scullery or sometimes next to the kitchen range (see chapter 12). This later developed into the coal-fired back boiler supplying hot water to bathrooms, sometimes via a storage cylinder built into the

Regular hot water cylinder, now superseded by....

Modern unvented pressurised hot water cylinder supplied direct from mains

## Hot water systems

### Low-pressure hot water

These systems are 'gravity fed' from a cold water tank in the loft (at least 114 litres). This is linked via a pipe to the hot-water cylinder below (usually in the airing cupboard). When you open a hot-water tap the cold water in the storage tank runs down the pipe into the base of the hot water cylinder, pushing hot water out the top of the cylinder to the tap. It's the cold-water tank that provides the pressure to push the hot water out, so to stop the flow of hot water at the taps you need to close a 'gate valve' on the supply pipe from the cold tank to the cylinder.

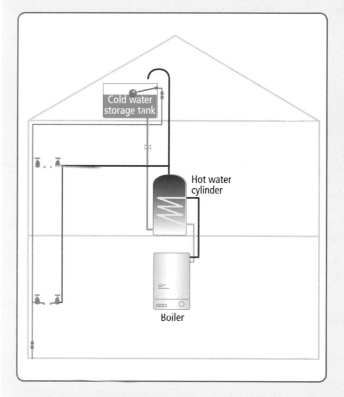

Cold water storage tank

Hot water cylinder

Boiler

brickwork at high level within the chimney breast above the kitchen range or more rarely a galvanised metal tank insulated with felt and encased in wood located in the bathroom.

Today, unless your house has a modern high-pressure sealed system (ie an unvented 'Megaflow' type cylinder or a combination boiler), the hot water will normally be stored in a conventional cylinder in the airing cupboard, usually incorporating an electric immersion heater to boost the hot water when needed. Cold water from the storage tank in the loft refills the cylinder as hot water is used. You may see a spaghetti of copper pipes in the vicinity, one of which is an emergency expansion pipe from the cylinder up to the feed and expansion tank. If the hot-water system is fed through old steel or cast-iron pipes, these could be subject to corrosion, causing the water to become rust coloured, and will need to be replaced. Modern cylinders are built with integral foam insulation: those with old lagging jackets are likely to need replacing shortly.

### CENTRAL HEATING

Central heating is a relatively recent addition to British homes, only becoming a standard feature in new properties from the 1960s. However, pioneering central-heating systems had been installed in some larger houses as far back as the late Victorian period with varying degrees of success. These comprised large-bore cast-iron pipes fed by gravity and the natural circulating force of hot water. Surviving pipes and bulky column radiators (made from hoops of cast iron) can often be adapted for use with modern pumped systems. Old cast-iron radiators are very

### High-pressure hot water

In most modern systems the hot water is either delivered instantaneously via a combination boiler, or via an unvented hot-water cylinder.

### Unvented hot-water cylinder

Modern unvented hot-water cylinders are very large, and made of steel with a flat top and (usually) white with numerous pipes leading in and out, plus an expansion vessel nearby (normally white or blue). In contrast, conventional (vented) hot-water cylinders are shorter and made from copper and covered with a red lagging jacket or yellow/green foam insulation, with a single pipe emerging from the top.

Unvented cylinders don't need a cold water tank in the loft (although your central heating system might).

### Combination boiler

'Combi' boilers deliver hot water almost instantly. Again, there is no need for a tank up in the loft storing water as the boiler will detect when you turn on a tap (or when a radiator 'calls' for hot water) and immediately start heating some mains cold water to supply it.

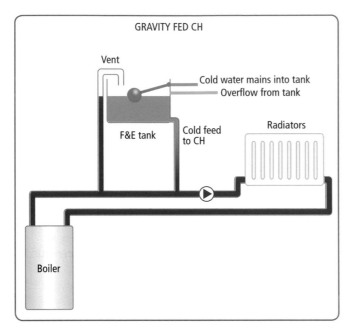

GRAVITY FED CH

Vent

Cold water mains into tank
Overflow from tank

F&E tank

Cold feed
to CH

Radiators

Boiler

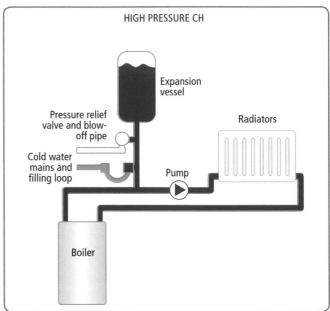

HIGH PRESSURE CH

Expansion
vessel

Pressure relief
valve and blow-
off pipe

Cold water
mains and
filling loop

Pump

Radiators

Boiler

# What type of central heating system do you have?

There are two types of central heating (CH) systems – gravity fed and high pressure.

## Gravity-fed CH

As with conventional cold-water systems, here the mains cold supply instead fills the smaller sister tank in the loft – known as the 'feed and expansion' tank (which is fairly small – about 500 x 300mm – and only about half full with distinctly murky-looking water). Gravity powers this water on its way down to the boiler (and hence the radiators) although relatively little water is needed. Once the system is full the ball valve closes, leaving the water within your boiler and radiators to circulate round and round for years at a time. So if you turn off the mains at the stopcock when you go on holiday, the CH system will still operate and keep your home warm.

Before draining the CH system you need to stop new water supplying the system. But because turning off the cold water at the mains stopcock also cuts the supply to the kitchen sink, a better method is to tie the ball valve on the tank to stop water replenishing the tank as you drain the CH system. This can be done by laying a strip of wood over the tank with a piece of string tied to the ball valve holding it up. Better still, if there's an isolation valve on the cold supply to the tank, simply close it off. Once the F&E tank is isolated, you can then drain down the CH system, starting at the drain cock on the radiator pipework near the front door.

## High-pressure CH

Most new CH and boiler installations are of this type. Here the tanks are dispensed with and the system fills directly from mains cold water via a 'filling loop' (a flexible hose with two taps connecting the mains supply to the CH system). The boiler will normally have a built-in pressure gauge, plus there's a red coloured expansion vessel nearby.

much back in fashion and are a visual asset well worth preserving. New replica period radiators, or salvaged originals, can look just right. Thankfully, old labour-intensive solid-fuel stoves and bad-tempered furnaces have long been consigned to history.

Modern central heating systems pump hot water from the boiler to steel radiators in each room via pipework that comprises separate parallel 'flow and return' circuits, usually in 15mm copper pipe. The 'flow' circuit feeds the hot water to each radiator and the 'return' circuit takes the old cooler water back to the boiler. The temperature output is controlled by thermostats – see 'controls' below.

Before fitting radiators, you need to calculate the required size of radiator to heat each room in BTUs (British Thermal Units) and to be sure that the boiler is man enough for the job. Radiators tend to be located on outer walls, often by windows where the room is coldest.

Common defects to look for are: DIY workmanship with irregular 'bendy' pipework (typically in thin 8mm microbore copper pipe which can be prone to blockage); pipes unsupported with insufficient clips; leaks at radiator valves; rusted radiators; unprotected pipes run in concrete floors; and butchered joists under badly refitted floorboards.

Central-heating systems need periodic maintenance, including flushing through to reduce limescale, and should be checked annually under a service contract. Pressurised combination-type systems require more frequent 'bleeding' to release built-up air.

Electric heating is quite common in rural areas without a gas supply, typically comprising fixed storage heaters that take cheaper off-peak electricity at night, store it in special bricks, and release the heat the next day. They need separate electric circuits with switched fused outlets to the heaters. Older ones are very bulky and there's relatively little control – once the bricks have been heated up you can't then decide you don't want them 'on'. Popular alternative fuels in areas without a mains gas supply are oil, bottled LPG gas and timber for use in wood-burning stoves.

**Above and below:** The modern AGA is descended from traditional Victorian cast iron solid fuel ranges

## COOKING

Even the cheapest Victorian kitchen had a cast-iron coal-fired range for cooking, with a primitive integral oven to the side of the fire, plus a hot plate and fire-grate. They combined the traditional British custom of open-fire roasting with room heating, oven baking and water heating, some even boasting a hot-water tap served by a tank on the other side of the fire. Due to their considerable size, ranges needed large fireplace openings, and brick or fireclay linings to cope with the huge amount of heat generated. But coaxing a temperamental black-leaded iron range to life every morning was not much fun in a

servant-less age, and few Victorian working kitchens survive. A deep recess where the range once stood can still be seen in many Victorian kitchens. The nearest modern equivalent is the Aga, powered either by solid fuel, electricity, or gas.

The widespread introduction of gas ovens was long delayed through fear of explosions and health scares about eating food impregnated with harmful fumes, so they did not begin to replace solid-fuel ranges in quantity until the end of the century.

## BOILERS

Many homes have boilers that date back a good 20 years or more. These may be floor-standing or located behind the living room fireplace (back boilers) with vertical flues snaking through chimneys, up to a metal terminal at the top of the stack. Older boilers may take their air for combustion from the room, so vents must be provided in the walls as described for open-flued gas fires.

Today, efficient wall-mounted condensing boilers designed to extract extra heat from the flue exhaust gases are standard issue. These are normally 'room sealed', taking air from outside and expelling exhaust gases externally through the same balanced flue projecting through the wall, sometimes fan assisted. Most new boilers are 'A' rated, which means they achieve efficiencies of at least 90% (ie they turn 90% of the fuel they use into heat). You can see how efficient your existing boiler is at www.sedbuk.com. Fitting an new boiler should significantly boost your home's Energy Rating.

Replacing a boiler or carrying out alterations to heating systems requires Building Regulations consent. In practice, however, this is usually undertaken by specialist contractors who can self-certify their work and provide the Completion Certificate when the job is done. As noted earlier, the installation of heat-producing gas appliances (eg boilers or fires) must only be carried out by a Gas Safe registered engineer. Boilers are available that run on a wide variety of fuels as well as mains gas, such as oil, bottled gas (LPG / propane), and even solid fuel (coal, wood etc.).

Coal range c 1890

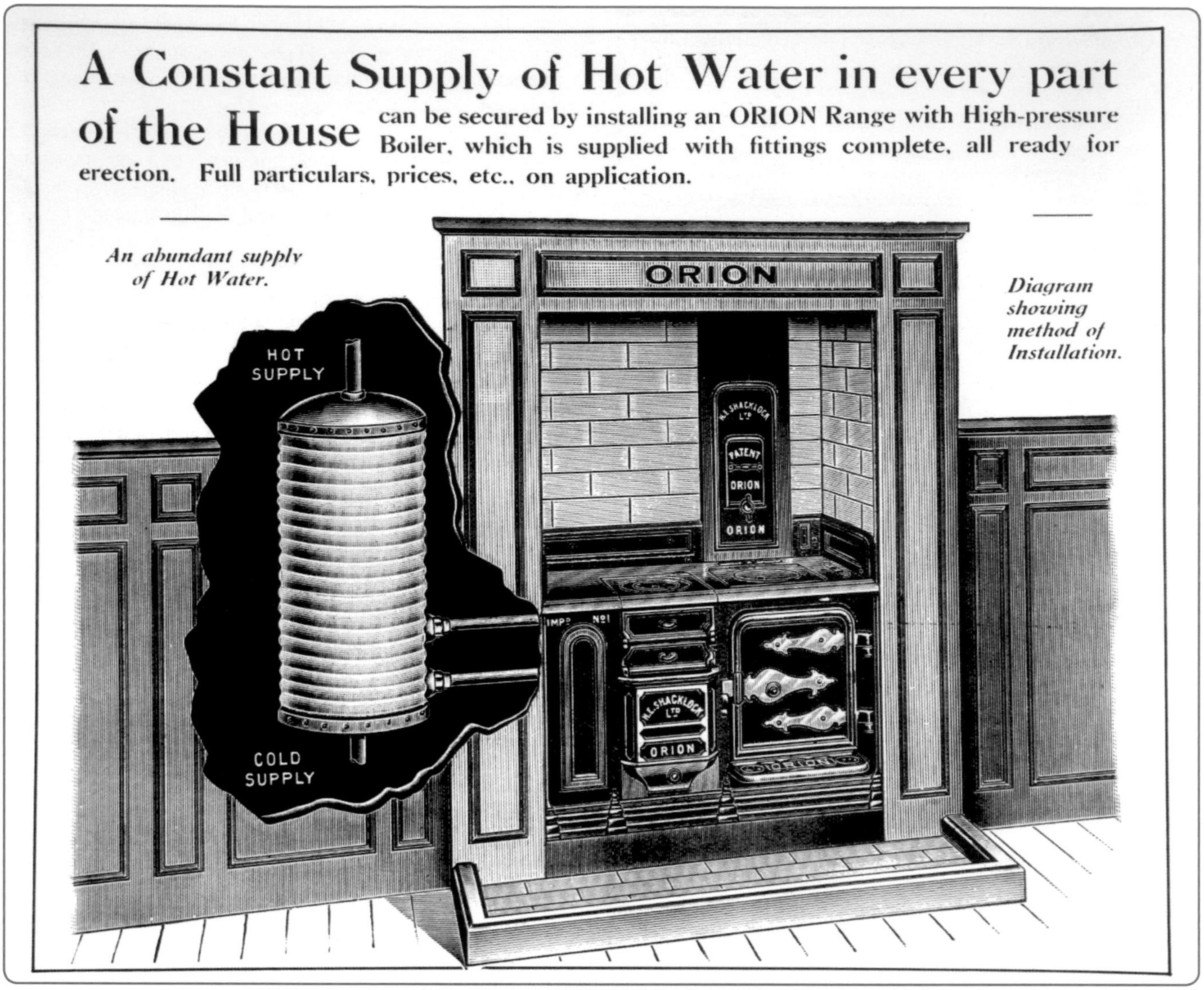

A Constant Supply of Hot Water in every part of the House can be secured by installing an ORION Range with High-pressure Boiler, which is supplied with fittings complete, all ready for erection. Full particulars, prices, etc., on application.

*An abundant supply of Hot Water.*

HOT SUPPLY

COLD SUPPLY

ORION

*Diagram showing method of Installation.*

The size of boilers and flues has reduced considerably in recent years, but installations still need to be carefully planned. One of the most important considerations is the position of the flue, as without due care this can easily damage the fabric of an old building and scar the exterior. The rules governing the location of balanced-flue terminals are complex, but generally they should be at least 300mm away from gutters, windows, doors, eaves, airbricks etc. (though fan-assisted flues can be closer), and should obviously not discharge into enclosed areas such as side passages. Although normally mounted on a main wall, there is some freedom about where to 'hang' the boiler since flues can be extended in length if they are fan assisted. If a conventional wall flue is inappropriate, vertical steel flues can be an option, venting via a roof slope or utilising a redundant chimney. The preferred location for boilers is within kitchens, utility rooms or garages, rather than in bedrooms, reception rooms or bathrooms, although sometimes they can be installed in lofts.

Modern systems are either 'conventional' with a small feed and expansion tank in the loft, or mains-fed 'sealed systems' such as combination boilers that don't require separate tanks. Combination boilers provide both heating and instant hot water (without the need for a hot water cylinder) but typically have a shorter lifespan than conventional boilers, depending on maintenance.

Like cisterns, boilers have overflow pipes to the outer wall, the difference being that boilers may need to expel boiling water under pressure. So regulations now require these pipes to be channelled down to ground level as a safety measure. In addition, a plastic condensate pipe needs to be connected to a waste pipe (preferably run indoors) to disperse surplus moisture from condensation in modern condensing boilers.

## CONTROLS

The aim of any heating system is to provide heat where there is a demand for it and avoid wasting money and energy where there is not. Room thermostats sense the air temperature in the room where they are installed but need to be positioned 1.5m above floor level to work accurately. A central heating system with modern programmable controls will be up to 30% cheaper to run than a standard system. The simplest and cheapest way to set different temperatures for each room is to fit thermostatic radiator valves (TRVs) to radiators. These sense and deliver the correct amount of heat on a room by room basis and are standard for new systems. But TRVs should not be

Gainassociation.org

back up the pipe – smells can't get past the water 'trapped' in the bottom of the U, which forms a seal. Simple, but highly effective.

This so called 'grey' waste water from upstairs bathrooms in older houses normally discharges via pipes through the wall that either link to the main soil stack via a branch pipe or simply into a hopper – a small, open collection box on top of a downpipe over a gulley. But hoppers tend to be prone to blockage, since being at high level they tend to be neglected, leading to overflowing and damp problems.

At ground level, waste water from kitchens and bathrooms usually discharges through pipes run out through the wall and down into a gulley. Gulleys should be enclosed within a protective surround with gratings to stop old hair, fat, bits of vegetables, leaves and general yuck from clogging them, and they need to be cleared periodically. Modern back-inlet gulleys have pipes that connect below the grating and traps that can be easily cleared. See chapter 4.

In older properties the waste pipes serving WCs, bathrooms, kitchens and sculleries were originally made of lead or cast iron. and may still be doing good service unless split or leaking. But many have subsequently been replaced or supplemented with new pipework. Because of their relatively large diameter (typically 30 to 50mm diameter and 100mm for WCs) waste pipes should not be run within timber floors – cutting huge chunks out of

installed in the same room as any existing wall-mounted thermostat, and if a TRV is fitted to every radiator it will be necessary to fit an automatic bypass valve at the boiler (unless it already has one internally). To save energy and keep bills down TRVs can be turned off completely when rooms aren't in use and room thermostats can be set to about 19°C.

You need to be able to control heating and water separately and most makes of boiler now have built-in clocks and programmers as well as a thermostat control that shuts off when the water gets to a certain temperature. A simple programmer that can set the heating to come on in the morning and again in the evening is essential to save energy, especially where a household is unoccupied during the day. More sophisticated programmable room thermostats and zoning controls allow different temperatures to be set depending on the level of occupation over a 24-hour period. Some controls have intelligent weather compensators that measure the temperature outside and self-adjust for greater efficiency making it possible to shut off the boiler earlier with consequent savings.

Traditionally, older systems relied on gravity and convection to move water around, but modern boilers use integral pumps. So if you need to extend your existing system, it may require an additional electronic 'slave' pump to improve flow.

# Drains

Drains have enormous potential to wreak havoc. Hidden away underground, persistent pipe leaks often go unnoticed for years on end – until disaster strikes. Defective drains are one of the main causes of subsiding foundations. Yet this is a subject we rarely give much thought to until accosted by foul stenches or overflowing toilets. So it's worth taking an interest to prevent such disasters arising in the first place.

### ABOVE GROUND FOUL WASTE
When you pull out the plug, the waste water from baths, sinks, and showers discharges via a U-shaped trap directly underneath the plughole. Traps cleverly prevent nasty sewer odours coming

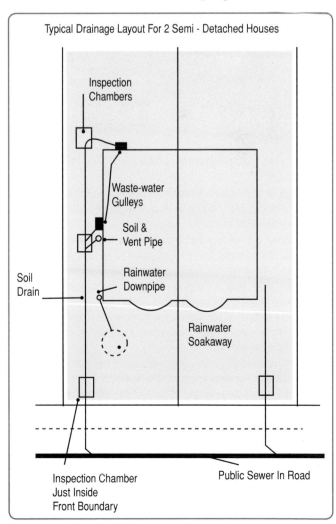

Typical Drainage Layout For 2 Semi - Detached Houses

Inspection Chambers

Waste-water Gulleys

Soil & Vent Pipe

Rainwater Downpipe

Soil Drain

Rainwater Soakaway

Inspection Chamber Just Inside Front Boundary

Public Sewer In Road

**Left:** Not ideal; spaghetti of original cast iron waste pipes, SVP and hopper – rust risks leakage even eventual collapse

**Right:** Many old cast iron SVPs now replaced in lightweight black PVC. This one has been massively extended to for new bathrooms, scarring architecture

**Below:** Modern gulley

**Right, lower left:** Original lead waste pipe, very rare; should be preserved

**Right, lower right:** The sweet aroma of foul drains! Looking down from back bedroom – SVPs should extend above eaves level

floor structures to accommodate them obviously isn't a good idea and hidden leaks can result in outbreaks of rot.

Modern plastic pipes are relatively trouble free if properly installed with sufficient support clips and suitable falls – (the fact that water needs to run downhill in order to discharge effectively is sometimes overlooked). External pipes should be of UV-resistant grey plastic. Internal white plastic fittings are push-fit, being either 32mm for basins, or 40mm for baths, showers and sinks (the quoted sizes are the internal pipe dimensions).

## SOIL AND VENT PIPES (SVPS)

Soil stacks, or 'stench pipes', are large-bore (100mm diameter) vertical waste pipes with branches for upstairs WCs and bathrooms. The job of the SVP is to channel foul waste safely down into the underground drainage system. These were traditionally manufactured from cast iron. Like funnels, the tops of SVPs are always open in order to ventilate the drainage system and alleviate pressure. This means foul odours can be emitted, so the top needs to project well above the level of upstairs windows. Normally terminating above eaves level is sufficient to disperse any malodorous wafts (unless there are roof windows at higher level). If it wasn't for the ventilation provided by the SVP, every

time you pulled out the plug, the waste water rushing down the pipe would create a vacuum in its wake, sucking out the water sitting in the traps, allowing foul sewer gases to rise up and pervade the house. This curious phenomenon known as 'siphonage' can also be caused by branch pipes that are too long, or set to incorrect falls. Problems can occur where SVPs become blocked, for example if birds are nesting in their open tops. One solution is to fit small protective 'wire balloon' cages to the tops.

'Branch' pipe connections into the sides of SVPs should be no longer than 3m, with provision for rodding access (to clear blockages), and must be properly supported. In contrast, old ground-floor WCs may have separate soil waste pipes that run down through the floor internally and out through the wall underground, at foundation level, where sufficient space should have been provided around the pipe to accommodate foundation movement and prevent cracking.

Original cast-iron SVPs are very heavy, so watch out for loose support brackets – it's not unknown for top lengths of pipe to come crashing down in stormy weather – a terrible way to end your days. Check for rust and limescale 'tide marks' at joints, indicating leaks or blockages. Unless regularly painted they tend to rust and may require replacement with modern plastic fittings.

Modern houses have internally-run SVPs to protect them from frost, but are boxed in, sometimes with very restricted access. Where you want to install additional bathrooms or WCs you don't have to go to the trouble of fitting another SVP. Instead 'air admittance valves' (aka 'Durgo' valves or stub stacks) terminating within the bathroom or loft can be used, as they have a one-way valve, but they must be located above the flood level of the highest sanitary fitting in the house.

## DRAINS AND SEWERS

Making alterations to drainage systems normally requires Building Regulations consent, so no work should commence until notification has been submitted to Building Control. Drains are pipes that carry waste water from just one property, and are the responsibility of the homeowner up to the point where they join a drain from another property or extend beyond your boundary. The drain then becomes a sewer and is maintained by the water authority.

Drainage is normally divided into two separate systems, foul waste and surface water. The relatively clean rainwater that collects as 'surface water' from roofs and hard surfaces is usually channelled to soakaways or into nearby watercourses. A separate foul waste system delivers 'greywater' from bathrooms and kitchens to mains sewers, or in rural areas sometimes to private drainage systems. Problems arise where the two get mixed up. One clue to this is where you see hoppers taking bathroom waste pipes as well as a rainwater downpipe. Foul waste

**Below:** CCTV camera reveals cracked sewer pipe, which can allow ingress of moisture-seeking roots

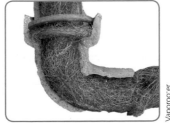

misconnections are illegal as well as a source of pollution. A simple example might be where a dishwasher is connected into a handy nearby rainwater downpipe causing the soakaway under the lawn to bubble with detergent. The same scenario with a WC macerator would have more disturbing consequences. Conversely, surface water is sometimes misconnected to foul-waste systems. This may not sound too serious, but it can be a recipe for disaster in storm conditions, as sewage works become deluged and the contents overspill. Either way the consequences are likely to be unpleasantly wet and sticky.

## GOING UNDERGROUND

The Victorians developed precision factory-made pipes of salt-glazed stoneware or cast iron that could be assembled on site using special cement for the proper jointing of pipes, a major advance in hygienic waste disposal. But old 'spigot and socket' mortar joints can be prone to cracking, being fairly rigid. In contrast modern plastic or clay pipes use flexible joint seals (collar couplings) that permit a degree of movement. Plastic underground pipes with flexible collar couplings were introduced in the 1960s and are still the most widely used type today. Adaptors are available for joining most types of dissimilar pipes.

Main underground waste pipes were typically 100mm (4 inch) or 150mm (6 inch) in diameter, laid in a trench to a gradient (fall) no flatter than 1:40 and bedded within 100mm depth of gravel or concrete, and then covered with about 300mm of soil. However, in most districts it was common practice to bed pipes directly in the earth, so that over time settlement in the surrounding ground has imposed stresses on the rigid pipes causing the cement joints to fracture.

Today, often the first challenge is to actually locate old drains. They may be hard to trace, disappearing under buildings or weaving around obstacles in a way that modern regulations wouldn't permit.

Many old drain runs in clay pipework are surprisingly shallow and can be fractured by ground movement or foundation settlement. Damage is sometimes also caused by heavy vehicles parked on top or by excavation work in close proximity.

Over time, ground movement sometimes affects the fall of underground pipe runs, which traditionally were laid to a slight incline. And where waste water can't flow easily, blockages are likely to occur. The most common weak points are at pipe joints, which are sometimes infiltrated by roots of trees and shrubs aggressively seeking out moisture, even through tiny cracks, ultimately choking the pipes with giant root balls. Another very common cause of damage to shallow pipes is from metal fence posts being hammered into the ground.

The simplest way to check

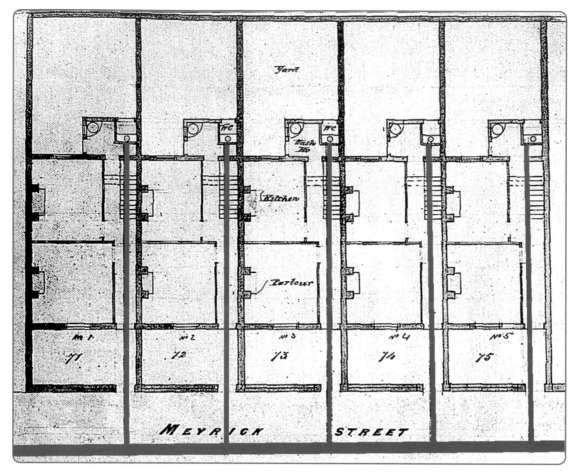

Yard

WC

Wash Ho.

WC

Kitchen

Parlour

No 1
71

No 2
72

No 3
73

No 4
74

No 5
75

MEYRICK STREET

Victorian terraces commonly have foul drain runs to the rear joining a main sewer running across the back gardens. Here they are run to the front under the entrance halls out to the sewer in the street

repaired internally using pressure grouting equipment or by inserting a new pipe inside the old one, but localised excavation and replacement may ultimately be the only solution.

Although Victorian builders generally tried to avoid laying drains directly below the house, in terraces drained to the front

whether drains are flowing freely, is to lift inspection chamber covers and then run the taps and flush WCs. Adding coloured dye can help identify the path of waste water from different locations.

All parts of the drainage system must be watertight otherwise effluent will leak out, contaminating the ground. Conversely, if the water table is high, water from the surrounding soil can enter the system so the pipes act in reverse, like land drains.

Leaking drains are a common cause of structural movement in older houses. Over time, the wet ground below the footings becomes more and more marshy and unable to support the foundation. This may go unnoticed for years as the wall may span a small 'soft spot'. But if it happens at a corner, cracking and localised collapse can eventually occur. Leaking drains can be checked with CCTV cameras, or you can hire equipment to test if the system is airtight. The procedure involves sealing off sections of pipe at inspection chambers using inflatable 'bungs' and pumping in air; a pressure gauge then measures if the air pressure drops (it shouldn't). Pipes can sometimes be

Inspection chamber with cast iron cover serving adjacent SVP and waste pipe

it could not always be avoided. In better-quality developments, the portion that passed beneath the house was formed of more expensive stronger cast-iron pipework that was extremely resistant to distortion. Most, however, used standard glazed stoneware pipes set within about 150mm of concrete. Where pipes ran under the walls, a lintel in the form of a brick arch allowed a protective space of about 50mm above the pipes.

Inspection chambers are needed for clearing blockages and there should be one at every change of direction of the pipe run, typically near the corners of buildings. But many older houses have no access at all, and construction of a new chamber around an existing pipe is a complex operation. Chambers are normally built of brickwork – either 230mm thick, like main walls (but often only of 115mm thick rendered brick), and a base of concrete covered with mortar benching. Chambers are rarely less than 750 x 600mm. The original covers were of thick cast iron with an airtight seal (more modern covers may be of concrete, steel, or plastic). By lifting the cover it should be easy to spot any damaged benching or render, which must be restored to a smooth finish to prevent blockages. In order to facilitate emergency access inspection chambers should not be obstructed or obscured. Unfortunately, however, many are now hidden under patios or paving, or may even have had extensions built over them, for which special consent should have been obtained.

Where an old system is defective, replacement with new pipework is normally necessary. Modern plastic pipes and ready-made inspection units are easier to install than in past. Even where an old system appears to be operating satisfactorily, if access is poor, it's a good idea to add new inspection chambers at major bends in pipe runs. Modern ready-manufactured chambers are of concrete or pre-formed plastic, although smaller units can also be used with screwed down covers through which drain rods can be inserted ('rodding eyes'). But as always with old houses, deep trenches mustn't be dug too close to shallow old foundations because of the risk of destabilising the wall. And as noted earlier, changes to drainage systems normally require Building Regulations consent.

Victorian houses sometimes have a projecting fresh-air vent pipe located by the last chamber nearest the road. This was to solve the problem of siphonage from the 'interceptor traps' that were sometimes fitted to the main drain leading to the sewer, to prevent smells passing via the drains into the house. These should be checked for blockages and damage because any gaps can allow ingress of rats; and some more intrepid members of the species have been known to make an unwelcome appearance in toilets! More conventional drainage afflictions are blockages from nappies flushed down WCs and from hot liquid fat poured down kitchen sinks, which then solidifies. Fortunately clearing pipes with drain rods is a normally fairly straightforward operation – see step-by-step.

## PRIVATE DRAINAGE

In rural areas houses may not be connected to a main sewer. Instead, you may encounter cesspits – large watertight underground chambers – usually made of brick and covered with a concrete slab. Cesspits need to be pumped out periodically into a mobile tanker to prevent overflowing: this was originally carried out by a horse-drawn van fitted with a manually operated suction pump. Septic tanks are similar to cesspits but have two or more interconnected chambers like mini sewage works with outlets in nearby fields to disperse treated liquids, so they need less frequent pumping out. Rainwater pipes must not be connected. These tanks can often leak with age, requiring lining, rebuilding or replacement with a modern pre-formed vessel.

## COMMON DRAINAGE PROBLEMS

If your waste water is not running away when you pull out the plug there are several possible causes listed below. This is often accompanied by smells or gurgling noises from plugholes.

- Waste pipes may not be laid to a sufficient slope, or may be sagging because not enough support clips have been fitted. The solution is to re-lay the pipes set to a decent fall and fit clips every 750mm.
- Hoppers or gulleys may be blocked or overloaded and need to be cleared.
- Blocked traps need to be cleared and refitted, or replaced with a new trap
- Siphonage can occur where the small amount of water that should remain in the trap as a seal is getting sucked out by a vacuum further down the pipe, allowing sewer smells into the house. If the pipe is very long or too small in diameter, or is serving several sanitary fittings at once, the sheer volume of water can cause a vacuum. The solution is to replace with larger bore pipes, or fit a special 'anti-siphon' trap or 'bottle trap', which retains more water in its seal and has a one-way

**Above:** Cast iron is tough but brittle, and can crack if subjected to extreme loadings
**Above left and right:** Low level sewer vents helped relieve pressure

air valve that lets air in to prevent a vacuum but stops smells getting out
- Underground drains may be blocked and need to be cleared. Light blockages can sometimes be cleared with a solution of caustic soda (wear goggles and gloves), or try flushing with a high-pressure water jet. Builders' rubble often gets washed down during works, as does loose render from inspection-chamber walls. Where there is a build up of grease or solid matter in a chamber or drainage pipe it can be often be cleared with drain rods – see step-by-step. If the neighbours' drains are also blocked, the problem may be in the main sewer in the road or beyond your boundary, which is the water company's responsibility. Where there are tree roots in the pipes or debris from collapsed underground pipes a specialist CCTV inspection will confirm the extent of any structural damage, but first notify your insurers as the cost may be covered.

Privy waste was collected from back alleys via small doors by 'nightsoil men'

# Clearing a blocked drain

If your bath water doesn't drain away, or the loo fills up when flushing, or if a gulley outside the house starts to overflow, it's likely that you have a blocked drain.

Most blockages can be cleared quite easily if you have access via inspection chambers. You will need a set of drain rods, which can be purchased from DIY stores. An alternative method is to use a high pressure water cleaner with a special drain cleaning attachment. Some have retrojets so that a cleaning unit will advance along the pipe unaided. The pressure created by one of these machines may be sufficient to clear some blockages.

Underground drainage pipework in older houses is normally 100mm or 150mm (4in or 6in) diameter pipes made from salt-glazed stoneware, laid in straight lines. Where pipes change direction there should be an inspection chamber.

Most houses should have more than one inspection chamber. The first one nearest the house is the highest. If this is full of waste water, the blockage is further out. It is always the chamber directly upstream of the blockage that will be full, whereas the one downstream of the obstruction (away from the house) will be empty. It is better to start rodding the blockage downstream from the empty inspection chamber. The water company is normally responsible for maintenance where blockages occur in pipework beyond your boundary.

Photos: Ian MacMillan

> ## TOOLS REQUIRED
> - ■ **Set of drain rods**
> - ■ **Pressure washer (optional)**
> - ■ **Jemmy and large screwdriver or drain keys**
> - ■ **Claw hammer**
> - ■ **Trowel**
>
> ## MATERIALS
> - ■ **Protective clothing, gloves etc**

**1** Locate the empty downstream inspection chamber. The inspection chamber upstream of the blockage will be full of water and solids. Clear the area around the cover to prevent any stones etc. lodging in the rim and stopping the cover from sitting back in the frame properly. Clean out the key holes and use a drain key to lift up one edge of the cover, or else use a jemmy and a large flat screwdriver to prise the cover up and a claw hammer to finish lifting – and to prevent damaged fingers if it slips. Put the cover safely to one side. Place a piece of board against the far-side outlet (downstream) so that when the blockage comes loose it will not then enter the next length of pipe.

**2** Connect your rods. These are a series of slightly flexible sections that are screwed together to make a much longer rod that can reach the obstruction.

**3** There are a number of different end fittings. The first to be used is the 'worm' or 'corkscrew'. (N.B. if your pipes are of more modern, pitch-fibre material, these fittings could potentially damage the soft surface of the pipes).

**4** In this instance one of the branches is blocked. Push the rods up the drain, remembering to constantly turn them in a clockwise direction – otherwise you risk unscrewing the rods and leaving one up the pipe. Push and turn the rods until you meet resistance, then keep going for a short distance further.

**5** Pull the rod back, and keep turning clockwise (resist the temptation to change to anti-clockwise turning when you pull the rod back). Some of the obstruction should now be cleared, coming into the chamber – be prepared for a sudden gush of effluent when the blockage is shifted. A flush of the WC can help wash the remaining soil further down and out of the pipe.

**6** The second attachment can now be used, the flexible scraper. This is pushed into the drain and passes over the remaining obstruction. When pulled back, it should bring the solids with it. Having dislodged the obstruction, remove it from the manhole with a trowel.

**7** The third tool to use is the rubber plunger. Locate the upstream chamber (if there is one) and push the plunger back down the drain to ensure the obstruction is well and truly clear. As a precaution, check any other branch pipes that enter both manholes with the rods and rubber plunger. The scraper can also be used to clear any sediment that may have solidified on the bottom surface of the chamber and pipes. Wash the drain through, flush the toilets, and make sure that all is working properly.

**8** Wash down the drain and the benching (the concrete surfaces).

**9** Replace the cover after greasing the edges to prevent rusting and to make it easier to gain access next time. Thoroughly clean all the equipment used.

# CHAPTER 14 Victorian House Manual
# THE SITE

Although your original cast-iron railings may have been carted away many years ago, it is surprising how many period features often remain within gardens, such as original tiled paths and quaint coal-hole covers. But it is here, in the site around the property, that many of the more serious risks to the house originate. Gardens can be minefields of unstable marshy ground, collapsing outbuildings, teetering retaining walls, dangerous electric cables and aggressive shrubs. They are also a potential legal quagmire, with adverse rights of way, ill-defined boundaries and unforeseen liabilities. In this chapter we show how to seek out and eliminate the dangers in your own back yard.

It's a jungle out there. Should you fail to maintain that all-important area immediately surrounding the property, the wellbeing of your home could be at stake. Many of the defects in the preceding chapters originate from neglect here, and even the most innocent-looking suburban gardens often contain dangerous structures and other potentially fatal pitfalls waiting to ambush the unwary.

But then your house might not have had any garden space at all, had it not been for the 1858 Building Act which stipulated '150 square feet at the back, free from any erection except a toilet block' for all new houses. For the better-off, this was the age of the tropical glasshouse, the conservatory and the new-fangled 'machine for cropping or shearing vegetable surfaces of grass plots and pleasure grounds' – aka the lawnmower. The Victorians set the pattern for most modern gardens – a refuge of lawns and flower beds, a haven from the world outside.

# Garden history

By the close of the century most new houses had a status-enhancing front garden, or at least a small paved forecourt around

the bay window. Front gardens were normally enclosed by heavy railings, perhaps set off by a small privet hedge. Early Victorian railings were mostly made in hand-crafted wrought iron, later superseded by moulded cast iron, with individual components bolted

together on site. These were sometimes set directly into the pavement, or more commonly into a low boundary wall built of spare stone or brick, sometimes with a facing of stucco applied to look more expensive. Railings were bedded with load into stone copings on top of the walls, or fixed into place with iron bolts passed through lugs incorporated into the design. Decorative lengths of railing were contained by horizontal bars supported every few feet by a small post with brackets.

In some front gardens you come across small square stumps of metal set into the rendered top of low level garden walls – the

remnants of the original railings patriotically sawn off for melting-down as part of the war effort in 1940 (except, it would seem, for those of the wealthiest homes, many of which remain intact to this day). Today, most off-the-shelf new metal railings and gates are made from mild steel, often replicating original patterns. However, where authenticity is important, it should be possible to have replicas made to order that match period pieces.

Various colours of paint have been found on historic ironwork. 'Invisible' greens (blending in with the background foliage) were widely applied to railings, gates and balcony balustrades. A fake 'bronze' finish, made with powdered copper dust to achieve a greenish hue, was popular, along with dark blues, maroon reds and chocolate browns.

Unless the house directly faced the street, there would

normally be an iron or timber front gate with a tiled path leading to the house. This could be clad with quarry tiles or encaustic tiles with coloured patterns, some incorporating an ornamental cast-iron cover to the coal hole (so coal could be delivered to the basement without disfiguring the hallway), the covers sometimes chained in place to deter thieves and practical jokers.

The level of the pavement does not usually equate to the original level of the plot. The street was normally built up during construction work, so the original ground level can generally be found at the back.

After the 1858 Act even the cheapest terraced houses would have their own back yard, complete with outside toilet and coal shed. Later terraced house designs with joined-up rear extensions meant that your back door faced that of your neighbour across a

shared yard, but most have since been separated by a dividing wall.

In more expensive mid-Victorian town houses the back garden was not always particularly valued and was often uncared for, in sharp contrast to the carefully laid-out ornamental front garden. Anything at the back of the house meant 'service', so the family rarely went there. Later in the century things changed, as the fashion for private garden retreats meant back gardens were rediscovered and small lean-to conservatories became popular. Towards the end of the century, the influence of the Arts and Crafts movement saw the revival of traditional cottage-style gardens.

# No home is an island

Your home might be at risk from things going on in the garden around it. Common danger signs include:

■ Over-enthusiastic garden makeovers that have heaped piles of earth and flowerbeds against the walls.
■ Paths and drives built up in layers over many years, submerging the damp-proof course.
■ Subsiding paving slabs sloping drunkenly towards the main walls instead of draining away.
■ Rainwater downpipes that discharge pools of water next to the house.

Previous chapters have stressed the need for external ground levels to be at least two courses of bricks below the damp-proof course (or about 200mm below floor level) to prevent damp getting inside or damaging brickwork. Airbricks must be kept clear of accumulated earth, plant growth and paint, which can block them and thereby hasten rot and beetle attack to floor timbers.

If your garden slopes down towards the house or is relatively high, the earth should be cut back so that there's a metre or more of permeable paving or gravelled ground next to the walls. This

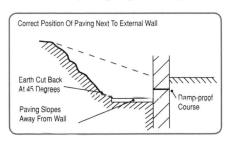

may require building a short retaining wall to the garden. The paving should tilt away slightly towards the garden. But in severe weather,

rainwater from a sloping site can come cascading down in the direction of your living space and needs to be chanelled away. So paths and drives need proper drainage, with channels or gulleys leading to soakaways to prevent surface water ponding near the building.

Where you have concrete paths abutting the main walls, the section closest to the house should be cut out to at least 300mm width and removed. Gravel or natural stone chippings can instead be placed in the space next to the house walls. As we saw earlier, with a small amount of excavation, a simple 'French drain' can be formed comprising a shallow gravel-filled ditch. The gravel should drain freely and allow the walls to breathe again.

Bear in mind also that paving over previously bare earth next to a building can cause the ground to dry out and clay subsoils to shrink, so it's worth leaving permeable gaps with sandy joints between slabs. In fact, if you want to pave more than 5m$^2$ of your front garden you need to use a permeable material such as gravel, permeable paving or reinforced grass otherwise planning consent is required. Known in planning terms as 'SUDS' (Sustainable Urban Drainage Systems) the object is to minimise the amount of surface water finding its way into public drains, rather than soaking harmlessly into gardens.

Downpipes should discharge into gulleys protected with gratings and small surrounding kerbs to ensure rainwater is safely channelled away so it can't cause trouble by seeping back to the walls or pouring through airbricks.

Defective drains are a major cause of structural movement to houses – look for the telltale signs of leakage, like marshy ground, extensive moss on paths and tide marks up side walls.

# Here be danger

Ah, the joy of the garden. The last place you'd expect disaster to strike you personally – until you consider the surprising number of potential dangers lying in wait to accost unwary gardeners:

- ■ Unprotected DIY electric extension cables run to outside lighting and outbuildings (often half-buried in flowerbeds or dangling merrily in the breeze like high-voltage washing lines).
- ■ Rusted ice-thin drain cover 'mantraps'.
- ■ Concealed stagnant ponds or forgotten wells.
- ■ Slimy, slippery paths and timber decking, and ponding water turning to black ice in winter.
- ■ Decrepit rotten sheds and garages, and tumbledown greenhouses awash with broken glass.
- ■ Unstable ornamental brick arches, and loose high-level coping stones balanced precariously on dodgy parapets.

But Number One on the danger list is the innocent-looking garden wall. The problem is, many garden walls are self-built with little or nothing in the way of foundations. As we saw earlier, in order to clear seasonal ground movement even the smallest structures need foundations at least half a metre deep. The fact is, pushing thin, tall garden walls over can be lethal child's play.

If a garden wall is leaning or you can make it move by pushing, it

**Top right:** Solid, chunky and eroding

**Centre right:** Holding it back – short-term solution

**Right:** Very common retaining walls to semi-basements' front gardens

**Left:** Bowing out - may need localised strengthening or rebuilding to safeguard passers by

will need to be rebuilt with decent foundations before a gust of wind blows it over and angry neighbours unleash a pack of slavering lawyers in your direction. Long stretches of wall also need expansion joints every 6m or so, as modern cement mortar is not tolerant of thermal movement and will crack. If a wall is low enough so you can step over it, it should be OK being built of single 115mm thick brick, otherwise it should be 230mm thick. And the cost of wall construction can be surprisingly steep. But remember, if you are contemplating carrying out building work to boundary walls with adjoining properties, legal notices first need to be served under the Party Wall Act 1996, even if you're on chummy terms with the neighbours (see www.victorian-house.co.uk).

Replacing a dodgy old boundary wall with new fencing would be a cheaper, although relatively short-lived alternative (the concrete post and base type will last longer), and fence posts require simple borehole foundations. But a word of warning before you wield the mallet: as noted earlier, serious damage to underground drains is frequently caused by enthusiastic fence-erectors banging in metal spikes along boundaries. So first spare a thought for the location of any hidden underground drain runs, which can be worked out from the position of inspection chambers or from plans at the water company/local drainage authority.

Now we come to the real killers – retaining walls. Fortunately, the Victorians knew a thing or two about how to build them. Giant railway embankments of beautifully laid dark engineering brick, built at a precise angle sloping back to restrain the ground beyond, are still going strong after more than a century. But in residential gardens some were cheaply and thinly constructed. In hilly towns, like Ashbourne and High Wycombe, retaining walls are commonplace, but many are now so old or poorly built that they are a liability, an avalanche waiting to happen. The place to check first is the upper side – on top of the ground that's being held back. Look for tension cracks and dips in the ground, which may be the first sign of subsidence in the retained earth, leading to bulging or overturning where the foot of the wall is pushed out, causing the top to collapse backwards. The thing that finally pushes them over is usually a build-up of water pressure, sometimes aggravated by the weight of cars parked above. So weep holes near the base are essential to relieve pressure, and these should be checked, as they may well have become blocked. For sufficient strength, retaining walls need to be constructed to a height:thickness ratio of about 3:1 (but often aren't).

It is at this point you may rue the day that you bought a listed building or a house in a posh conservation area. It would be bad enough should Local Authority Building Control decide to slap a 'dangerous structure' notice on your rickety old stone retaining wall, requiring it to be rebuilt at your expense. But the cost could triple if the Planners demand reconstruction to exactly the same appearance as the original wall, requiring the same bricks or stones to be re-laid in exactly the same positions with traditional materials.

## OUTBUILDINGS
Even the most superbly renovated properties can conceal guilty secrets in the form of gruesome old structures festering in their back gardens. All manner of noxious horrors can make their unwelcome presence felt: rotten timber shacks, corroding concrete panelled garages and, very commonly, roofs of

**Right:** The classic Victorian outside toilet
**Below:** Watch out! Collapsing outbuilding

corrugated asbestos sheeting.

Old outbuildings are naturally a magnet for children, who will inevitably be drawn to anything potentially lethal or toxic. The best option is often demolition and complete removal, but the cost involved should not be underestimated. Asbestos cement materials are very common and generally not dangerous if left alone (see chapter 8). However, such roofs are notoriously fragile, and could be a legal liability should some daft person unwisely choose to walk on top of one and then fall through, possibly injuring themselves on the razor sharp shards. Council tips can be reluctant to accept asbestos materials and specialist contractors know how to charge.

Beware also of primitive lean-to conservatories and attached garden walls built without any foundations whatsoever, just placed over old paving slabs. They will move around and pull away in tune with seasonal changes in the ground and the vagaries of the weather.

## TREES AND SHRUBS
The finishing touch to many a late Victorian or Edwardian suburb was the planting of rows of neat forest trees. Avenues lined with plane

**Right:** Adjoining tree pushes out wall – collapse could kill a child

**Above:** Garden walls with shallow foundations often show signs of movement due to near trees

**Right:** The dreaded Japanese knot weed

trees, limes and flowering cherries helped create the archetypal leafy suburb. But many have now grown to 20m or more.

All was well until the arrival of long, parched summers, when houses started displaying symptoms of apparent subsidence – cracks in the internal plaster, or to the brickwork, particularly to bay windows, and doors or windows that suddenly decided not to close properly. At least 15 million houses in Britain are built on shrinkable clay, which, as we have seen, is very prone to movement. Insurance companies get an average of 40,000 subsidence claims a year, the majority of which are tree related, with an annual cost of around £400 million, the peak in claims corresponding to periods of drought.

Any fast-growing tree or shrub and thirsty broadleaf trees will extract a lot of moisture from the soil and can upset ground conditions. Roots can extend more than one-and-a-half times the height of the tree, so any structure within that 'influencing distance' can potentially be affected. On the other hand, mortgage lenders' legal advisers often insist that surveyors carrying out valuations insert standard tree-warning phrases at the slightest hint of a shrub nearby, and this often alarms homeowners into a slash and burn mentality and unnecessary bouts of pre-emptive tree felling. The resulting acres of naked concrete may be less threatening technically, but can significantly detract from a property's 'kerb appeal', which in turn can impact negatively on its market value.

You may not necessarily be at liberty to cut down trees even in the privacy of your own garden. Sometimes they are legally protected by a Tree Preservation Order – although this normally applies to more substantial specimens. Similarly, those in Conservation Areas enjoy automatic legal protection and it is an offence to fell or prune such a specimen without permission. If you want to fell, top or lop such a tree, six weeks' notice has to be given to the Local Authority so they can consider the contribution the tree makes to the character of the area. There are some exceptions, such as trees with a trunk diameter less than 75mm at a height of 1.5m above ground level, and ones that are dead or dangerous.

Depending on the specific circumstances, a suitable remedy might involve pollarding (severe pruning) to restrict their growth and moisture uptake, or complete removal of the offending tree, or insertion of a root barrier into the ground. The best advice is to first obtain an arboricultural report from a tree specialist before jumping to conclusions. Today, insurance companies are much more likely to monitor cracking over several months and may well conclude that it is due to settlement from cyclical movement, albeit influenced by nearby trees.

Another tree-related issue is that of the infamous leylandii (cypress conifer), which can grow nearly a metre per year, reaching over 25m unless regularly trimmed, blocking natural light to all around. If not maintained, the Control of High Hedges and Anti Social Behaviour legislation allows Local Authorities to issue remedial notices enforcing action where trees are in excess of 2m (6.5ft) high.

The proverbial back alley – to millions of workers' terraces

Back-to-back houses with 'tunnel' passageway to shared back yard

Typical side gate for right of way across back gardens to terraced houses

We are not amused - terracotta Disraeli

But it's not just trees that can cause trouble. Some less lofty forms of garden life can be surprisingly pernicious and aggressive. One shrub to be especially wary of is fast-growing Japanese knotweed. So seriously is this highly invasive weed regarded that mortgages can be declined where infestation is evident. It has become fairly widespread in recent years and left untreated can become strong enough to damage foundations, concrete and tarmac. Complete eradication is difficult because the roots can penetrate to depths of 3m, requiring a mix of weeding and pesticide. Knotweed-contaminated soil is classed as controlled waste that has to be removed by a licensed operator to a designated landfill site. Control and eradication is an expensive, specialist job.

# Flying freeholds and rights of way

Some older terraces have small passageway 'tunnels' that allow shared access from the street to a courtyard of gardens at the rear. Where there are bedrooms located over the passageway below it is known legally as a 'flying freehold', a phenomenon that solicitors like to mull over at some length, with regard to rights of access and the liability for maintenance, in a similar way to flats. There may also be practical issues to consider, such as a lack of insulation (gales blowing under your bedroom floor).

Victorian terraces also often have peculiar rights of way that may not be obvious on site, frequently running across a series of back gardens. There may be clues from the position of nearby gates in fences, or perhaps the odd group of militant ramblers trekking over your prized tomato crop. Only the land registry deeds can clarify the true position. You may even be the proud owner of a separate 'non-contiguous' back garden somewhere down the road. In industrial cities many older terraces were built with a back lane so that coal could be delivered, and to allow for collection of refuse and toilet contents.

So when buying a Victorian property, all such rights of access should be thoroughly investigated by your solicitor along with the usual pre-purchase checks, such as confirming which boundaries are your liability and enquiries concerning any planning applications, any history of flooding, and the location of wells and old mineshafts.

# LIVING FOR TODAY – MAKING MORE SPACE AND BOOSTING INSULATION

**Above:** Solar panels generate electricity, but add loadings to old roofs
**Above right:** The Victorian Terrace Project by BRE proved that old buildings can be retro-fitted to match or exceed newbuild thermal efficiency standards

Most Victorian and Edwardian houses have scope for a spot of upgrading to suit modern lifestyles. There are two main areas where improvement can be especially beneficial:

■ Thermal efficiency: Older solid-wall buildings tend to leak more heat compared to new houses, so boosting insulation levels can have a significant impact with lower energy bills.
■ Making more space: Most homes have good potential for extending or converting lofts. Kitchens can be bit cramped by today's standards, so filling in the side return is a popular improvement that can reap major dividends.

There is one other area where older buildings can sometimes benefit from beefing up – home security. However, we sometimes forget that crime is nothing new, and most Victorian houses were built with effective defensive measures. Front gardens were commonly fortified with legions of spear-like cast-iron bars to deter potential intruders; if yours are missing, this is an original feature that's well worth reinstating. Unlike boundary walls, hedges or fences, railings cleverly combine the obstacle of a physical barrier with visual exposure – visibility through railings is a powerful burglar deterrent. A second line of defence was provided with stout bolts on sturdy main entrance doors, and internal shutters sometimes fitted to windows. However, to comply with current insurance requirements there are some fairly straightforward improvements that can be made to upgrade locks to doors and windows – see chapter 12.

# Making your home warm and comfortable

Compared to most modern buildings, Victorian houses are considerably more 'sustainable'. They were built using locally sourced, natural materials, and are easily repaired. Plus of course they've lasted a very long time. The one area where they sometimes fall short is in 'hanging onto their heat'. Fortunately, it's normally possible to upgrade them to meet modern thermal efficiency standards.

## ENERGY PERFORMANCE

All residential properties being sold or rented must by law have an Energy Performance Certificate (EPC). These measure the building's

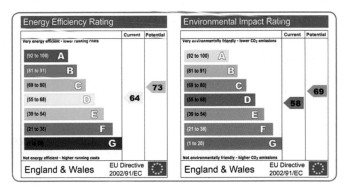

energy efficiency, from a cosy A down to a cold and draughty G. But EPCs aren't always terribly accurate for older solid wall properties. And the 'recommended improvements' churned out are computer generated and shouldn't be taken as gospel.

Repair is always a better option than replacement. Yet current energy policy is designed to steer homeowners towards wastefully replacing sound old windows and doors for example with expensive new short-life products on the grounds that they take less energy to run. This, of course, completely ignores the energy and resources used to manufacture these new products as well as the embodied energy in the old ones that are thrown away.

Thermal imaging: 'warm' colours (red, orange, yellow) show surfaces leaking most heat

Expanding foam is useful for larger sealing gaps that aren't visible

and doors, and checking letter boxes, key holes, cat flaps and loft hatches can pay big dividends.

Some gaps are blindingly obvious, such as spaces between old floorboards or where a door or window doesn't close properly. Others require a bit more detective work. Using special 'smoke puffers' can show how big the draughts are and where they go. Or if there's a smoker in the family, invite them to light up! If funds permit, you could go one better and buy or hire a thermal-imaging camera to highlight the white, yellow and orange 'hot spots' where valuable warmth is seeping out of your property.

Alternative solutions, such as draught-proofing and secondary glazing, are much cheaper and can be highly effective at improving comfort and reducing heat loss, while consuming far less energy.

## DRAUGHT-PROOFING

Of all the possible energy-saving measures you can carry out, draught-proofing will normally give the fastest payback, and is a straightforward DIY project. The problem with draughts is that they make our homes feel cold and uncomfortable – so the natural reaction is to turn up the heating to compensate, wasting energy and pushing up fuel bills. As if that wasn't bad enough, the same gaps and cracks that invite cold air in, also allow expensively heated warm air to escape. So sealing gaps in floors, windows

Draughtex.co.uk

Special artificial rubber filler strips can seal gaps up to 11mm without adhesive 'Draughtex' rubber

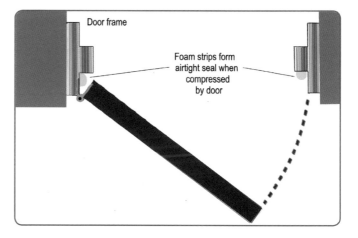

Door frame

Foam strips form airtight seal when compressed by door

## U-values

The Building Regulations set targets to minimise heat loss, not just in new homes but also for upgrading existing properties. Heat loss is expressed in terms of 'U-values'. These describe the amount of heat leakage (measured in Watts) transmitted through 1 square metre of a wall or roof etc. The lower the figure, the less heat should escape.

U-values are expressed in $W/m^2K$ – that is watts per square metre for each degree of difference between the indoor and outdoor air temperature. So if the walls have a stated U-value of $0.30W/m^2K$ it means that 0.30 watts of heat can pass through each square metre of wall for every degree of temperature difference. This is a useful way to compare performance; for example, a wall with a U-value of 1.0 will lose heat twice as fast as a wall with a U-value of 0.5 $W/m^2K$. A typical uninsulated 230mm-thick solid-brick wall in a Victorian house has a relatively poor U-value of $2.1W/m^2K$, compared to about $0.25W/m^2K$ in a super-insulated new home.

# Insulation

Many households are paying for warmth lost through cold walls, roofs, floors, windows and doors. So it obviously makes sense to stem this waste of valuable heat and reduce energy consumption, thereby saving money and making your home more comfortable.

Some improvement works can be done any time the fancy takes you – such as laying loft quilt or draughtproofing. Other more disruptive jobs, like insulating timber floors, are better suited to being done on a room by room basis or as part of a major refurbishment. So if you're planning to redecorate or re-fit the kitchen, bathroom or cloakroom any time soon, it could be a good opportunity to line the walls internally with insulation.

As with all home improvements, it's always a good idea to take a few minutes at the outset to plan how life can go on around the works with the minimum of disturbance. As a rule, the more you can do to reduce dust, mess and power cuts the more harmonious your home life will be!

Thermafleece

Natural sheep's wool insulation

## WHAT CAN GO WRONG?

Probably the biggest risk from insulating your home is a hidden time-bomb – dampness caused by condensation becoming trapped in walls. Any timber in contact with damp insulation will be more at risk of decay. And once insulation becomes wet, its effectiveness is massively reduced, and can even suck heat out of a property.

Cold spots attract condensation and mould

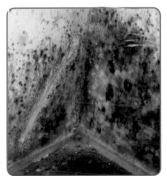

Extreme mould – cold, unvented cupboard in house with steamy, humid air

## CONDENSATION AND THERMAL BRIDGING

It's a fact that warm air can hold more water vapour than colder air. So a warm, well-insulated room can be surprisingly humid – especially steamy bathrooms and kitchens. This isn't necessarily a problem unless the warm, humid air finds a way to leak through gaps in walls, floors, or ceilings. Then as soon as it comes into contact with a cold surface it will cool and condense back into water. Lining walls internally makes the wall concealed behind the new insulation colder, as it's now cut off from the warmth of the room. So any moist air that gets through gaps in the insulation and hits the cold wall surface will condense, causing damp inside the structure of a property. This is known as 'interstitial condensation'. Being out of sight, this can go unnoticed until serious decay has taken hold – for example if it drips down the wall to affect hidden floor timbers.

Dampness from condensation is also attracted to any areas where insulation is thin or has been omitted. This is down to 'thermal bridging' – where cold from outside is able to physically penetrate into the house. Notoriously 'cold spots' include the reveals around openings for windows and doors, and sloping

Extractor fan neatly ducted to outside via roof vent, in well insulated loft

upstairs ceilings near the eaves. When moist, warm air encounters these cold surfaces, it will condense into water – eventually attracting an ugly patina of black mould. So it's important to make the insulation as continuous as possible without any gaps.

## VENTILATION AND AIR-TIGHTNESS

To minimise condensation the obvious thing to do is reduce the amount of moist air that's generated in the first place, *eg* by ensuring clothes driers are properly vented to outdoors, and not

Some types of extractor fan can recover up to 75% of the heat from a room's air normally lost through extraction

Envirovert.co.uk

boiling food. The next best thing is to get rid of it before it has a chance to cause any problems. Ventilation is important to prevent damaging levels of damp building up from airborne moisture. The aim is to have *controllable ventilation* by fitting extractor fans to kitchens and bathrooms, and trickle vents to windows. Or simply get into the habit of opening a window for 15 minutes after showering. But when fitting new extractors you need to consider where to locate them on the external walls so the character of the house isn't spoilt.

The next line of defence is to prevent moist air entering the structure of a building. This can be done by sealing newly insulated surfaces with vapour-control membranes laid over the insulation on the (warmer) room side. These are basically polythene sheets designed to stop the passage of airborne water vapour.

However, for older buildings with solid walls there's an alternative approach that can be more appropriate. Instead of blocking all moist air from getting through, it's accepted that a certain amount may get into the fabric. This is OK as long as it's free to evaporate out again through 'breathable' surfaces – emulating the way old buildings deal with moisture. By using natural insulation materials that work in tune with traditional materials, such as sheep's wool and wood-fibre boards, the moisture doesn't get trapped and is free to escape.

# Insulation materials

There are different types of insulation made from different materials:

## Foamed plastics

*Eg* Polyurethane and polystyrene.

These are mostly supplied as rigid boards designed for insulating walls, roofs and floors.

Polyurethane boards (often referred to by brand names such as 'Kingspan' or 'Cellotex') offer the best performance of all the mainstream products. These are about twice as effective as polystyrene (*eg* white Styrofoam similar to packaging material) but are also more expensive.

## Mineral based

*Eg* glass fibre and rock/stone wool.

Best known as mineral quilts used for loft insulation. Some types are now made from recycled glass bottles with a soft texture similar to cotton wool.

## Natural materials

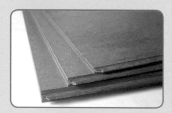

*eg* Sheep's wool, wood-fibre, hemp, cellulose fibre

Natural materials tend to be a little dearer but are generally much nicer to work with. Not surprisingly they score best in terms of environmental impact. Being 'breathable' they suit older buildings with solid walls.

## Reflective foils

'Multi-foils' comprise thin sheets of aluminium, rather like *Bacofoil*, and multiple layers of tissue-like insulant or 'bubble wrap'. Also known as 'radiant heat barriers', these work primarily by reflecting solar energy. Relatively thin strips of multifoil can insulate as effectively as thicker layers of mineral wool but need a clear air space either side for optimum performance.

## Hi-tech

*Aerogel and vacuum panels*

These recently developed 'space-age' materials offer extremely high

performance, and only need very thin layers to match the best mainstream materials. Although very expensive, they can be useful in areas where space is limited, such as window reveals.

**Different materials** are sold in different physical forms making them better suited to insulating specific parts of a building:

## Loose materials

These are handy for reaching awkward spots in lofts etc. and can be laid by hand or blown-in using special machines. One of the best known is cellulose fibre (recycled newspaper). Loose tufts of mineral wool or small beads of polystyrene are also widely used as cavity-wall insulation.

## Fluffy quilt

Soft, fluffy (and sometimes itchy) materials sold in rolls of quilt include mineral wool, glass-fibre wool, as well as natural materials such as sheep's wool and cellulose. Widely used for insulating lofts.

## Rigid boards

Mostly made from petrochemicals, such as polyurethane and polystyrene, although some natural materials like wood-fibre boards are also available. Ideal for lining solid walls, floors, ceilings and flat roofs.

## Semi-rigid batts

Flexible batts sold in handy chunk sizes are a 'halfway house' between fluffy quilts and rigid boards. Made from mineral wool, sheep's wool and hemp etc., these can be wedged between studwork, joists or rafters.

Levels of heat loss in a typical house through different parts of its 'thermal envelope'

## APPLYING INSULATION

The percentage heat loss attributed to different parts of a typical property's envelope are widely quoted as:

- Roof 25%
- Walls 35%
- Floor 15%
- Doors and Draughts 15%
- Windows 10%

Some of these areas are easier and more cost-effective to insulate than others. So let's take a look at each in turn.

# Roofs

It's a well-known fact that nearly half of our body heat escapes via the tops of our heads – hence the importance of wearing a hat in cold weather. The same logic applies with buildings because heat rises. So applying generous layers of insulation at roof level can pay big dividends.

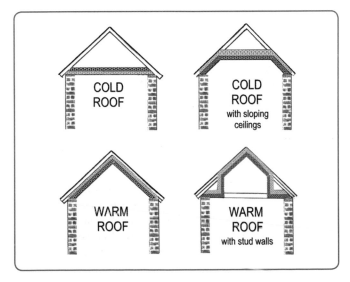

## LOFT INSULATION

The loft is one of the most effective places to insulate a house as well as usually the easiest. Loft insulation can be laid in rolls or as loose lay. A minimum depth of 270mm is generally considered sufficient to meet Building Regulations heat-loss targets. The most widely used

materials are rolls of mineral wool quilt (either glass-fibre or rockwool), which also have good fire resistance and sound insulation qualities. However, these materials are irritants to the skin and respiratory systems, so some are now produced in pre-wrapped form. New 'eco' loft quilts claimed to be free from floating fibres. Natural insulation materials such as sheep's wool quilt or loose cellulose (made from recycled newspaper) are more pleasant to work with and are well suited for use in older buildings. To reach awkward spaces, loose lay cellulose fibre insulation can be sprinkled by hand or blown in by machine.

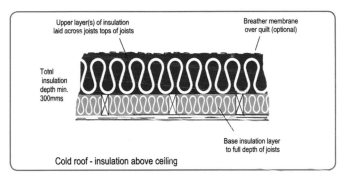

There are a few key points to bear in mind when laying loft insulation. First of all, try to avoid pushing it into the eaves because ventilation paths at the edge of the roof are important for keeping roof timbers aired and dry, to avoid timber decay. Victorian slates and roof tiles were generally laid without any roofing felt underneath, which allowed plenty of air to enter via lots of very small gaps, keeping roof spaces nice and airy. But many old roofs have subsequently been lined with underfelt, hence any remaining ventilation paths are all the more important. To maintain a clear gap between the underfelt and the top of the insulation quilt, purpose-made ventilation trays can be inserted between the rafters at the eaves.

Where there are recessed lights to ceilings below, gaps are normally left in the loft insulation so the bulbs don't overheat. But this blows a hole in the your home's defences, allowing steamy air and valuable heat escape into the cold loft, causing dampness and wasting energy. One solution is to replace the bulbs with modern LEDs, which don't emit any significant heat so recessed lights can safely be boxed and insulated on top. Better still, recessed light fittings can be replaced with sealed shower lights. LEDs consume a fraction of the electricity and are claimed to last for 25 years. Bear in mind also that if electric cables serving power sockets or showers are buried deep within insulation they can potentially be prone to overheating. Instead cables should be run on the surface or within conduits, or higher capacity cable used.

## RAFTER INSULATION

Roof spaces can also be insulated higher up at rafter level – although to justify the cost of keeping it heated this normally only makes sense where you want to make use of your loft space. To achieve sufficient thermal performance, you might need to install approximately 50mm depth of rigid insulation between the rafters, with another 70mm or so lining the undersides of the rafters.

To comply with Building Regulations, an air gap of at least 50mm must normally be left beneath the roof tiles for ventilation. However, this may not be necessary where the roof tiles aren't underfelted or where there's a modern breather membrane underlay.

Alternatively, modern multifoils can be used to provide 'under rafter insulation' by being stapled in place to the rafters and secured from inside the loft with horizontal timber battens. These sheets of silver foil insulation comprise several thin layers of reflective and insulating material. They are heat-reflecting and although only 30mm or so thick, claim performance equal to much thicker insulation quilts. However, to achieve optimum efficiency, a clear air space of about 20mm needs to be maintained around the material. Multifoils can also come in useful for insulating awkward, narrow spaces in old houses.

Insulating from above is technically the most effective solution. Where insulation boards are placed above the rafters it forms what is known as a 'warm roof'. This is obviously a major undertaking, and because it will raise the height of the roof there may be planning implications (at least for main roofs). However,

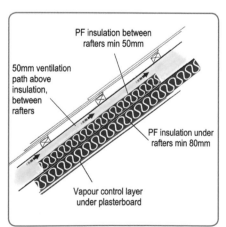

where an old roof needs to be stripped anyway, or you want to insulate smaller subsidiary roofs, it can be a good opportunity to strip and re-lay the tiles over new insulation placed on top of the rafters, such as sheets of wood-fibre sarking board.

## SLOPING CEILINGS

Many larger Victorian homes were built with attic rooms for servants and nurseries, with the lower rafters around the edges of the rooms lined in lath and plaster. But because no insulation was installed, these short, sloping ceilings close to the eaves can be very cold. Insulating sloping ceilings is not usually practical from above, as stuffing insulation down them will block ventilation to the loft. The optimum method is to strip the roof and lay insulation boards externally before recladding it, but this will raise the height of the roof. Alternatively you could take down the old lath and plaster and replace it with thicker insulation and plasterboard, but this destroys old fabric and is incredibly messy. It may be a better solution to fix sheets of insulated plasterboard over the existing surface secured with dry wall screws into the rafters, although there is a small penalty in terms of loss of headroom.

## LAGGING TANKS AND PIPES

To protect against frost, cold water tanks should be fully insulated, and any pipework run through lofts should be lagged (as should pipes in other cold spaces such as floors and larders). This should prevent pipes bursting due to water freezing in winter and will also reduce the potential for condensation forming on cold surfaces and dripping down causing damp patches.

Loft insulation is usually omitted to ceilings under water tanks, the theory being that warmth from the rooms below can seep upwards, keeping frost at bay. This relied on tanks being draped with insulation forming a 'tent' so that warm mist air from below can't enter

*Diagram labels:*
PF insulation between rafters min 50mm
50mm ventilation path above insulation, between rafters
PF insulation under rafters min 80mm
Vapour control layer under plasterboard

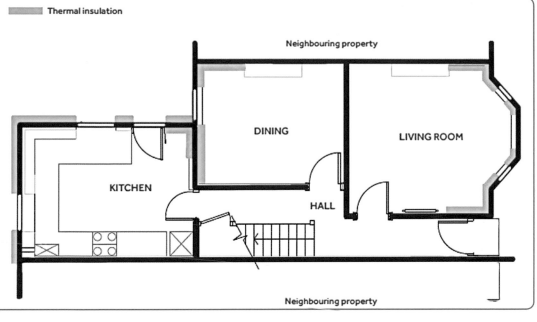

Victorian terrace showing mix of internal and external solid wall insulation. Party walls don't need lining.

Thermal insulation

Neighbouring property

DINING

LIVING ROOM

KITCHEN

HALL

Neighbouring property

the roof space causing condensation and heat loss. In reality, of course, tank lagging is patchy at best. So a better approach is to line the underside of the tank or deck with rigid polyurethane insulation. One potential danger to watch out for is fluffy white asbestos lagging, which, although very rare, is sometimes found on pipes and tanks (particularly in larger Edwardian houses). These can emit dangerous fibres and require specialist removal.

# Walls

Solid walls can be insulated either by lining them on the inside or the outside. The main concern is that adding layers of insulation can potentially act as a barrier to the natural breathing process of old walls, trapping moisture. More obviously, there is also the danger of ruining the character of a classic old building by obscuring period features or damaging fragile historic fabric. And trashing the kerb appeal of a Victorian house means it will be worth less when you come to sell.

## EXTERNAL WALL INSULATION (EWI)

Cladding the walls externally with thick insulation boards can give excellent thermal performance. But external cladding will obviously have a radical effect on the appearance of an old house. So it makes more sense on less visible side or rear walls and is likely to be more appropriate in houses with rendered walls. The works typically involve fixing rigid insulation boards to the outer face of the walls with special adhesives and fixings, followed by two coats of render. However, making walls thicker raises

Special super-wide gutters to eaves, and verge extension, provides neat solution avoiding the need to extend the roof to accommodate fatter externally insulated walls

Prewett Bizley Architects

End terrace with external insulation to flank wall

External insulation can be clad, eg with traditional tiles

Simmonds Mills Architects

Kingspan

Frewett Bizley Architects

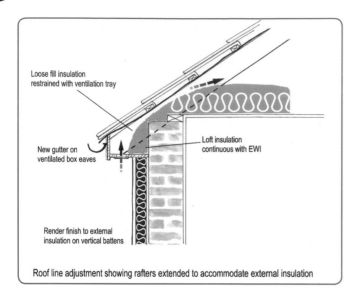

Loose fill insulation
restrained with ventilation tray

New gutter on
ventilated box eaves

Loft insulation
continuous with EWI

Render finish to external
insulation on vertical battens

Roof line adjustment showing rafters extended to accommodate external insulation

**Above:** Lining internal walls with rigid boards secured with adhesive dabs

**Below:** Slim high performance aerogel pre-bonded boards applied to timber battens (right) or direct to wall for minimal loss of floor space

Kingspan

Proctorgroup.com

a number of practical problems, such as extending roof overhangs and window sills, modifying window and door openings, and integrating it with any adjoining walls next door, plus the headache of trying to re-route downpipes and flues etc. Dealing with these sorts of issues, together with the need for scaffolding, makes EWI a relatively expensive option.

But probably the biggest worry is the fact that sealing up old solid-wall buildings by sticking giant slabs of polyurethane onto their outsides can store up trouble by trapping damp. To reduce this risk, works are best carried out after a dry spell in the early autumn when solid walls should be relatively free of moisture (both from minimal rainfall and less risk of condensation indoors). Where this is a concern, natural permeable insulation materials can be used, such as wood-fibre board, with a lime render finish that allows the walls to breathe. Alternatively, where it matches the local vernacular, traditional tile cladding hung from battens can be a good solution for very exposed walls.

## INTERNAL WALL INSULATION (IWI)

Lining the inside of the main walls in your home can achieve very good improvements in thermal efficiency and can also be a suitable DIY project. The main drawback is the disruption involved in removing and reinstating skirting boards, shelving,

radiators, sockets, switches etc. Staircases and period features such as cornices need to be integrated with new wall linings, and getting the detailing right around windows etc. can be tricky. The increased thickness of insulated walls also means there will be a small (often imperceptible) loss of room space.

Internal insulation can be an appropriate solution for many Victorian houses, where their historic visual appearance and 'kerb appeal' is important. A good compromise can be to insulate externally to the narrow rear addition kitchen area, and internally to the main house. Terraced properties have the advantage of fewer main walls that need upgrading. Lining the walls internally has the

Lining walls internally with rigid polyurethane insulated plasterboard, fixed to battens

Kingspan

Mineral wool quilt applied to timber framework before plasterboarding

Knauf

added advantage of providing a smooth new surface to decorate.

There's some debate about whether it's appropriate to fit standard non-breathing insulation boards to the inside of the main walls. This is less of an issue than with external insulation, and as long as the walls aren't damp and are free to breathe on the outside, it may be an option. One solution is to use natural breathable insulation materials. It can also help if the works are carried out after the summer when the walls should normally be at their driest.

The works typically involve fixing rigid insulation boards to the walls with special adhesives and screw fixings. Alternatively, batts of mineral-wool quilt can placed within a timber framework, with a skimmed plasterboard finish. As noted above, the main worry when insulating the inside of solid walls is the risk that warm, moist air from indoors could penetrate through any gaps in the new insulated lining and condense on the cold wall behind – hence the importance of good detailing to prevent 'interstitial' condensation causing damp and rot problems. The provision of a plastic sheet vapour barrier (with joints overlapped and taped) before plasterboarding will act as a barrier to water vapour.

An alternative solution that allows the wall to breathe (eg where there's any history of dampness in the wall) is to build an independent new insulated inner wall with a ventilated space left between it and the main wall. For example, a new 100mm timber studwork insulated inner leaf can be constructed leaving a 25mm air gap between the studwork and the main wall. Any moisture in this cavity should be able to escape via vents to the

Woodwool boards are breathable and can be lime rendered - here shown hammering oakum caulking into gaps at edges

outside, or via the floor or a void above. The studwork can be filled with natural sheepswool, cellulose or hempwood batts and lined on the room side with 20mm woodfibre board for a traditional lime plaster finish.

## HOLLOW WALLS

As we saw earlier, a small number of early cavity wall houses were built in the late Victorian and Edwardian periods. These had relatively thin cavities with simple cast-iron wall ties, and the ventilation through cavities helps disperse moisture. Injecting such walls is not normally advisable, as the cavity is less than 50mm and there is increased risk of rusting wall ties. These walls should be treated as if they were solid and are best lined internally.

# Windows

As we saw in chapter 7, the 'payback period' for replacement double glazing is generally very poor. It can take a lifetime to cover the cost of installation calculated from the resulting savings in energy bills. Also, it's a bit of a myth that UPVC windows 'last forever'. Their lifespan is typically 30 years or less. UPVC is vulnerable to UV degradation and cannot be adapted and repaired like timber. And sealed units commonly fail and become 'misted'

Restoration of period windows is often the best option

Timberrepairs.co.uk

**Below:** New double glazed sashes installed; refurbishment of bay window, including draught seals and repairs
**Below right:** High performance secondary glazing

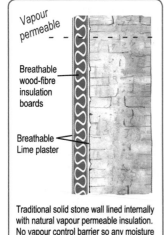

Vapour permeable

Breathable wood-fibre insulation boards

Breathable Lime plaster

Traditional solid stone wall lined internally with natural vapour permeable insulation. No vapour control barrier so any moisture in the wall can evaporate out.

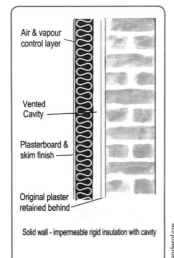

Air & vapour control layer

Vented Cavity

Plasterboard & skim finish

Original plaster retained behind

Solid wall - impermeable rigid insulation with cavity

Slenderglaze

SelecaGlaze.co.uk

One fact the double-glazing industry doesn't go out of its way to advertise is that much of the benefit gained from replacement double glazing is down to the reduction in draughts by fitting new window frames. So where you still have the original windows, overhauling them and draught-proofing can reduce air leakage by as much as 50%, slashing the amount of energy needed to heat the room.

The Victorians made their homes cosy at nightime by enclosing windows behind thick curtains (and sometimes with shutters). Entrance doors might be draught-proofed with curtains in the daytime as well. Studies by English Heritage have shown that thick curtains can reduce heat loss through a window by 41%, and shutters by up to 50%, while also helping to reduce noise. Furthermore, air infiltration through a sash window in good condition can be reduced by as much as 86% by draught-proofing.

Fitting internally opening secondary glazing on the inside of your existing windows can achieve excellent results. Secondary glazing can even be double glazed, with performance equivalent to many replacement windows. Installing good-quality, low-emissivity, secondary glazing, can reduce heat loss by 58% achieving U-values as low as 1.7 – more than meeting current Building Regulations targets. A number of different designs are available, including some that operate very discreetly like shutters, or simple DIY versions where the frames are held in place with magnets facilitating easy removal in summer.

The big risk with replacing windows is that it could seriously damage the historic nature of the property, if carried out insensitively. So rather than replacing entire widows, a good compromise can be to overhaul and retain the existing frames but reglaze them with special slim double-glazed units. Made to exacting historical likenesses, featuring replica period glazing bars, these are sometimes so authentic that they can be hard to discern from the originals. Yet high-performance krypton-filled sealed units, incorporating warm edge spacers can achieve impressively low U-values. However, new double-glazed panes will inevitably be far thicker than the original glass, and may not always fit existing rebates. It will also be heavier, putting additional strain on hinges and sash balances.

# Floors

In a typical home the amount of heat lost through the ground floors accounts for around 15% of the total. This is less than the amount lost through roofs and walls, but is still significant.

### TIMBER FLOORS
Ground floors constructed of suspended timber are almost ubiquitous in Victorian and Edwardian properties. These can sometimes result in draughts through gaps, which can be sealed as follows:

■ Gaps of a centimetre or more are often found around the edges of the room to the skirting boards. Strips of artificial rubber draught seal can be compressed and inserted, or expanding foam carefully applied. Smaller cracks along the top and bottom edges of skirting boards can be sealed with silicone mastic.

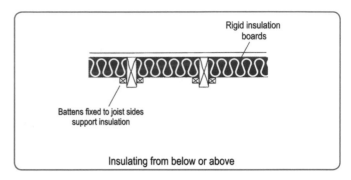

Insulating from below or above

■ Larger gaps between floorboards can be sealed with purpose-made slivers of timber bedded in place with sealant (where exposed floor surfaces are visually important). For smaller gaps compressible foam rods or beads can be pressed into place (available in a range of diameters). Silicone sealant can also be very effective, either clear or colour matched. Where there are extensive gaps or the floor coverings are in very poor condition, draughts can be reduced by screwing hardboard sheets over the boards and sealing the edges. Carpet or laminate etc. laid over rubber underlay can be very effective at preventing draughts.

Where you're planning to strip and polish the floorboards it's a good opportunity to insulate them. This can either be done from above, by temporarily removing all the boards, or from below by lifting a few boards to get access to the void below. In most cases there's a generous amount of 'crawl space' to the sub floors. The ideal situation is where there's a cellar or basement below permitting easy access to the exposed floor structure.

There are a number of different ways floors can be insulated, for example rigid insulation can be wedged between joists and secured on

battens, or mineral-wool quilt batts can be held in place with plywood strips or sheets of breather membrane stapled underneath to the joists.

Upstairs floors generally have no need for insulation unless they're located above cold passageways. However, where improved acoustic insulation would be beneficial, mineral-wool loft quilt laid between joists can be very effective at sound deadening.

### SOLID FLOORS
Solid concrete floors are relatively difficult to insulate. This is because laying a new insulated 'floating floor' on top of the existing surface raises the floor level, with all kinds of knock-on

Side return kitchen extension with enlarged rear opening for bi-fold doors

effects to doors and especially to stairs. Also, many Victorian houses have beautiful original floor tiling that is important to preserve. But rigid insulation boards can sometimes be laid on top, polyurethane foam boards generally providing the most thermally efficient floor insulation. But in practice, all the

associated disruption to internal joinery, fittings and pipework can make this an unrealistic proposition, unless you are carrying it out as part of a major refurbishment.

*[For detailed practical guidance see The Haynes Home Insulation Manual].*

# Making more space

When it comes to adding more living space to a Victorian house, there are a number of possible options. Much will depend on whether the property is terraced, or has already been extended. The least expensive options are usually adding a small conservatory, or taking out internal walls to change the layout (which is not always advisable – see chapter 9). But to gain significant new habitable space you normally need to build an extension or convert the loft.

## CONSENTS AND DESIGN

Before making any major alterations it's important to have the necessary consents in place, so everything's above board and legal. This may involve Planning, Building

Regulations, and the Party Wall Act. And if your building is Listed or located in a Conservation Area, complying with Planning requirements will be more onerous. When it comes to designing your project an architect or a specialist building surveyor can provide professional advice. Then there's the all-important question of picking the right builders and managing the project.

New cavity walls laid in Victorian style solid wall. Flemish bond with 'split header' bricks

Unusual end terrace with corner bay; extension with nice matching brickwork, windows and (artificial slate) roof

Carefully excavated trench with exposed drain runs

## PLANNING

Most renovation work isn't likely to concern the Planners. Under the Permitted Development Rules (PDRs) even quite sizeable extensions can often be acceptable without the need for a formal application. But in most cases an application will need to be made.

Planning rules are far stricter in Conservation Areas, where any significant alterations that are visible to the 'principle elevation' (normally the front) will need consent, including new windows or doors. Where a property is Listed it will enormously restrict your freedom to carry out all but the most minor changes. Any proposed works, including internal alterations will need Listed Building consent. However, like-for-like repairs do not usually require consent. If in doubt, consult the Conservation Officer at your local authority.

The application process usually takes around eight weeks. If

Mix of trad and modern works well.

it's ultimately refused you have a right to appeal within six months. But the best advice is to carefully run through your plans in advance with the local planning team to pre-empt any possible problems. Once planning consent has been granted it normally remains valid for a period of three years, so if work has not been 'substantially commenced' during that time, a fresh application will need to be made.

## BUILDING REGULATIONS

A Building Regulations application is required for more types of work than you might imagine. In addition to regulating matters such as structural alterations, fire safety, drainage and new building work (*eg* extensions), the Building Regulations are increasingly focused on thermal insulation and energy efficiency. Approval of works such as electrical wiring and replacement windows is normally delegated via approved installers where a 'competent person' can certify that work has been carried out in compliance with the Building Regulations.

For any structural alterations or new building work the application will normally require detailed drawings accompanied by an engineer's calculations. Unlike simpler planning drawings, these need to include detailed information specifying precise details of openings and the strengths and thermal performance of materials.

When it comes to submitting your application, there are two options: a 'proper' 'Full Plans' application, or the short-cut method known as a Building Notice. The fee is the same in both cases, but with a Building Notice you can normally commence work 48 hours after giving the minimum two day's written notice. With this method you are basically making a promise up front that you will comply with the Building Regulations on site, rather than submitting detailed drawings to prove it in advance. But the big risk is that a site inspection could later uncover something that doesn't comply, which could prove highly disruptive and expensive.

So for all but the smallest works, it's normally best to make a Full Plans application, not least because it encourages you to think through the details in advance, thereby reducing the risk of problems arising later on site. It also means your builders will

## Building Regulations – self certification

Approval of works such as electrical wiring, replacement windows and heating installations is normally delegated via approved installers registered as *'competent persons'* who can certify that work has been carried out in compliance with the Building Regulations. Certain trade bodies are allowed to 'self-certify' their members' work and issue completion certificates:

1. FENSA – Replacement windows should be installed by a company registered with Fenestration Self Assessment (FENSA) who will be able to self-certify the compliance of the works (not required where only the glass is being replaced).
2. GAS SAFE – registered contractors can issue certificates for installations and alterations to gas, hot water and heating systems. But the contractor must be a registered member, not just a service engineer.
3. OFTEC – the equivalent of Gas Safe for oil fired boilers and appliances
4. HETAS – the equivalent of Gas Safe for solid-fuel-burning boilers and appliances
5. Electrical Contractors – must be registered under one of the 'Part P' schemes in order to issue certificates for domestic electrical work.

have an approved set of plans to work to. The application itself should take no more than five weeks, but you can still choose to take a chance and start work on site prior to approval. Once approval has been granted you need to use it or lose it, and start work within a three-year period. You need to notify Building Control at least two days prior to start on site, and then at key stages so their surveyor can inspect the works as they progress.

## PROFESSIONAL ADVICE

For larger jobs you are likely to require the services of a professional advisor. A building surveyor or architect can be appointed to draw up plans, write a specification and tender the job, as well as advising on complying with relevant legislation (never assume your builder 'knows the law'). For larger projects they should be able to perform the role of project manager and administer the contract fairly between the employer and building contractor on site up to completion. A structural engineer is normally required to design structural alterations, and Building Control are likely to request engineer's calculations to be submitted with your application.

## THE PARTY WALL ACT

If you need to carry out building work to a wall shared with neighbours, the Part Wall Act comes into play, as a legal requirement. This primarily applies to works to party walls in terraced and semi-detached houses. But it also applies where you want to excavate foundations within 3m of an adjoining property (in some cases within 6m), and can equally relate to works affecting walls on boundary lines between gardens. You have to serve a formal notice on the owners of the adjoining property at least two months before the work commences, which normally means having to appoint a Party Wall Surveyor.

## CONSERVATORIES

Conservatories are a relatively inexpensive way of adding extra space to your home and can often be built without the need for a planning application under the Permitted Development Rules (PDRs). But to qualify as a 'conservatory' as opposed to a highly glazed extension or garden room (which are considerably more expensive to build) there are a number of very specific rules that

**Left:** Side return extension with enlarged kitchen

**Right:** Original Vic conservatory overlooks neighbours – planners would never allow this today!

# Top tips for employing builders

- Only select builders who have good experience of Victorian buildings.
- Get at least three written quotes (not estimates). For larger jobs this should be priced against an itemised specification that you've provided. Get the cost broken down into as much detail as possible, so you can see what you're getting and use it as a guide for any additional works.

- Check if VAT is included in the quote. Individual trades my have a turnover that's below the VAT threshold and quite legitimately not need to charge VAT.
- Ask for references from previous jobs, and go to see them.
- When comparing quotes, check for hidden extras, and don't be tempted to automatically choose the lowest price.
- Agree payment stages in advance and only pay for completed work. Never pay up front.
- Check that the builder is suitably insured for risks to persons and property. Ask for copies of certificates for full public liability insurance and (for main contractors) employers' liability cover.
- Ask who will actually be doing the work. Will any subcontractors be able to produce work of a high standard?
- Be very clear about precisely what work you want done and don't keep changing your mind later on site or you will be charged lots of expensive extras.
- Use a written contract with firms or detailed letters with trades
- Confirm the start and finish dates in writing, along with the agreed price and arrangements for payment.
- Confirm exactly what's included in the price – does it for example include lifting and relaying carpets, moving any furniture, scaffolding, clearing all rubbish from the site, skip hire and cleaning up?

must be met to make them exempt from the Building Regulations:

- The internal floor area has a limit of 30m².
- Not less than 75% of the roof and 50% of the walls should comprise a translucent material – either glass or polycarbonate sheeting. All critical and low-level glazing must be toughened or laminated safety glazing and window sill heights should be a minimum 800mm above floor level.
- Conservatories must be thermally separated from the main house with exterior quality doors (including patio or French doors) to prevent heat loss. Simple polycarbonate roofs can allow more than 15 times the amount of heat to escape than conventional tiled roofs.
- Where the conservatory blocks the windows/doors to a habitable room, it should be able to provide sufficient ventilation for both rooms. So conservatory doors and opening windows should in total be no smaller than 1/20th of the combined floor area of both rooms. Plus you need background ventilation of at least 8,000mm². The closable openings separating the conservatory from the house need to meet similar size and ventilation requirements.
- No fixed heating system should be installed – ie not linked to your existing central heating system. Portable heaters are acceptable.

The Building Regulations will however come into play if the structure is built over a shared drain run, and for any new openings cut into the walls from the house.

When it comes to designing your conservatory, there are some general points to bear in mind.

- Better-quality designs have base walls of cavity masonry construction laid to normal foundation depths, upon which the superstructure is fixed – as opposed to simply being erected upon a thin concrete slab (which can lead to problems with structural movement).
- Flues serving wall-mounted boilers are often located so they clash with the proposed conservatory. This normally means having to relocate the boiler so the flue is well clear of the building.
- Ventilated roof ridges should be incorporated to relieve air pressure and prevent 'wind uplift' that can push out lightweight roof panels.

## EXTENSIONS

A good extension should enhance the original building. But that doesn't mean you can replicate Victorian building methods. Traditional solid-wall construction is too thermally inefficient for new construction to comply with today's Building Regulations. Paradoxically, however, the Planners may insist on a design that faithfully replicates the original. To resolve this apparent conflict, modern insulated cavity construction can be disguised so that,

Lovely job – skilled period detailing complements and enhances original house

viewed externally, it resembles traditional solid walls in Flemish bond pattern. To achieve this, each lengthways brick is interspersed with a brick cut in half and laid across the wall ('split headers').

From a practical point of view, a 'sympathetic' Victorian style is usually the best option. But extensions to Victorian houses don't always need to replicate the original. Good modern design can sometimes work extremely well alongside period buildings. A highly glazed extension can create a large open plan kitchen/diner with a pleasant, bright feel. The transparent nature of pure glass extensions can make them almost invisible, nicely complementing the original architecture. However, this can involve a fair amount of additional work insulating the main house to compensate for the extent of new glazed areas (new glazing is normally limited to the equivalent of 25% of the extension's floor area plus an additional area to compensate for any existing doors or windows that are covered up due to the works).

One key point to bear in mind when designing extensions to old buildings is the risk of differential settlement. When an extension with deep foundations is built next to old house with shallow foundations, the old structure will still want to move slightly in accordance with changes to ground conditions, while the new one remains rock solid. This inevitably leads to cracking at or near the point where the two structures meet. So it's important to design a flexible junction that allows them to move independently.

Modern' glass box' (left) can add style, but in most cases trad (right) works best

# Terraced house side return extensions

The majority of Victorian houses are terraced, and the kitchens to the rear addition can sometimes be a little narrow and cramped (estate agents sometimes describe them as 'galley style'). So 'filling in' the adjoining alley space to the side

the kitchen, a major structural alteration. One challenge is to retain as much natural light as possible to the old dining room once it is 'hemmed in' by the extension. This is easier to achieve in single storey extensions where a glazed roof over a single-storey side return can bring in plenty of light, although off-the-shelf rooflights or solar light tubes can be a less-expensive solution.

In some circumstances you may be allowed to build right up to a

can transform the property. A side return extension usually adds only 1 or 2 metres to the kitchen width, but this should be sufficient to create an attractive kitchen/diner or family room.

Because of the limited building space, the likely effects on adjoining properties must be carefully considered. It's worth asking the neighbours if they want to extend too, as there are savings to be made sharing the same designer and builder, and it would make building the new party wall between the properties a lot easier.

Building a side return extension involves chopping out a large part of the main side wall to

boundary or party wall, although planners sometimes like to see them set back at least 900mm to allow access for future maintenance. End terrace houses normally offer more scope for extending although where a large multi-storey side extension brings you within range of a highway you may only be permitted to build up to 1m from the road (for a single storey you may be allowed to build up to the boundary). In many areas there is a maximum permitted distance that a rear extension can be built out from a terraced property.

One of the toughest dilemmas when extending terraces is where to put the new windows. There are minimum guideline distances that should be maintained between your new extension windows and next door's existing windows, *ie* where the sides of two properties face each other. For two-storey extensions, so as not to overlook a neighbouring property, any side-facing window may need to be obscure glazed and fixed shut, or any opening must be 1.7m above the floor level. But the reality for most urban properties is that space is always going to be limited, so the planners may agree to them being closer than recommended in the guidelines, especially if there is a substantial boundary fence or garden wall between the buildings.

In order to preserve privacy some Planning Authorities insist that the new rear windows on your extension must be no closer than 18 metres to the widows of any houses at the bottom of the garden (or 14 metres if it's just a ground floor extension).

noticeable change, making this perfectly feasible even for properties in Conservation Areas.

The necessary structural work normally involves introducing steel beams to take loadings diverted from the roof and to support new loft floors. However, old buildings have relatively shallow foundations and often fairly weak lintels and thin or missing firebreak party walls in the loft. So the best advice is to consult a structural engineer to design a suitable method of supporting the new loadings and produce the necessary calculations and drawings.

Converting a loft will also have a major impact on the interior of the existing house, and particular care is needed to protect fragile historic plasterwork to ceilings below.

The new loft stairs can usually be installed above the existing main stairs, although a new rear facing large dormer may be needed to create sufficient headroom. Also, complying with fire regulations can mean having to replace all the internal doors to habitable rooms en route down to the main entrance with new fire doors. However, to conserve the original doors Building Control may accept upgrading them instead. As you might expect, the fire regulations become more draconian in houses of more than two storeys, and may require a new sprinkler system, an external fire escape, or enclosing the stairs with a fire door at the top or bottom.

## CELLARS AND BASEMENTS

A cellar is an underground storage space, which is often windowless and has minimal headroom. They were often originally used as coal stores.

### LOFT CONVERSIONS

Most Victorian loft spaces are pretty generous and well suited to conversion, with minimal visual impact externally. The exceptions are early townhouses with parapet roofs, and some mid-Victorian Italianate-style houses with shallow pitched, hipped roofs. But in most cases a couple of conservation skylights to the rear, or a discreet window to a side gable may be the only

A basement, on the other hand, is a full-sized room commonly taking the form of a 'semi-basement' with part above ground level. Basements are generally found in more expensive early to mid-Victorian townhouses, where it was probably once the 'downstairs'

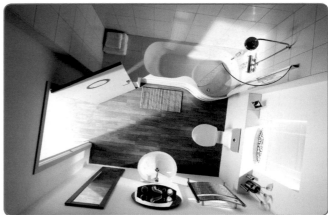

Heritagebathrooms.com

occupied by servants (along with rooms in the attic) and the location of the kitchen and scullery.

In terms of potential for conversion to liveable space there's usually not a lot you can do with cramped, windowless, cellars. The nuclear option – excavating and rebuilding at enormous cost – is rarely advisable. One thing cellars and basements have in common is the tendency to suffer from damp. In more severe cases, such as where standing water has accumulated, it can normally be dispersed with a pump. However, if the water table is higher than the level of the cellar floor, specialist advice will be required as it may be necessary to install an external land drain.

Converting an existing basement into habitable space can often be a realistic proposition. However, there are still likely to be significant limitations, not least the cost. Issues such as damp-proofing, light and ventilation, access, headroom and potential for flooding all need to be carefully weighed up.

The best-known method of damp-proofing basements is 'tanking', where thick coats of render are applied to seal the walls and floor. But this is now considered to be the least-effective system. The cold wall surfaces can be prone to condensation, and cracks can develop in the render as old buildings move, making them potentially vulnerable to leakage from external water pressure.

Today ventilated dry-lining systems are the preferred method used in basement conversions. These work by isolating the original wall surfaces behind a new inner lining, leaving a ventilated air gap in between. This is achieved by sandwiching a special dimpled plastic sheet between the basement walls and a new interior wall, usually of dry-lined timber studwork. To the floor, a screed is laid over a damp proof membrane (DPM). This method has the advantage of not taking up much space and also allows walls to breathe, but good ventilation is vital. However, provision should be made to collect any water that may penetrate and drain or pump it away.

In basements where damp is likely to be more excessive, a more robust solution can be to build a 'room within the room'. This is achieved by constructing a new floor and blockwork walls, which are isolated from the main structure by DPMs and a drained cavity. The main disadvantage is that the blockwork takes up valuable space.

## ADDING NEW KITCHENS AND BATHROOMS

Installing new sanitary fittings or kitchen units in older properties with uneven walls and sloping floors can be fairly challenging. So

rather than opting for conventional fitted units, it's sometimes preferable to select free-standing kitchen furniture and a range cooker, and a classic Victorian 'roll top' bath. A traditional kitchen dresser can provide generous amount of storage space whilst imparting suitably authenticity (and may cost significantly less than installing conventional fitted units).

One of the key considerations when designing new bathrooms or kitchen is to carefully plan all the services in advance. The position of sinks and appliances must be integrated with pipework for the hot and cold water, waste plumbing, electric supplies, gas or oil pipes.

Bear in mind also that cooking and bathing are the biggest contributors to indoor water vapour. To extract steam and dispatch stale cooking smells to the outdoors the Building

Regulations require extractor fans to be installed. This is an essential part of your armoury in the war against damaging condensation. But any holes for vents that need to be cut in external walls must be done carefully so as not to disfigure the building.

architectyourhome.com

# GLOSSARY

**Access tower** A movable scaffolding platform allowing access for high-level work on roofs etc.

**Aggregate** Gravel, shingle, or pebbles, etc, used in the manufacture of concrete.

**Airbrick** A perforated brick or metal/plastic grille used for ventilation, typically in external walls to suspended-timber ground-floor voids.

**Apron** A metal strip, usually of lead, fitted at the base of a chimney or under window sills above tile-hung bays, to provide a waterproof joint.

**Architrave** Moulded wood strip covering the joint at the edge of a door or window frame and the surrounding wall.

**Asbestos cement** Cement with 10–15% asbestos fibre as reinforcement. Hazardous fibres may be released if it is cut or drilled.

**Ashlar** Finely dressed natural stone, a superior grade of masonry.

**Asphalt** Black tar-like substance used as an adhesive and impervious moisture barrier on flat roofs and floors.

**Back gutter** A metal flashing strip forming a waterproof seal between the back of a stack and its roof slope.

**Balanced flue** Metal vent that allows gas appliances both to draw air in from outside and to expel exhaust fumes. Some are fan assisted.

**Balusters** Vertical spindles supporting the handrail of a staircase.

**Balustrade** A row of balusters/spindles joined to a horizontal rail, typically to a staircase or landing.

**Bargeboards** Boards placed along the verges of a roof, usually at gable ends, often of decorative timber. Also known as 'vergeboards'.

**Battens** Thin lengths of timber to which tiles or slates are nailed or fixed.

**Beetle infestation** Larvae of various species of beetle that tunnel into wood causing damage, usually evident as small boreholes. The generic term is 'woodworm'.

**Benching** Smooth layer of concrete alongside the drainage channel in an inspection chamber. Also known as 'haunching'.

**Binder** A horizontal timber beam laid across ceiling joists in some roof spaces to help strengthen the structure.

**Bitumen** Black, sticky substance related to asphalt. Used in sealants, mineral felts and DPCs.

**Bond** The pattern in which bricks are laid in mortar to form a wall.

**Bonding timbers** Small timbers that help tie the party walls in a terrace to the front and rear walls.

**Box section gutter** Square-shaped valley gutter or parapet gutter.

**Bressummer** Old word for a large timber lintel, now often used to describe timber lintels over wall openings to single-storey bays.

**Butterfly roof** M-shaped roof found on Georgian and some early Victorian houses, hidden from street view by parapet walls. It comprises two 'lean-to' roof slopes, each leaning against one of the party walls and meeting at a central valley.

**BTU** British Thermal Unit.

**Buttress** A brick or stone support to a wall designed to resist thrust movement and to give added stability.

**Carriage** A thick length of timber running along the underside of a staircase.

**Casement** A window hinged at one edge and designed to open inwards or outwards.

**Cavity wall** Standard modern (post 1930) main-wall construction comprising two leaves of brick or blockwork about 280mm thick, separated by a gap (cavity) of about 50mm, which can be insulated. The leaves are secured together with wall ties.

**Cement fillet** The covering over a junction (eg between roofs and walls) made from mortar instead of a metal flashing.

**Cesspit** or **cesspool** An underground tank to hold sewage and foul waste, needing regular emptying.

**Chimney breast** That part of the chimney flue that projects into a room. The 'stack' is the part above roof level.

**Cill** Variant spelling of 'sill'.

**Cistern** A water storage tank (usually to a WC or water tank in loft).

**Coke breeze** Waste ashes from coke- or coal-fired furnaces, often used as a crude covering to the earth under timber floors, sometimes mixed with sand and cement or lime.

**Collar** A horizontal timber member that joins and restrains opposing roof slopes.

**Combination boiler** Modern 'sealed system' gas boiler that activates on demand for hot water or central heating and does not require water tanks or cylinders, being supplied direct from the mains.

**Coping** Masonry covering laid on top of a wall to stop rain soaking into it, usually of stone or concrete.

**Corbel** Projection of brick, stone, timber, or metal jutting out from a wall to support a load such as a beam.

**Cornice** and **coving** Ornamental plaster around the joint of a wall and ceiling. Coving is a curved strip covering the joint.

**Cowl** A cap to a chimney or vent pipe.

**Dado** Protective wooden or tiled horizontal strip running along internal walls, about 1m above the floor.

**Damp-proof course (DPC)** An impervious layer (eg slate, felt, PVC) built into a wall to prevent the passage of dampness.

**Damp-proof membrane (DPM)** An impervious layer (polythene sheeting, bitumen, etc.) within a concrete ground floor slab to prevent rising damp.

**Distemper** Solution of ground chalk and animal glue size used in decoration.

**Dormer window** A window that projects out from a roof slope.

**Dowel** Thin timber plugs that hold jointed sections of timber together.

**DPC(s)** Damp-proof course(s).

**DPM(s)** Damp-proof membrane(s).

**Drip groove** Groove cut beneath sills, doorsteps and coping stones to disperse rainwater.

**Eaves** The overhanging edge of a roof near the gutter.

**Encaustic tiles** Tiles moulded from one clay, usually red, and inlaid with white or coloured clay patterns that fuse when fired. Used mainly for floors.

**English bond** Bricklaying pattern in which courses laid lengthways alternate with courses laid crossways.

**Fall** The slope or gradient, typically of a pipe run or flat roof, to ensure water run off.

**Fanlight** A small window above a door or casement.

**Fascia** Horizontal timber boards that run along the eaves at the base of roof slopes. They often cover the ends of the rafters. Gutters may be fixed to them.

**Fibreboard** Lightweight board for ceilings or internal walls, made of compressed wood pulp. Now superseded by plasterboard.

**Fillet** A small strip of cement/lime mortar, timber, plastic, etc., used to cover or seal the junction between two surfaces.

**Finial** A small 'spire' type ornament, often of terracotta or iron, fixed to the roof ridge at a gable end.

**Flagstones** Large paving stones used on floors in older houses.

**Flashing** A thin strip, usually of lead or zinc, used to cover roof joints to prevent leakage (eg to chimney stacks).

**Flaunching** Smooth contoured cement mortar around the base of chimney pots.

**Flemish bond** Bricklaying pattern in which the bricks in each course are alternately laid lengthways and crossways.

**Flue lining** Flues are the 'exhaust ducts' for gases from fires or appliances. Flue linings are long tubes fitted within flues, usually of stainless steel, clay pipework, or concrete.

**French drain** Shallow gravel-filled drainage ditch.

**Frieze** The part of a wall above the picture rail.

**Gable** The triangular upper part of a wall under the verges at the edge of a pitched roof ('gable end').

**Gather** Alternative name for the throat of a chimney flue.

**Granolithic** A mixture of cement and stone chippings.

**Gulley** An opening into a drain, receiving water from downpipes or waste pipes.

**Hair mortar** Lime mortar incorporating horse or ox hair as a binding agent.

**Half-round guttering** Guttering that is nearly semi-circular in section. The most widely used modern type.

**Hangers** Vertical beams under the roof ridge supporting the ceiling binders.

**Haunching** See 'benching'.

**Header** The end of a brick, visible in solid walls when laid crossways (see 'stretchers').

**Heave** The opposite of subsidence. It occurs when the ground swells with water.

**Hip** The external junction where two roof slopes meet.

**Hip iron** Protruding galvanised metal straps found on hipped roofs, screwed to the base of hip rafters to help prevent hip tiles slipping off.

**Hip roll** See 'ridge roll'.

**Hopper** or **hopperhead** An open-topped box or funnel at the top of a downpipe. It collects rainwater or waste water from one or more pipes.

**Inspection chamber** A 'manhole' with a removable cover providing access to the drainage channel at its base.

**Jamb** Vertical side part of a door frame or window frame.

**Joists** Horizontal structural beams used to construct ceilings, timber floors and flat roofs.

**Kingposts** Traditional name for hangers.

**Lantern light** A kind of small greenhouse structure built on some flat roofs to allow light through to a stairwell.

**Lap** The overlap of courses of slates or tiles.

**Lath** Thin strips of wood traditionally used as a backing to plaster.

**Leaded light** Window divided up into small, often diamond-shaped panes by strips of lead.

**Ledged and braced door** A 'ledged door' is made from vertical timber boards fixed to thick horizontal cross-timbers called ledges. Some are strengthened with diagonal braces.

**Limewash** A solution of powdered lime and water used in decoration.

**Lintel** Horizontal structural beam over a window or door opening. Normally made of timber, concrete, stone, or steel.

**Louvres** Glass or timber slats laid at an angle or hinged so that they can be opened to allow ventilation.

**Mansard roof** A roof constructed so each slope has two different pitches – a shallow upper part and a steeper lower part – so as to provide a top floor of usable space within a roof structure.

**Mastic** A generic term for any sealant used in the building process, eg for sealing joints around window openings.

**Matchboarding** Vertical tongue-and-groove timber cladding, traditionally used to cover the lower part of walls internally.

**MCB** Miniature circuit breaker.

**Mortise** A slot cut in a section of wood for a corresponding 'tenon' 'tongue' of another section to fit into.

**Mullion** An upright division of a window such as a vertical bar dividing individual lights.

**Newel** A stout post at the bottom or top of a stair to which the handrail is fixed.

**Nib** The projecting lug on the back of a tile that hooks over the supporting batten.

**Nogging** or **noggin** A short timber batten that fits between a pair of joists or timber studs to add strength.

**Nosing** The rounded projecting edge of a stair step.

**Ogee guttering** Wavy-fronted guttering, usually screwed directly to the fascia.

**Oriel window** A window projecting from an upper floor.

**Padstone** A stone or robust block laid under the end of a beam or steel joist, to help distribute the load.

**Pantiles** Large curved roofing tiles with lugs that hook over the battens.

**Parapet** Low wall along the edge of a flat roof or balcony.

**Parapet roof** A pitched roof with the main walls built up to a parapet that hides the gutters.

**Parging** Lime and sand mortar, traditionally mixed with cow dung and ox hair for added strength.

**Parquet floor** Small strips of wood usually laid on a solid floor to form a pattern.

**Party wall** The wall that separates, but is shared by, adjoining properties.

**Pier** A vertical column, usually built in brickwork, used to strengthen a wall or support a weight.

**Pitch** The angle or slope of a roof, technically the ratio of span to height.

**Place bricks** Cheap bricks used in the construction of hidden walls.

**Plasterboard** Large thin sheets made of plaster sandwiched between coarse paper, used for ceilings and internal walls.

**Plinth** The projecting base of a wall, usually of brick or render.

**Pointing** The smooth outer edge of mortar joints between bricks, stone, etc.

**Purlins** Horizontal beams in a roof upon which the rafters rest.

**Quarry tiles** Plain single colour 'geometric' floor tiles made from clay, often red or brown, usually unglazed. From the French word *carré*, meaning square.

**Queenpost** Traditional method of framing a roof to cover a larger span. See illustration in chapter 2.

**Quoin** Projecting bricks or stone blocks traditionally used at corners of walls.

**Rafter brackets** Metal brackets screwed to the ends of rafters to support guttering where there is no fascia board.

**Rafters** The main sloping roof timbers to which the tiles/slates, battens and felt are fixed.

**Rails** The horizontal framing members of a door or window, usually at the top or bottom (see also 'stiles').

**RCD** Residual current device.

**Rebate** A recess, groove, or rectangular step cut in the edge of a piece of timber or stone, etc. ('rebated' or set back) to receive a mating piece.

**Red rubbers** Soft red bricks that could be cut and rubbed to shape.

**Render** or **rendering** General term for the finish applied to external wall surfaces of sand and cement/lime (or the first coat of internal plastering). It may be smooth, or finished in roughcast, pebbledash, etc.

**Retaining wall** Usually a garden wall built to hold back or retain a large bank of soil, rubble, etc.

**Reveals** The vertical sides of an opening cut in a wall (typically of brick or stone), *eg* between a door or window frame and the front of the wall (see also 'jambs').

**Ridge** The top or 'apex' of the roof where two slopes meet, formed from a timber board (the 'ridge plate') joining the tops of the rafters and covered with shaped 'ridge tiles'.

**Ridge roll** or **hip roll** A strip of lead or zinc flashing wrapped round a wooden pole with side wings, sometimes used in place of tiles to seal a ridge or hip.

**Riser** The vertical portion between treads of stairs.

**Rising damp** Moisture soaking up a wall from the ground, by capillary action, or through a floor (see 'damp-proof course').

**Rodding access** Removable covers at bends in drainage pipes, gulleys, etc., allowing access for clearing blockages.

**Roof spread** The outward thrust of a poorly restrained roof causing a wall to bow out. (See 'collar'.)

**Room sealed appliance** One that takes its combustion air from outside via a 'balanced flue' and expels exhaust fumes via the same flue. Most modern boilers are room sealed.

**Roughcast** A rough render finish to external walls, usually incorporating gravel.

**RSJ** Rolled steel joist used for structural support (*eg* to walls or floors), usually spanning relatively wide openings.

**Saddle piece** Middle part of the flashing for a centrally located chimney stack.

**Sarking boards** Alternative name for battens.

**Sarking felt** A layer of bituminous felt used for covering roofs before laying battens as a secondary defence against rain. Not normally fitted in pre-war houses.

**Sash window** Two-part, vertically sliding window that can be opened at both top and bottom.

**Screed** A smooth finish coat on a solid concrete floor slab, usually of mortar, concrete, or asphalt.

**Scrim tape** A special weaved material for reinforcing the filler concealing joints between plasterboard panels.

**Secret gutter** Soaker at the junction of a roof obscured by the lap of the slates or tiles.

**Semi-basement** Rooms in a lower floor not set in the ground to their full height, often with a 'trench' separating the front wall and the higher garden level. Windows are normally visible from ground level.

**Septic tank** Private drainage system comprising underground tanks where sewage decomposes through bacteriological action. Can require periodic emptying.

**Settlement** General disturbance in a structure showing as distortion in walls etc. Usually the result of initial compacting of the ground due to the loading of the building.

**Sill** The lower horizontal member at the bottom of a door or window frame. Externally it should throw water clear of the wall below (of stone, concrete, brick, or timber). Internally it is a shelf at the bottom of a window.

**Skylight** A window in a roof slope or ceiling to admit daylight.

**Sleeper wall** A dwarf wall supporting the joists under a suspended timber ground floor.

**Snapped headers** Facing bricks cut in half, generally used in a cavity wall to imitate the appearance of a solid 'Flemish bond' wall.

**Soakaway** A rubble-filled pit for rainwater dispersal.

**Soakers** Strips of metal (usually lead, zinc, or copper) fitted beneath tiles to provide a waterproof joint at the junction of a roof with a wall or a chimney. Normally overlain with flashings.

**Soffit** The underside ('external ceiling') below eaves, balconies, etc.

**Spalling** The crumbling of masonry or tiles as a result of weather damage.

**Spouting** Moulded timber guttering found on some terraced houses.

**Squareline guttering** Modern guttering with a rectangular section.

**Stiles** The vertical framing members of a door or window (see also 'rail').

**Stocks** Quality bricks used to face the main walls.

**Stretcher** The side of a brick, visible in walls when laid lengthways (see 'header').

**String** Long length of timber running from top to bottom of a staircase.

**String course** A course of brickwork that projects beyond the face of an external wall (or band course).

**Struts** Timber props supporting purlins, found in roof spaces.

**Stucco** A hard external plaster used to imitate stone, superseded by modern cement render.

**Stud partition** Lightweight internal wall, usually of a timber framework faced with plasterboard or lath and plaster, usually non load-bearing.

**Subsidence** Ground movement, often as a result of clay shrinkage, drainage problems, or mining activities.

**Subsoil** Soil lying immediately below the topsoil, upon which foundations usually bear.

**SVP** Soil and vent pipe. Vertical stack taking 'soil' waste from WCs and bathrooms, etc., typically of plastic or iron, and vented at the top, normally terminating at roof level.

**Swept valley tiles** Purpose-made curved tiles used to cover joints between roof slopes (valleys).

**Tanking** Waterproof treatment to basements.

**Terrazzo** Marble chips set in mortar.

**Throating** Alternative name for a drip groove.

**Tie bar** Large metal bar passing through a wall to brace a structure suffering from structural instability.

**Tilting fillet** A timber fillet fixed under the roof coverings above the eaves, in order to raise the edge of the first row of slates.

**Tingles** Small folded strips of lead or copper used to keep loose slates in place.

**Tongue-and-groove boarding** Close-fitted boards where the edge of one board fits into a groove of the adjoining board.

**Torching** Mortar traditionally applied on the underside of slates or tiles to help prevent moisture penetrating.

**Transom** Horizontal bar of wood or stone across a window or the top of a door.

**Tread** The horizontal 'flat' part of a step or stair.

**Trimmer** A small section of timber joist run at right angles to the ends of the main floor joists to form an opening, *eg* for stairs or a fireplace.

**TRV** Thermostatic radiator valve. An adjustable sensor valve next to a radiator allowing its temperature to be set.

**Undercloak** Extra line of tiles at verges, placed underneath the end slates or tiles to tilt them up and create a neat edge.

**Underfelt** or **underlay** Alternative names for sarking felt.

**Underpinning** A method of strengthening weak foundations where a new stronger foundation is placed beneath the original.

**Valley** The junction of one roof with another at an angle, typically where a bay window roof meets the main roof. Often formed by a strip of lead or zinc sheeting over a timber board base. 'Open valleys' leave the metal strip exposed; 'mitred valleys' have slates over the edges; and 'laced valleys' have rows of tiles over the trough.

**Valley gutter** Gutter at the bottom of the 'V' at the junction of two roof slopes.

**Verge** The edge of a roof, especially over a gable.

**Vergeboards** See 'bargeboards'.

**Wainscot** Wood panelling or boarding on the lower part of an internal wall.

**Wall plate** A timber beam placed on a wall, *eg* at eaves level for the roof rafters, or to receive floor joists.

**Wall ties** Strips of metal built across cavity walls to join the inner and outer skins.

**Water bar** A strip of metal fixed to a door sill to prevent water flowing under the door.

**Weather board** A board fixed to the bottom of a door on the outside to prevent rain driving underneath.

**Weatherstruck** Style of mortar pointing to brickwork. It slants downwards, projecting slightly from the wall surface, to help disperse rainwater.

**Woodworm** General term for beetle infestation.

# BIBLIOGRAPHY

Anon. *DIY and Home Improvements*, Colour Library 1986.

Barrett, Helena, and Phillips, John. *Suburban Style: The British Home 1840–1960*, Macdonald 1987.

Collings, Janet. *Old House Care and Repair*, Donhead 2002.

Johnson, Alan. *How to Restore and Improve Your Victorian House*, David & Charles 1984.

Marshall, John, and Willox, Ian. *The Victorian House*, Sidgwick & Jackson 1986.

Melville, I.A., Gordon, I.A., and Borwood, A. *Structural Surveys of Dwelling Houses*, Estates Gazette 1974.

Morris, Ian. *Bazaar Property Doctor*, BBC Books 1989.

Muthesius, Stefan. *The English Terraced House*, Yale University Press 1982.

Reyburn, Wallace. *Flushed with Pride: The Story of Thomas Crapper*, Pavilion 1989.

Seeley, Ivor H. *Building Maintenance*, Palgrave Macmillan 1987.

Wedd, Kit. *The Victorian Society Book of The Victorian House*. Arum Press 2002.

Wickersham, John. *Repair and Home Renovation*, Haynes 1995.

**Websites**

www.bricksandbrass.co.uk

www.chimneybuster.com

www.lafargeplasterboard.co.uk

www.periodproperty.co.uk

www.propertybooks.co.uk

# INDEX